Current Topics in Membranes, Volume 68

Membrane Fusion

Current Topics in Membranes, Volume 68

Series Editors

Dale J. Benos
Department of Physiology and Biophysics,
University of Alabama,
Birmingham, Alabama

Robert Balaban
National Heart, Lung and Blood Institute,
National Institute of Health,
Bethesda, Maryland

Sidney A. Simon
Department of Physiology and Biophysics,
University of Alabama,
Birmingham, Alabama

Current Topics in Membranes, Volume 68

Membrane Fusion

Edited by

Leonid V. Chernomordik
Laboratory of Cellular and Molecular Biophysics
National Institute of Child Health and Human Development
National Institutes of Health
Bethesda, MD, USA

Michael M. Kozlov
Department of Physiology and Pharmacology
Sackler Faculty of Medicine
Tel Aviv University
Tel Aviv, Israel

ELSEVIER

AMSTERDAM • BOSTON • HEIDELBERG • LONDON
NEW YORK • OXFORD • PARIS • SAN DIEGO
SAN FRANCISCO • SINGAPORE • SYDNEY • TOKYO
Academic Press is an imprint of Elsevier

Academic Press is an imprint of Elsevier
225 Wyman Street, Waltham, MA 02451, USA
525 B Street, Suite 1900, San Diego, CA 92101-4495, USA

First edition 2011

ISBN: 978-0-12-385891-7
ISSN: 1063-5823

For information on all Academic Press publications
visit our website at elsevierdirect.com

Printed and bound by CPI Group (UK) Ltd, Croydon, CR0 4YY

11 12 13 10 9 8 7 6 5 4 3 2 1

Dedication

We dedicate this 68th volume of the Current Topics In Membranes (CTM) book series to our longtime friend and colleague Dr. Dale J. Benos Ph.D. (1950-2010). Dale was the series, and often a contributing, editor whose untimely death at the age of 60 has been a huge loss for his family, his friends, biomedical research community as well as the CTM editorial team.

Dale earned his Bachelor of Science degree in Biology in 1972 from Case Western Reserve University. He earned his Ph.D. in Physiology and Pharmacology from Duke University in 1976 with Dr. Daniel Tosteson. His Ph.D. work was well before his time using genetic mouse models in studying the ion transport in red blood cells. He followed this work with a post-doctoral fellowship in epithelial physiology at Duke in 1978 with Dr. Lazaro Mandel beginning his long standing interest in sodium transport and amiloride- sensitive ion channels. It was at Duke University where both of us met, socialized and worked with Dale on a variety of projects. Upon completing his postdoctoral work, he moved to Boston as an Andrew W. Mellon Scholar in the Laboratory of Human Reproduction and Reproductive Biology at Harvard Medical School, where he rose to the position of Associate Professor in the Department of Physiology and Biophysics.

In 1985, Dale joined the University of Alabama faculty an Associate Professor in the Department of Physiology and Biophysics. Two years later was promoted to Professor and in 1996 became department chair, a position he retained until his untimely death.

The scope of Dale's scientific interest and contributions were remarkably broad and insightful. He made major contributions to the understanding of the function of the amiloride-sensitive Na channel in epithelia as well as studies on the metabolism and function of the reproductive system from blastocysts to sperm as well as his most recent work on the role of Na transport in the functioning of the brain in health and disease. His studies were always complete and meticulously conducted challenging competitors as well as collaborators to move to the next level of understanding of any problem he engaged.

In addition to his own research, Dale also played a major role in the dissemination of scientific information as well as teaching. In addition to his work at CTM in which not only was the series editor, but also edited several volumes, Dale, also had major editorial chores on the Journal of Biological Chemistry, American Journal of Physiology, Biochem Biophys Acta (Biomembranes) and many other periodicals. He was also the President of the American Physiological Society and served on the boards of numerous scientific societies dedicated to the dissemination of biomedical science as well as the teaching both science and ethics of the next generation of scientists. Based on personal accounts and the number of teaching awards he received at the University of Alabama, Dale was a spectacular teacher getting students engaged and excited about the most fundamental aspects of physiology and membrane biophysics. Clearly this enthusiasm was also imparted on the contributing editors to CTM generating the remarkable series of scholarly contributions to the field of membrane biology over the last ~15 years.

We cannot replace Dale's friendship, enthusiasm, intellect and remarkable positive energy. Many of these traits are revealed in this picture of Dale from the 1970's at Duke where most people will recognize the enthusiasm and positive attitude but also the remarkable fact that he looked almost the same on his 60th birthday. Clearly, it was much too soon for him to leave us with so much more to contribute to unraveling nature and teaching the next generation the lessons learned by this generation. His scientific legacy is remarkable; however his real long term impact is the training of future generations. Moreover, his role in generating numerous volumes of CTM will be one of the many monuments he has left all of us. We pledge to carry on the tradition of scientific excellence in the future volumes of CTM series.

Sidney A. Simon and Robert S. Balaban

Co-series editors

Contents

PART 2 VIRAL FUSION

PART 3 INTRACELLULAR FUSION

PART 4 DEVELOPMENTAL CELL FUSION

PART 5 THEORETICAL MODELING

Contributors

Numbers in parentheses indicate the pages on which the authors' contributions begin.

Aurélie Albertini (49) Centre de Recherche de Gif, Laboratoire de Virologie Moléculaire et Structurale, CNRS (UPR 3296), Gif sur Yvette, France

Ori Avinoam (209) Department of Biology, Technion, Israel Institute of Technology, Haifa, Israel

Julie Boutilier (107) Department of Microbiology and Immunology, Dalhousie University, Halifax, Nova Scotia, Canada

Stéphane Bressanelli (49) Centre de Recherche de Gif, Laboratoire de Virologie Moléculaire et Structurale, CNRS (UPR 3296), Gif sur Yvette, France

Elizabeth H. Chen (235) Department of Molecular Biology and Genetics, Johns Hopkins University School of Medicine, Baltimore, Maryland, USA

Roy Duncan (107) Department of Microbiology and Immunology, Dalhousie University, Halifax, Nova Scotia, Canada

Yves Gaudin (49) Centre de Recherche de Gif, Laboratoire de Virologie Moléculaire et Structurale, CNRS (UPR 3296), Gif sur Yvette, France

Andreas Herrmann (1) Department of Biology, Faculty of Mathematics and Natural Sciences I, Humboldt-University Berlin, Berlin, Germany

Meyer B. Jackson (187) Department of Neuroscience, University of Wisconsin School of Medicine and Public Health, Madison, Wisconsin, USA

Dae-Hyuk Kweon (161) School of Life Science and Biotechnology and Center for Human Interface Nanotechnology, Sungkyunkwan University, Suwon, South Korea

Jean Lepault (49) Centre de Recherche de Gif, Laboratoire de Virologie Moléculaire et Structurale, CNRS (UPR 3296), Gif sur Yvette, France

Albert J. Markvoort (259) Institute for Complex Molecular Systems & Biomodeling and Bioinformatics Group, Eindhoven University of Technology, Eindhoven, The Netherlands

Siewert J. Marrink (259) Groningen Biomolecular Sciences and Biotechnology Institute & Zernike Institute for Advanced Materials, University of Groningen, Groningen, The Netherlands

Sascha Martens (141) Max F. Perutz Laboratories, University of Vienna, Vienna, Austria

Harvey T. McMahon (141) MRC Laboratory of Molecular Biology, Cambridge, United Kingdom

Gregory B. Melikyan (81) Department of Pediatrics, Infectious Diseases, Emory University, Atlanta, GA, USA

Marcus Müller (295) Institut für Theoretische Physik Georg-August-Universität, Göttingen, Germany

Jörg Nikolaus (1) Department of Biology, Faculty of Mathematics and Natural Sciences I, Humboldt-University Berlin, Berlin, Germany

Ben O'Shaughnessy (1) Department of Chemical Engineering, Columbia University, New York, New York, USA

Benjamin Podbilewicz (209) Department of Biology, Technion, Israel Institute of Technology, Haifa, Israel

Thomas J. Pucadyil (33) Indian Institute of Science Education and Research, Pashan, Pune, Maharashtra, India

Michael Schick (295) Department of Physics, University of Washington, Seattle, Washington, USA

Yeon-Kyun Shin (161) Department of Biochemistry, Biophysics, and Molecular Biology, Iowa State University, Ames, Iowa, USA

Jason M. Warner (1) Department of Chemical Engineering, Columbia University, New York, New York, USA

Tae-Young Yoon (161) Department of Physics and KAIST Institute for the BioCentury, KAIST, Daejeon, South Korea

Preface

Cell boundary and intracellular space are organized by biological membranes. The continuity and structural stability of the membranes are essential to their primary function as dividers between different environments. At the same time, countless normal and pathological processes involve topological rearrangement of the membranes consisting in merger of two lipid bilayers into one (fusion) or division of a single continuous membrane into two (fission). Research on the mechanisms of membrane fusion is vital for understanding synaptic transmission and viral diseases, fertilization and protein trafficking.

Ultimately, any fusion event boils down to rearrangement of membrane lipid bilayers and the pathway of these rearrangements appears to be conserved between disparate fusion processes. Due to intrinsic stability of lipid bilayers of conventional compositions, drastic changes in their structure require application of forces and input of energy, which are thought to be provided by dedicated proteins often referred to as fusion proteins.

Most studies suggest that fusion starts with a stage of hemifusion where the apposed leaflets of the fusing membranes merge whereas the inner leaflets remain separate. Subsequent and usually most energy demanding stages of the fusion pathway involve transition from this hemifusion intermediate to an expanding fusion pore that allows mixing of the contents of the fusing compartments. The specific mechanisms by which conformational energy released by restructuring of fusion proteins is used to drive the membrane rearrangements remain elusive. However, the major features of these mechanisms are likely to be shared by different membrane systems as suggested by a striking similarity of postfusion conformations of the fusion proteins of diverse enveloped viruses. Recent works emphasize the essential role of the bilayer elastic stresses, and, specifically, the curvature stresses, as mediators between refolding of the fusion proteins and restructuring of the lipid bilayers leading to merger of the latter. At the same time, along with elucidation of the shared properties of fusion processes, it is equally important to uncover their specific features in diverse biological systems characterized by dramatic variations of the regulatory mechanisms, rates, and compositions of the membranes.

This volume brings together reviews on different aspects of membrane fusion. The leading experts present both the broad picture of corresponding research fields and the in-depth discussion of the most recent developments.

A section on fusion and fission of lipid bilayers is followed by three sections discussing different families of processes involving fusion of biological membranes: fusion in viral infections, intracellular fusion, and cell-to-cell fusion in development. The last section on theoretical modeling reviews some numerical approaches used to simulate the fusion reaction and to understand the underlying physics.

The first section presents two chapters that discuss mechanisms of remodeling of lipid bilayers in fusion (J. Nikolaus et al.) and fission (T. J. Pucadyil). J. Nikolaus et al. concentrate on the pathway of lipid bilayer restructuring in fusion with emphasis on the hemifusion stage. They describe a recent direct observation of formation of an extended hemifusion diaphragm between giant lipid vesicles and discuss theoretical approaches providing insight into energetics and kinetics of hemifusion. T. J. Pucadyil focuses on a specific protein, dynamin 1, which drives bilayer fission. In addition to an overview of structural and biochemical properties of dynamin, he discusses a recently developed experimental system, which, for the first time, made possible investigation of the dynamin-driven fission of tension-free membranes under most natural conditions of constant GTP turnover.

We then proceed to three chapters covering fusion mediated by viral proteins. Most viruses including very important human pathogens deliver their genetic material into cells by fusing viral envelope to cell membranes (different enveloped viruses such as vesicular stomatitis virus and HIV discussed in A. Albertini et al. and G. Melikyan, respectively) or fusing cell membranes with each other (nonenveloped viruses, J. Boutilier and R. Duncan). The review by A. Albertini et al. focuses on the pre- and postfusion structures of viral fusion proteins and ultrastructural analysis of viral particles that suggests that fusion requires the cooperative action of a large number of fusion proteins. G. Melikyan explores the distinct stages of viral entry and fusion elucidated at the level of individual virions and describes his recent paradigm-shifting finding that HIV fuses with endosomal membrane rather than with cell's plasma membrane. J. Boutilier and R. Duncan discuss a unique family of fusion proteins of nonenveloped viruses strikingly different from other well-studied fusion proteins and thus reminding us that our current models of how proteins fuse membranes will have to be revised and extended.

The intracellular fusion section of the book focuses on exocytosis, the process of cellular secretion in which substances contained in intracellular vesicles are discharged from the cell. S. Martens and H. McMahon discuss the protein machinery that drives calcium-dependent exocytosis in chromaffin cells and neurons with the special emphasis on the mechanistic role of C2 domain-containing proteins that serve as calcium sensors. M. B. Jackson reviews the electrophysiological data on exocytosis of norepinephrine from chromaffin cells and comes to the conclusion that SNARE-mediated fusion can proceed by a pathway that starts with an opening of a gap-junction-like proteinaceous fusion pore rather than with hemifusion. The chapter by T.-Y. Yoon et al. reviews recent studies of the molecular mechanisms of SNARE-dependent fusion explored using proteoliposomes. The chapter focuses on the small molecule inhibitors of SNARE blockers and state-of-the-art assays for dissection of fusion pathway for a single vesicle fusion event.

Cell-to-cell fusion in development is represented in the volume by two reviews. O. Avinoam and B. Podbilewicz discuss two families of eukaryotic cell–cell fusogens: fusion failure proteins identified in *Caenorhabditis elegans* and syncytins derived from proviruses in mammals. The authors analyze the similarities and distinctions between these cell fusogens and the experimental approaches used to identify and characterize them. E. H. Chen's chapter concentrates on fusion between myoblasts and discusses invasive actin-supported podosome-like structures that initiate this fusion process, essential for the development and repair of skeletal muscles. A detailed view of the fusion site at the ultrastructural level will undoubtedly stimulate significant new insights into fusion mechanism.

Overviews of numerical modeling of membrane fusion are given in two chapters. A. J. Markvoot and S. J. Marrink review the whole range of computational methods used for simulation of membrane fusion and fission. Interestingly, while some of the computations substantiate the hemifusion stalk to fusion pore pathway of fusion suggested by phenomenological theories and supported in experiments on fusion of black lipid membranes and giant lipid vesicles (see the review by J. Nikolaus et al.), other simulations predict alternative scenarios of lipid rearrangement which await their experimental verification. M. Müeller and M. Schick describe the results of a specific computational method where lipid molecules are modeled as amphiphilic block copolymers and water as a hydrophilic homopolymer. This method identifies one of the alternative fusion pathways and the related field-theoretical analysis suggests energies of the structures emerging at the intermediate stages of the fusion reaction.

In general, we believe the book both illustrates the current knowledge on membrane fusion and emphasizes that many important aspects of this exciting research field remain to be understood. For many fusion processes the proteins that mediate membrane rearrangements are yet unidentified. Even for the best characterized fusion proteins we still need to figure out the specific mechanisms by which they initiate and drive lipid bilayer restructuring. The pathway of this restructuring is also an area of active research as illustrated by conflicting data on what comes first in exocytotic fusion, hemifusion, or fusion pore (T.-Y. Yoon et al. vs. M. B. Jackson). Another example is the discrepancy between the currently available experimental data on the hemifusion structures and phenomenological theories of the fusion pathway on one hand (reviewed in J. Nikolaus et al.) and alternative pathways of lipid rearrangements uncovered in the computer simulations on the other (M. Müeller and M. Schick and A. J. Markvoort and S. J. Marrink). Only future work will clarify the universality and biological relevance of the specific fusion motifs discussed in this book.

Leonid V. Chernomordik and Michael M. Kozlov

Previous Volumes in Series

Current Topics in Membranes and Transport

Volume 23 Genes and Membranes: Transport Proteins and Receptors* (1985)
Edited by Edward A. Adelberg and Carolyn W. Slayman

Volume 24 Membrane Protein Biosynthesis and Turnover (1985)
Edited by Philip A. Knauf and John S. Cook

Volume 25 Regulation of Calcium Transport across Muscle Membranes (1985)
Edited by Adil E. Shamoo

Volume 26 Na^+–H^+ Exchange, Intracellular pH, and Cell Function* (1986)
Edited by Peter S. Aronson and Walter F. Boron

Volume 27 The Role of Membranes in Cell Growth and Differentiation (1986)
Edited by Lazaro J. Mandel and Dale J. Benos

Volume 28 Potassium Transport: Physiology and Pathophysiology* (1987)
Edited by Gerhard Giebisch

Volume 29 Membrane Structure and Function (1987)
Edited by Richard D. Klausner, Christoph Kempf, and Jos van Renswoude

Volume 30 Cell Volume Control: Fundamental and Comparative Aspects in Animal Cells (1987)
Edited by R. Gilles, Arnost Kleinzeller, and L. Bolis

Volume 31 Molecular Neurobiology: Endocrine Approaches (1987)
Edited by Jerome F. Strauss, III, and Donald W. Pfaff

Volume 32 Membrane Fusion in Fertilization, Cellular Transport, and Viral Infection (1988)
Edited by Nejat Düzgünes and Felix Bronner

Volume 33 Molecular Biology of Ionic Channels* (1988)
Edited by William S. Agnew, Toni Claudio, and Frederick J. Sigworth

**Part of the series from the Yale Department of Cellular and Molecular Physiology.*

Volume 34 Cellular and Molecular Biology of Sodium Transport* (1989)
Edited by Stanley G. Schultz

Volume 35 Mechanisms of Leukocyte Activation (1990)
Edited by Sergio Grinstein and Ori D. Rotstein

Volume 36 Protein–Membrane Interactions* (1990)
Edited by Toni Claudio

Volume 37 Channels and Noise in Epithelial Tissues (1990)
Edited by Sandy I. Helman and Willy Van Driessche

Current Topics in Membranes

Volume 38 Ordering the Membrane Cytoskeleton Trilayer* (1991)
Edited by Mark S. Mooseker and Jon S. Morrow

Volume 39 Developmental Biology of Membrane Transport Systems (1991)
Edited by Dale J. Benos

Volume 40 Cell Lipids (1994)
Edited by Dick Hoekstra

Volume 41 Cell Biology and Membrane Transport Processes* (1994)
Edited by Michael Caplan

Volume 42 Chloride Channels (1994)
Edited by William B. Guggino

Volume 43 Membrane Protein–Cytoskeleton Interactions (1996)
Edited by W. James Nelson

Volume 44 Lipid Polymorphism and Membrane Properties (1997)
Edited by Richard Epand

Volume 45 The Eye's Aqueous Humor: From Secretion to Glaucoma (1998)
Edited by Mortimer M. Civan

Volume 46 Potassium Ion Channels: Molecular Structure, Function, and Diseases (1999)
Edited by Yoshihisa Kurachi, Lily Yeh Jan, and Michel Lazdunski

Volume 47 Amiloride-Sensitive Sodium Channels: Physiology and Functional Diversity (1999)
Edited by Dale J. Benos

Volume 48 Membrane Permeability: 100 Years since Ernest Overton (1999)
Edited by David W. Deamer, Arnost Kleinzeller, and Douglas M. Fambrough

Volume 49 Gap Junctions: Molecular Basis of Cell Communication in Health and Disease
Edited by Camillo Peracchia

Volume 50 Gastrointestinal Transport: Molecular Physiology
Edited by Kim E. Barrett and Mark Donowitz

Volume 51 Aquaporins
Edited by Stefan Hohmann, Søren Nielsen and Peter Agre

Volume 52 Peptide–Lipid Interactions
Edited by Sidney A. Simon and Thomas J. McIntosh

Volume 53 Calcium-Activated Chloride Channels
Edited by Catherine Mary Fuller

Volume 54 Extracellular Nucleotides and Nucleosides: Release, Receptors, and Physiological and Pathophysiological Effects
Edited by Erik M. Schwiebert

Volume 55 Chemokines, Chemokine Receptors, and Disease
Edited by Lisa M. Schwiebert

Volume 56 Basement Membranes: Cell and Molecular Biology
Edited by Nicholas A. Kefalides and Jacques P. Borel

Volume 57 The Nociceptive Membrane
Edited by Uhtaek Oh

Volume 58 Mechanosensitive Ion Channels, Part A
Edited by Owen P. Hamill

Volume 59 Mechanosensitive Ion Channels, Part B
Edited by Owen P. Hamill

Volume 60 Computational Modelling of Membrane Bilayers
Edited by Scott E. Feller

Volume 61 Free Radical Effects on Membranes
Edited by Sadis Matalon

Volume 62 The Eye's Aqueous Humor
Edited by Mortimer M. Civan

Volume 63 Membrane Protein Crystallization
Edited by Larry DeLucas

Volume 64 Leukocyte Adhesion
Edited by Klaus Ley

Volume 65 Claudins
Edited by Alan S. L. Yu

Volume 66 Structure and Function of Calcium Release Channels
Edited by Irina I. Serysheva

Volume 67 Advances in Adrenergic Receptor Biology
Edited by Qin Wang

PART 1

FUSION AND FISSION OF LIPID BILAYERS

CHAPTER 1

The Pathway to Membrane Fusion through Hemifusion*

Jörg Nikolaus[1], Jason M. Warner[2], Ben O'Shaughnessy[2] and Andreas Herrmann[1]
[1]Department of Biology, Faculty of Mathematics and Natural Sciences I, Humboldt-University Berlin, Berlin, Germany
[2]Department of Chemical Engineering, Columbia University, New York, New York, USA

I. OVERVIEW

Membrane fusion is ubiquitous in life and requires remodeling of two phospholipid bilayers. Many experimental results and theoretical studies support the view that bilayer fusion proceeds through a hemifusion intermediate. This hemifusion diaphragm (HD) is believed to be a radial extension of a first

*Jörg Nikolaus and Jason M. Warner have contributed equally to this chapter.

Current Topics in Membranes, Volume 68 0065-230X/10 $35.00

DOI: 10.1016/B978-0-12-385891-7.00001-5

stalk-like connection between the two contacting bilayers. The HD separates the aqueous contents and is formed by the distal (inner) leaflets, while the proximal (outer) leaflets are fused. Growth of the HD may promote the formation of a fusion pore within the HD, which completes the fusion process. In an overview, we present evidence supporting the proposed fusion through hemifusion pathway for fusion of artificial lipid bilayers as well as for biological membranes, driven by synthetic fusogenic molecules or viral or intracellular fusion proteins, respectively. Theoretical approaches are presented and discussed, including continuum models and computer simulations, which provide a rationale to understand the energetics and kinetics of the formation of the HD as well as their dynamic, size and stability. In particular, a quantitative model of the kinetics of the growing HD and the hemifusion equilibrium was developed, showing that bilayer tension drives HD expansion. Interleaflet tension was found to act as an opposing force, because HD growth involves a compression of the outer leaflets. The model and its predictions fit nicely with observations on giant unilamellar vesicles (GUVs) with reconstituted fluorescently labeled transmembrane peptides. Using fluorescence microscopy it was possible to follow the inevitable sequestering of these peptides upon HD formation.

II. INTRODUCTION

A. Membrane-Separated Cellular Units

An early step in evolution was the self-assembly of simple amphiphilic molecules to supramolecular structures yielding a bilayer-like assembly (Apel, Deamer, & Mautner, 2002). Since then these first basic cell-like units have much evolved. Ever since nature came up with lipid membranes separating cellular processes, it became also necessary to forge a rapid and regulated membrane fusion mechanism required to maintain cellular functions. Membrane fusion occurs in cell–cell fusion during fertilization and also in cell growth and tissue genesis, intracellular vesicular trafficking in exocytosis and protein trafficking, neurotransmitter release and hormone secretion, and also in tumorgenesis and infection by enveloped viruses (Earp, Delos, Park, & White, 2005; Jahn, Lang, & Sudhof, 2003; Kielian & Rey, 2006; Marsh & Helenius, 2006; Mohler et al., 2002; Shemer & Podbilewicz, 2003; Sudhof & Rothman, 2009; Ungermann & Langosch, 2005).

These membrane fusion events are mediated by a specialized set of proteins overcoming the repulsive force of the hydration energy between the polar heads of two approaching lipid bilayers (Rand & Parsegian, 1984). The proteins seem to be able to supply the energy for the lipid rearrangement necessary to merge the two apposing membranes. They are anchored in the membrane via a transmembrane domain (TMD) and establish a connection to the opposing bilayer either by insertion of a hydrophobic anchor peptide as in viral fusion (Harrison, 2008) or by zippering up with yet another protein anchored already in the opposed membrane by a TMD as in intracellular fusion events mediated by SNARE (soluble *N*-ethylmaleimide-sensitive factor attachment protein receptor) proteins (Sudhof & Rothman, 2009). For a detailed introduction of viral and intracellular fusion proteins, see chapters 3 and 7 within this book. Although the fusion proteins are presumably present close to the emerging fusion pore, their membrane anchors do not seem to exclusively line the pore edges (Xu, Zhang, Su, McNew, & Shin, 2005). The fusion pore connecting the two membrane enclosed compartments was found to be dominated by lipids (lipidic pore model) (Chernomordik & Kozlov, 2003). This is in contrast to the proposed proteinaceous fusion pore model. Here the entire connection between two fusing bilayers would be first established by transmembrane proteins building a pore in each of the juxtaposed membranes that subsequently align to establish a channel-like gap junction connecting both luminal reservoirs (Jackson & Chapman, 2006; Lindau & Almers, 1995; Tse, Iwata, & Almers, 1993). The edges of the pore would first be lined by proteins and only in a subsequent step lipids would invade the area in-between the protein subunits and remodel the two bilayers into one. The questions of whether and how lipids could enter this arrangement and how the double ring of proteins forming the gap junction-like pore would disassemble are not clear (Jahn & Grubmuller, 2002; Jahn & Scheller, 2006). Problematic about this model is also that either entirely hydrophobic TMDs would be exposed on one side to an aqueous polar environment in the open channel-like ensemble or an amphiphilic protein segment would be buried in the hydrophobic lipid environment before channel formation. Both situations would be energetically unfavorable. EPR data, however, showed that the transmembrane helix resides in the nonpolar bilayer environment (Xu et al., 2005).

B. One Common Fusion Pathway – Fusion through Hemifusion

Although the protein machinery accomplishing membrane fusion in the different systems is quite distinct, the underlying pathway concerning lipid rearrangements within the bilayers seems to be similar within the three classes of developmental, intracellular, and viral fusogen systems (Sapir, Avinoam, Podbilewicz, & Chernomordik, 2008). Membrane fusion of biological membranes as well as of artificial protein-free bilayers is primarily of lipidic character—following the lipidic pore model, which will be reviewed below. Many experimental insights about the lipid remodeling in the fusion process came from early studies of protein-free liposome and flat bilayer fusion in the presence of, for example, calcium ions or poly(ethylene glycol) (PEG) or from vesicle systems with high membrane curvature (Chernomordik, Melikyan, & Chizmadzhev, 1987; J. Lee & Lentz, 1997) since protein-free bilayers do not fuse when just brought in very close contact (Chernomordik & Kozlov, 2003). In biological membranes, fusion proteins bring the membranes into close apposition and prime the juxtaposed membranes for fusion (see Section V.A).

Many experimental findings support the view that lipid bilayers fuse via the hemifusion pathway (Fig. 1), which was proposed based on the theoretical work of Helfrich (1973). Kozlov and Markin (1983) were the first to suggest this mechanism. It was proposed that after close approach and the formation of a point-like protrusion (Fig. 1CII) (Chernomordik & Kozlov, 2003; Efrat, Chernomordik, & Kozlov, 2007) a subsequent deformation of two bilayers establishes a first connection between the two proximal lipid bilayer leaflets after dehydration of the initial site of contact (Chernomordik & Kozlov, 2003; Kozlov & Markin, 1983). This so-called fusion stalk (Fig. 1BI) would comprise the two highly curved proximal monolayers which may subsequently expand radially into an extended HD, with the distal membrane leaflets and the aqueous luminal contents remaining separated (Fig. 1AII, BII, and CIII) (Chernomordik et al., 1987; Chernomordik & Kozlov, 2003; Melikyan, White, & Cohen, 1995). Opening of an aqueous fusion pore within the HD may then establish a connection between the opposing, formerly separated compartments (Fig. 1AIII and BIII). However, once hemifusion is established, complete fusion is not inevitable. Besides a further, irreversible enlargement of a small fusion pore, also closing of early fusion pores was described. Even reversibility of hemifusion intermediates was found for some artificial systems, resulting again in the initial state of two separate bilayers (Chanturiya, Chernomordik, & Zimmerberg, 1997; Chernomordik & Kozlov, 2008; Giraudo et al., 2005; Lentz, Malinin, Haque, & Evans, 2000; Melikyan, Niles, & Cohen, 1993). For intracellular fusion events also a transient opening of a small pore releasing the vesicle content was observed by capacitance measurements (Alvarez de Toledo, Fernandez-Chacon, & Fernandez, 1993; Klyachko & Jackson, 2002; Neher & Marty, 1982) and amperometry (Alvarez de Toledo, Fernandez-Chacon, & Fernandez, 1993; C. T. Wang et al., 2003). This was termed "kiss-and-run mechanism", in which the vesicle would pull back and seal itself again instead of merging completely with the plasma membrane.

Much evidence supports the widely held view that hemifusion is a key intermediate, not only in protein-free model membrane systems but also in the biological fusion pathways of viral fusion (Armstrong, Kushnir, & White, 2000; Chernomordik, Frolov, Leikina, Bronk, & Zimmerberg, 1998; Chernomordik & Kozlov, 2003; Cleverley & Lenard, 1998; Kemble, Danieli, & White, 1994; Melikyan, Lin, Roth, & Cohen, 1999; Nussler, Clague, & Herrmann, 1997) and intracellular fusion (Chernomordik et al., 1993;

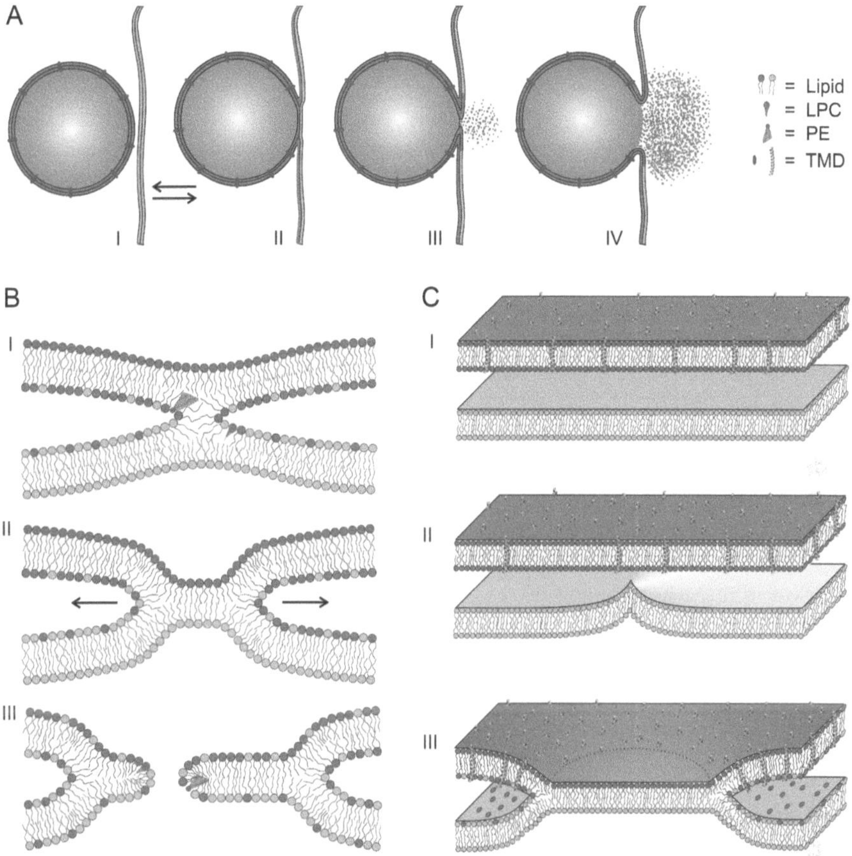

FIGURE 1 Fusion through hemifusion pathway. (A) Shows a sequence of membrane fusion with approach and reversible attachment of two bilayers (I), formation of a hemifusion intermediate, with continuous outer membrane leaflets allowing for lipid redistribution (II), reversible opening of a first small fusion pore (III) that irreversibly expands and allows for content mixing (IV). (B) Detailed view of the possible hemifusion intermediates: fusion stalk with a negative curvature (I), radially expanding hemifusion diaphragm (II), and opening of a fusion pore with a positive curvature (III). Note that a cone-shaped lipid like PE and the inverted-cone-shape of, for example, LPC either enhance or inhibit the formation of the respective intermediate due to the required curvature of these structures. Also, tilting of the hydrocarbon chains allows filling the hydrophobic interstices that would otherwise cause a huge energy deficit. (C) Illustrates the sequestering of transmembrane peptides from the forming HD. Note, that in the contacting membrane leaflet lipid mixing occurs, while distal leaflet mixing only takes place after HD rupture (shown in B.III).

Giraudo et al., 2005; Hofmann et al., 2006; Jahn & Scheller, 2006; Jun & Wickner, 2007; Liu, Wang, Chapman, & Weisshaar, 2008; Lu, Zhang, McNew, & Shin, 2005; Reese, Heise, & Mayer, 2005; Rizo, 2006; Xu et al., 2005; Yoon, Okumus, Zhang, Shin, & Ha, 2006), and recently also in developmental fusion (Podbilewicz et al., 2006) evidence for a hemifusion intermediate was presented. Since the hydrophobic effect drives the self-assembly of lipids and provides membranes with an immense stability (Tanford, 1980), it is thought that specialized fusion proteins are required to drive biological membrane fusion (Chernomordik, Zimmerberg, & Kozlov, 2006; Graham & Kozlov, 2010; Harrison, 2008; Kozlov, McMahon, & Chernomordik, 2010). Therefore, it was proposed that lowering this energy barrier for membrane merger is accomplished by fusogenic proteins (Chernomordik & Kozlov, 2008; Jahn et al., 2003) using either a hairpin or a zipper-like mechanism in the viral or intracellular fusion pathway, respectively (see Chapters 3 and 7 of this book). In SNARE-mediated fusion a zipper-like mechanism is thought to lead to the formation of a stable four-helix bundle, which brings the two membranes together (Jahn & Scheller, 2006). The free energy released in the formation of the helix bundle is believed to contribute significantly to membrane fusion (Sutton, Fasshauer, Jahn, & Brunger, 1998). These binding energies were recently determined to be about 35 k_BT for an intermediate SNARE complex that was only about 60–80% assembled. This roughly corresponds to the energy that was proposed to be necessary to initiate the fusion of the outer bilayer leaflets, that is, hemifusion (F. Li et al., 2007; Rizo & Dai, 2007).

III. CRITERIA AND ASSAYS TO IDENTIFY HEMIFUSION

A. Hemifusion Diaphragm Definition

The HD separates two membrane enclosed compartments only by a single bilayer formed by the two formerly inner membrane leaflets with the outer leaflets fused and pushed outwards by the forming HD. Whereas the definition of hemifusion is quite simple, the experimental verification and characterization of the HD turned out to be difficult due to its normally transient nature and very small size. Experimentally verifiable criteria for a hemifusion intermediate (a stalk or an extended HD) are lipid mixing before content mixing, redistribution of outer leaflet lipids before inner ones, dependence on lipids with a certain intrinsic curvature and a sequestering of transmembrane proteins upon formation of an HD (Fig. 1). Especially in biological membrane fusion HD detection is difficult. For example, lipid mixing due to outer leaflet merger may be restricted by a fence of membrane anchors of the fusion proteins leading to a so-called "restricted" hemifusion intermediate as it was proposed in case of the viral fusion protein hemagglutinin (HA) of influenza virus. Here, although a continuous membrane connection was established, no lipid mixing was observed (Chernomordik et al., 1998). Furthermore, hemifusion was difficult to detect since formation of an HD and its conversion into a fusion pore were found to be quite rapid events. To slow down the transition from hemifusion to complete fusion and thus increase the chance for the observation of the hemifusion intermediate the number of fusion proteins can be reduced, their membrane anchor can be modified, or temperature can be reduced (Chernomordik et al., 2006).

B. Hemifusion Assays

The existence of a hemifusion intermediate in biological and model membrane fusion is supported by a number of observations. Redistribution of lipids between two hemifused membranes can be followed experimentally by a fluorescence-dequenching assay (Struck, Hoekstra, & Pagano, 1981). This assay can, however, only provide indirect evidence for the formation of a hemifusion intermediate (Langosch et al., 2001). Incorporation of a water-soluble dye into a liposome enables to determine that lipid mixing (redistribution of lipid-bound fluorophores) precedes content mixing (J. Lee & Lentz, 1997; Wilschut, Duzgunes, Fraley, & Papahadjopoulos, 1980). Using such an assay detecting lipid mixing and content release, it could recently be shown in a single-vesicle fusion assay driven by reconstituted SNAREs that even very fast fusion events, previously thought to result from prompt full fusion, are in fact productive hemifusion events thus proceeding through a hemifusion intermediate (T. Wang, Smith, Chapman, & Weisshaar, 2009). Fusion in biological systems was investigated using similar lipid mixing assays studying fusion between viruses and liposomes (Stegmann, Hoekstra, Scherphof, & Wilschut, 1985), fusion protein (e.g., HA) expressing cells and red blood cells (RBCs) (L. V. Chernomordik, Leikina, Frolov, Bronk, & Zimmerberg, 1997; Frolov, Cho, Bronk, Reese, & Zimmerberg, 2000; Kemble et al., 1994; Melikyan et al., 1995) and cell–cell fusion (Cleverley & Lenard, 1998). An elegant lipid mixing assay was used by Nüssler et al. (1997) to show hemifusion between HA expressing cells and RBCs by specifically labeling the inner and outer membrane leaflet of RBCs solely with different fluorescent lipid analogues, thus able to determine hemifusion by an exclusive movement of the outer leaflet dye. Observation of single virus particles fusing to larger and therefore flat membrane patches provided information about distinct fusion intermediates like the time points of HD and fusion pore formation, but still not able to visualize the HD itself (Floyd, Ragains, Skehel, Harrison, & van Oijen, 2008; Imai, Mizuno, & Kawasaki, 2006; Joo, Tai, Lee, Wong, & Wang, 2010; Melikyan, Barnard, Abrahamyan, Mothes, & Young, 2005). A first observation of a hemifusion intermediate in biological fusion was reported by Zampighi et al. (2006) using conical thin sectioning and electron tomographic analysis of cortical synapses. A drawback of this method, which provides a high optical resolution, is that no kinetic information can be obtained.

IV. EXPERIMENTAL EVIDENCE FOR HEMIFUSION

A. Hemifusion and Fusion Depend on the Lipid Composition

In biological membranes and artificial bilayers, the lipid composition of the bilayer influences the formation of a hemifusion intermediate. As reviewed in detail by Chernomordik and Kozlov (2003), the energy required for stalk formation is predicted by models to be dependent on the spontaneous curvature of the fusing membrane leaflets. Addition of inverted cone-shaped lipids like LPC, with a positive spontaneous curvature arching in direction of the lipid head group when forming a lipid monolayer, to the contacting monolayer were found to inhibit the formation of hemifusion intermediates (Fig. 1BI) of protein-free bilayers (Chernomordik, Chanturiya, Green, & Zimmerberg, 1995; Chernomordik & Kozlov, 2003), of biological membranes in viral (Chernomordik et al., 1993; Gaudin, 2000; Melikyan et al., 2000; Russell, Jardetzky, & Lamb, 2001; Yeagle, Smith, Young, & Flanagan, 1994), and intracellular fusion (Chernomordik, 1996) as well as in a reduced SNARE-like model system (Robson Marsden, Elbers, Bomans, Sommerdijk, & Kros, 2009). In contrast, cone-shaped lipids with a negative spontaneous curvature like phosphatidylethanolamine (PE) arching toward the lipid fatty acids in a lipid monolayer is promoting the formation of the hemifusion intermediate in the contacting membrane leaflet (Fig. 1BI) again found in protein-free (Chernomordik & Kozlov, 2003) as well as in biological membrane fusion events (Chernomordik, 1996; Chernomordik et al., 1997; Gaudin, 2000). Interestingly, PE was very recently found to reduce the number of SNARE complexes needed to support membrane fusion from approximately eight to only three in a model system with reconstituted SNAREs observing single-vesicle fusion events, although PE reduced the probability of fusion (Domanska, Kiessling, & Tamm, 2010). For the formation of a fusion pore within the HD, which is lined with polar lipid head groups (Chernomordik et al., 1987) the lipid profile properties are reversed. Here, the inverted cone-shaped LPC was found to promote the pore opening when incorporated in the distal monolayer (Fig. 1BIII) in viral fusion for HA (Chernomordik et al., 1998; Melikyan, Brener, Ok, & Cohen, 1997; Razinkov, Melikyan, Epand, Epand, & Cohen, 1998) and G protein (Gaudin, 2000) and for SNARE-mediated fusion (Amatore et al., 2006; Grote, Baba, Ohsumi, & Novick, 2000).

B. Hemifusion in Biological Membranes

Stable transmembrane anchoring of the fusion proteins in the membranes is crucial for their functionality. Trans-SNARE pairs are both anchored in the apposed bilayers by TMDs. Replacing the TMD by a more flexible lipid anchor such as geranylgeranyl or glycosylphosphatidylinositol (GPI) residing in the contacting monolayer allows only for lipid mixing but impedes fusion pore formation. These observations suggest that the fusion pathway is arrested in the hemifusion intermediate. This was found for biological fusion in the intracellular fusion pathways (Grote et al., 2000; Y. Wang, Dulubova, Rizo, & Sudhof, 2001) and also in model systems for reconstituted SNAREs (McNew, Weber, Engelman, Sollner, & Rothman, 1999). Bilayer contact and HD formation might be driven by SNARE protein-induced bending stresses transmitted by a linker region from the four-helix bundle to the transmembrane anchor within the membrane. Note that fusogens can only do so if their bending rigidity exceeds the rigidity of the bilayer that has to be bent in order to establish membrane contact. Typically, lipid bilayer bending rigidity is in the order of 10 kcal/mol (Chernomordik et al., 2006), whereas protein stiffness of such a linker region has not been measured. By molecular dynamic simulations the partially unstructured linkers connecting the SNARE zipper region and the transmembrane anchor could be shown to provide a mechanical coupling that may reduce the energy barrier for fusion (Knecht & Grubmuller, 2003).

Additionally to mere anchoring leading to an apposition of the membranes destined to fuse, evidence is accumulating that the membrane anchors also ensure a destabilization of the bilayers (Abdulreda, Bhalla, Chapman, & Moy, 2008; Kesavan, Borisovska, & Bruns, 2007; McNew et al., 2000; Siddiqui et al., 2007) facilitating lipid rearrangement on the way to fusion pore formation within the HD. Isolated peptides derived from the TMD of SNARE proteins alone were shown to promote membrane fusion (Hofmann et al., 2006; Langosch et al., 2001; Siegel et al., 2006). Insertion of the TMD in a 35–55° angle relative to the membrane normal depending on the lipids in the bilayer might account for the lipid mixing activity by perturbing the lipid arrangement within the bilayer (Bowen & Brunger, 2006; Thomas & Brasseur, 2006). Finally, SNAREs seem to recruit certain lipids like diacylglycerol (DAG) interrupting the bilayer integrity or phosphoinositides and phosphatidic acid (PA) for a synergetic assembly of the fusion competent SNARE complexes (Das & Rand, 1984; Mima, Hickey, Xu, Jun, & Wickner, 2008; Siegel et al., 1989; Vicogne et al., 2006).

A rather detailed scenario how a HD intermediate can evolve can be drawn from dynamic interaction of viral fusion proteins with membranes along their fusion mediating conformational changes. In viral fusion the contact to the target cell membrane is established by insertion of the fusion peptide, a 10–30 amino acid long protein part, either N-terminal or within a protein loop. For influenza HA the fusion peptide was found to be a kinked, amphipathic helix, located at the interface between the lipid head groups and the hydrophobic hydrocarbon region upon insertion (Han, Bushweller, Cafiso, & Tamm, 2001; Q. Huang, Chen, & Herrmann, 2004; Vaidya, Huang, & Takagi, 2010). As the fusion peptide of HA is one of the most conserved regions of the protein it is not surprising that its boomerang-like structure is intimately connected with the fusion process. Combining nuclear magnetic resonance (NMR) and electron paramagnetic resonance (EPR) data it was found that at pH 5 a realignment of charged residues within the peptide perturbs the lipid packing and might thus facilitate formation of an HD between the fusing membranes (Han et al., 2001; Tamm, 2003). At pH 5 a conformational change within the HA protein is triggered that is believed to initiate membrane merger (R. T. C. Huang, Rott, & Klenk, 1981; Korte et al., 1997; Krumbiegel, Herrmann, & Blumenthal, 1994; Wilson, Skehel, & Wiley, 1981). How membrane fusion is able to relax the stress generated by the peptide-induced perturbations still has to be established (Chernomordik et al., 2006). The binding energy of the HA fusion peptide at neutral pH was estimated to be $\sim$15 kT resulting in a force of $\sim$20 pN to pull back a peptide of $\sim$3 nm in length (Ishiguro, Matsumoto, & Takahashi, 1996; Kozlov & Chernomordik, 1998). This provides a sufficient attachment of the viral fusion proteins for downstream fusion processes. Intercalation of the hydrophobic residues of the amphipathic fusion peptide into the contacting juxtaposed monolayer (Gibbons et al., 2004; Heldwein et al., 2006; Roche, Bressanelli, Rey, & Gaudin, 2006) probably introduces membrane curvature, bending the endosomal membrane toward viral membrane similarly to curvature mediated by BAR domains (Zimmerberg & Kozlov, 2006). Assuming a ring-like alignment of several fusion protein trimers, many fusion peptides would then be plunged into the target membrane interface upon fusion protein activation causing the formation of a protein-free membrane dimple stressed at its tip (Chernomordik & Kozlov, 2005; Gibbons et al., 2004; Kozlov & Chernomordik, 1998; Weissenhorn, Hinz, & Gaudin, 2007). An oblique insertion angle between 25° and 38° relative to the membrane normal was found for the HA fusion peptide by NMR and EPR (Han et al., 2001; Y. Li et al., 2005; Macosko, Kim, & Shin, 1997). Coarse-grained MD simulation found comparable peptide orientations of $\sim$30° relative to the membrane normal (Lague, Roux, & Pastor, 2005; Vaidya et al., 2010). It should be pointed out that the oblique insertion angle of the fusion peptide is matching the orientation found for the transmembrane anchor, emphasizing a possible similarity in function of inducing curvature and thus bending the membrane dimples toward one other and priming the opposing membranes for the formation of a hemifusion intermediate.

Similar to SNARE proteins also viral fusogens are anchored by a single transmembrane anchor in the viral membrane. Replacing the TMD by a lipid anchor, for example, by GPI, also arrests fusion at hemifusion (Kemble et al., 1994; Melikyan et al., 1995; Nüssler et al., 1997; Salzwedel, Johnston, Roberts, Dubay, & Hunter, 1993; Tong & Compans, 2000; Weiss & White, 1993). Also mutating of semiconserved residues or shortening of the TMD below a critical length that ensures contact with both bilayer leaflets inhibits complete fusion (Armstrong et al., 2000; Cleverley, Geller, & Lenard, 1997; Cleverley & Lenard, 1998; Kozerski, Ponimaskin, Schroth-Diez, Schmidt, & Herrmann, 2000; Melikyan et al., 1995, 1999; Schroth-Diez, Ponimaskin, Reverey, Schmidt, & Herrmann, 1998). However, in most viral families the short cytoplasmic tail (CT) being C-terminal of the TMD seems not to be required for fusion (Melikyan et al. (1999) detailed review see Schroth-Diez et al. (2000)).

Notably, studies using model peptides of the influenza HA TMD showed an increased acyl chain ordering associated with a dehydration of the bilayer surface (Tamm, 2003; Tatulian & Tamm, 2000) as it was also shown for the fusion peptide (see below) of HA (Han, Steinhauer, Wharton, & Tamm, 1999). Bilayer surface dehydration was proposed to facilitate the formation of a hemifusion intermediate as it reduces the hydration repulsion between the apposing bilayers (Tamm, 2003).

When the structural rearrangement within the trimer, for example, of HA is completed the linker regions of the fusion peptide to the coiled-coil and N-terminal of the TMD interact and form a N-cap structure (Borrego-Diaz, Peeples, Markosyan, Melikyan, & Cohen,, 2003; Chen, Skehel, & Wiley, 1999; Harrison, 2008; Park, Gruenke, & White, 2003). This interaction of transmembrane anchor and fusion peptide was proposed to support the hemifusion to fusion transition and open up an expanding fusion pore within the HD (Melikyan et al., 2000; Qiao, Armstrong, Melikyan, Cohen, & White, 1999; Weissenhorn, Dessen, Harrison, Skehel, & Wiley, 1997).

C. Hemifusion in Artificial Bilayer Systems

Artificial membrane systems seem to offer an easier way to control the parameters influencing the fusion pathway such as lipid composition, membrane curvature, lateral membrane tension, and intermembrane distance. Fusion experiments with artificial lipid membranes were carried out using a variety of bilayers systems. Formation of a fusion stalk-like structure was found after reducing the hydration of a stack of parallel phospholipid bilayers judged by the electron density distribution from X-ray diffraction (Yang & Huang, 2002, 2003). A nonlamellar stalk structure was also proposed by Leikin, Kozlov, Chernomordik, Markin, and Chizmadzhev (1987) supported by experiments using planar lipid bilayers (reviewed in Chernomordik et al. (1987)). Fusion of planar lipid membranes with liposomes was already demonstrated in 1980 (Cohen, Zimmerberg, & Finkelstein, 1980; Zimmerberg, Cohen, & Finkelstein, 1980) whereas only later the hemifusion intermediate could be detected using an advanced experimental setup combining simultaneous electrical and fluorescence measurements (Chanturiya et al., 1997). An analogue setup was used to demonstrate the regulatory influence of the lipid composition on the formation of an HD (Chernomordik et al., 1995). Studying PEG-mediated liposome–liposome fusion by using light scattering, fluorescence lifetime measurements of a fluorescent membrane probe, and an aqueous fluorescent content marker, Lee and Lentz observed vesicle aggregation and rapid lipid mixing of the two contacting outer membrane leaflets which was followed by the formation of a semistable intermediate separating the two vesicle compartments. The authors suggested that these findings might be explained with the formation of a fusion stalk allowing for rapid outer leaflet lipid mixing while the formation of stable vesicle aggregation would correspond to an extended HD (J. Lee & Lentz, 1997). Hemifused bilayers have also been observed as metastable or stable intermediates in fusion of large unilamellar vesicles (LUVs) (Kuhl et al. (1996) and ref. therein). A very recent study on LUV fusion comprising reconstituted SNAREs identified lipid merger without content mixing and thus hemifusion, but was also able to show that the SNARE proteins alone were able to expand the fusion pore within the HD to a size large enough to transmit ~11 kDa cargo reporter molecules (Diao et al., 2010).

Whereas stalk and HD are too small to be optically resolved directly in the above-mentioned artificial systems as well as in biological membranes, GUVs have proven to be an excellent tool to visualize the fusion process as shown previously (Haluska et al., 2006; Heuvingh, Pincet, & Cribier, 2004; Lei & MacDonald, 2003, 2008; Pantazatos & MacDonald, 1999). Due to their remarkably large size with a diameter of 1—100 μm (Walde, Cosentino, Engel, & Stano, 2010), giant vesicles are easily visible by light and fluorescence microscopy. The electroformation method is a reliable technique for the preparation of GUVs (Angelova & Dimitrov, 1986) and allows for the preparation of GUVs with a defined lipid composition (Veatch & Keller, 2003) and the reconstitution of transmembrane peptides (Stockl, Nikolaus, & Herrmann, 2010). Using GUVs labeled with about 1 mol% of fluorescently tagged transmembrane peptides, the formation and expansion of an HD could be investigated (Nikolaus, Stöckl, Langosch, Volkmer, & Herrmann, 2010). For these experiments GUVs were prepared from a mixture of unsaturated phospholipids DOPC:DOPE:DOPS (3:1:1, molar ratio) (Nikolaus et al., 2010). This mixture resembles the major fraction of phospholipids in intracellular membranes, in particular of the Golgi, and in the cytoplasmic leaflet of the plasma membrane of mammalian cells, the major sites for fusion processes in the cell (Holthuis & Levine, 2005; Holthuis, van Meer, & Huitema, 2003). Furthermore, this lipid composition has been shown previously to be optimal for TMD peptide-mediated fusion in studies using smaller liposomes (Gurezka, Laage, Brosig, & Langosch, 1999; Hofmann et al., 2004; Ollesch et al., 2007). Different rhodamine-tagged transmembrane peptides or the fluorescent lipid analogue N-NBD-PE were reconstituted into the GUV membrane during vesicle preparation. Peptide and lipid analogue were both homogenously distributed in the membrane of the respective GUVs.

Aggregation of GUVs containing PS was achieved by addition of divalent cations at 2 mM Ca^{2+} or Mg^{2+} (Papahadjopoulos et al., 1977; Papahadjopoulos, Nir, & Duzgunes, 1990). After raising the Ca^{2+} concentration to 6 mM, time resolved imaging revealed that fusion was preceded by displacement of peptides and fluorescent lipid analogues from the GUV–GUV adhesion region beginning at a single starting point and expanding from there on (Fig. 2A). A more detailed analysis of this microscopic visible structure being several micrometers in size revealed that peptides were completely sequestered as expected for the formation of an HD (Fig. 2C). Lateral distribution of lipid analogues also provided strong evidence for the formation of an HD but not for the presence of two adherent bilayers as shown in several different experimental setups (Fig. 2C and cf. Nikolaus et al. (2010)). A structure of rather high fluorescence intensity was formed at the rim of this growing region, which may correspond to a transient enrichment of sequestered molecules (cf. Nikolaus et al. (2010) Fig. 3). Finally, the diaphragm ruptures, very likely at the junction site of the three bilayers at the HD periphery and retracts to the other side (Fig. 2B, last image). These findings are in line with calculations that the lateral tension concentrates along the rim of the HD and is sufficiently strong to rupture the bilayer forming a crack-like fusion pore along the HD rim (Chernomordik & Kozlov, 2003; Kozlovsky, Chernomordik, & Kozlov, 2002). Sequestering of peptides was independent of a specific amino acid sequence as well as a secondary structure and even full length HA was sequestered upon HD formation (Nikolaus et al., 2010). A dependence of the HD size on peptide concentration was found. Increasing concentration decreased HD size, which can be rationalized by a 2D lateral pressure of the sequestered peptides outside the HD pushing onto the expanding HD boundary thus constricting HD growth (Nikolaus et al., 2010). Consisting with previous observations (Pantazatos & MacDonald, 1999), no HD formation was found without the negatively curved phosphatidylethanolamine (PE) (Nikolaus et al., 2010). Presence of lipids with sufficiently negative spontaneous curvature like PE is required since HD formation and growth increases the length of its rim where negative monolayer curvature is large (Chernomordik & Kozlov, 2003; Kozlov & Markin, 1983; Kozlovsky, Efrat, Siegel, & Kozlov, 2004; Kozlovsky & Kozlov, 2002; May, 2002). Driving forces of the HD formation in this model system are, first, the interaction of negatively charged phospholipids with divalent cations (Nikolaus et al., 2010; Pantazatos & MacDonald, 1999) and thus a reduced repulsion and a dehydration of lipid head groups (Feigenson, 1986; Mattai, Hauser, Demel, & Shipley, 1989). Second, divalent cation addition leads to an asymmetric surface area reduction only on the outer leaflet (Chanturiya, Scaria, & Woodle, 2000). HD formation seems to be able to compensate for the evolving surface area difference between the two

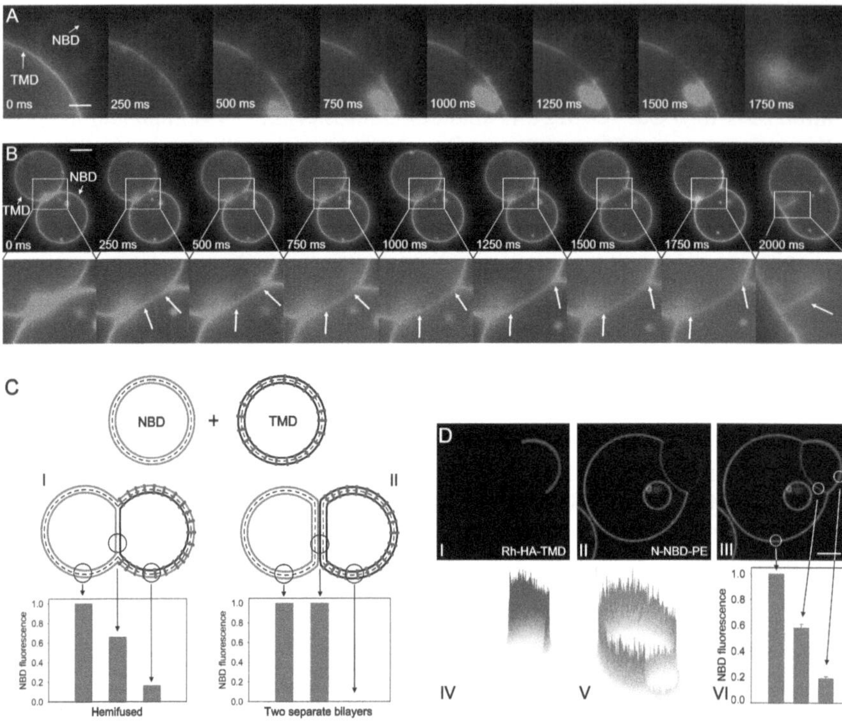

FIGURE 2 Hemifusion of GUVs made from a DOPC:DOPE:DOPS mixture (3:1:1, molar ratio), containing either 1 mol% rhodamine-labeled TMD peptide or 1 mol% N-NBD-PE. (A and B) Sequence of fusion triggered by addition of 6 mM Ca^{2+}. Pairs of GUVs were imaged by fluorescence microscopy. (A) Excitation filter 510/50 nm, emission filter >590 (long pass filter); bright spot in the lower figure part corresponds to fluorescent aggregates inside the large GUV. In the last image, the GUVs disintegrate. (B) Excitation filter 480/20 nm, emission filter >510 nm (long pass filter). Arrows in the magnifications indicate the borders of the growing HD. In the last images, the HD ruptures very likely at the junction site of the three bilayers at the HD periphery and retracts to the opposite side (indicated by an arrow). (C and D) To verify that the evolving structure is in fact an HD and not just two closely adhered membranes from which the TMD peptide is sequestered for steric reasons, the NBD fluorescence intensity of the lipid analogue is analyzed (left GUV labeled with N-NBD-PE; right GUV with reconstituted peptide). Intensity is shown for two possible different structures of the adhesion region: (I) HD; (II) two separate adherent bilayers. While for (II) no N-NBD-PE is found in the peptide-containing GUV, the outer leaflet of the peptide-containing GUV becomes labeled by the analogue for (I). However, NBD intensity is reduced by approximately about 50% due to FRET from NBD to Rh-labeled peptides. (D) GUVs containing the peptide Rh-HA-TMD and N-NBD-PE labeled GUVs were mixed. Mg^{2+} (2 mM) was added to trigger adhesion of GUVs. Distribution of Rh-labeled peptide (I) and N-NBD-PE (II) as well as an overlay (III) are shown. Fluorescence intensity profiles of rhodamine (IV) and NBD (V). N-NBD-PE fluorescence intensity in three different bilayer regions is given in (VI): Region of the NBD-labeled GUV outside the HD (intensity was set to 100%), HD, and region of the peptide-containing GUV outside the HD. Intensity profiles coincide with the fluorescence intensity distribution expected for a HD. Microscopy at 25°C. Bar corresponds to 5 μm. Figure modified from Nikolaus et al. (2010) with permission.

membrane leaflets (Nikolaus et al., 2010). Third, addition of cations results in an increase of membrane tension (Ohki, 1982; Sinn, Antonietti, & Dimova, 2006) also believed to drive hemifusion (Shillcock & Lipowsky, 2005).

The GUV model system allows studying μm-sized HDs, but these dimensions have not been observed *in vivo*. Although reconstituted full-length protein was sequestered in the GUV model system (Nikolaus et al., 2010), the situation in cellular membranes is different. Here membrane proteins are at a much higher density compared to any model system. The required sequestering of densely packed transmembrane proteins upon membrane merger is energetically unfavorable and probably confines HD dimensions. Cytoskeleton anchoring of some membrane embedded proteins might also account for limited HD growth in biological membranes. The addition of cations effectively reduces the outer membrane leaflet and provides thus a major driving force for HD formation in the GUV model system, but whether similar mechanisms are available to cells is not known. In particular, negatively charged phospholipids as PS with a high affinity to Ca^{2+} are essentially localized on the cytoplasmic leaflet of the plasma membrane of eukaryotic cells, while on the exoplasmic leaflet neutral phospholipids as PC and sphingomyelin are enriched.

V. MODELING THE BIOPHYSICAL MECHANISMS OF HEMIFUSION ON THE PATHWAY TO FUSION

Models developed over the past three decades have illuminated many aspects of the pathway to fusion. Most models have described protein-free membrane systems. Biological fusion machinery presumably operates within the constraints of the physical principles established by these models. Continuum models have invoked a small number of measurable physical variables or parameters such as membrane tension or leaflet spontaneous curvature to make simple, testable mechanistic predictions. Computer simulations have aimed to capture the entire fusion pathway including complex intermediate states where leaflets may be discontinuous. Since simulations require considerable computation time they have typically followed fusion pathways involving a small number of lipids (e.g., ~20-nm vesicles).

A. Membrane Contact and Destabilization Are Required to Nucleate the Hemifusion Connection

Membranes must come into contact to allow a hemifusion connection to be nucleated. Evidence suggests that considerable force may be required to drive membrane contact before hemifusion. In multilamellar membrane stacks, it was found that membrane separations less than ~2 nm are strongly opposed by repulsive hydration forces which arise from the affinity of hydrophilic lipid head groups for water (Rand & Parsegian, 1989). Pressures of ~100 atm were required to bring macroscopic mica-supported eggPC bilayers to within ~2 nm (Leckband & Israelachvili, 2001). Thus it has been proposed that membrane contact occurs through small membrane protrusions, which limit the surface area of membrane regions suffering energetic penalties from close proximity (Fig. 1CII) (Efrat et al., 2007; Kuzmin, Zimmerberg, Chizmadzhev, & Cohen, 2001).

Membranes create secure compartments because their disruption would expose hydrophobic lipid tails to the aqueous environment with high free energy cost. Thus, it is thought that membrane contact alone may be insufficient to induce hemifusion. Additional membrane stresses may destabilize and sever the outer monolayers and promote their reconnection in the new configuration of the hemifused state. Such stresses may be generated in supported bilayers by depletion of lipids, which decreases lipid packing density and may expose portions of the membrane hydrophobic layer to induce high surface tension (Leckband & Israelachvili, 2001). Indeed, depletion of bilayers strongly promoted hemifusion and fusion (Helm, Israelachvili, & McGuiggan, 1989). In synthetic vesicle fusion systems, divalent cations create large membrane tensions that are thought sufficient to destabilize bilayers to induce fusion (Arnold, 1995; Ohki, 1982; Parsegian, Rand, & Gingell, 1984). In other synthetic systems, inclusion of lipids of high spontaneous curvature, amphipathic molecules or peptides is thought to destabilize lipid packing and has been shown to promote fusion (Chernomordik et al., 1995; Dennison, Bowen, Brunger, & Lentz, 2006; Lentz, McIntyre, Parks, Yates, & Massenburg, 1992; Murata et al., 1993).

In biological fusion, a mechanism by which fusion proteins may destabilize bilayer structure to induce hemifusion is the local perturbation of lipid packing by the inclusion of protein TMDs or membrane-inserted fusion peptides. Viral fusion proteins and SNAREs have been proposed to use this mechanism (Dennison et al., 2006; Kozlovsky, Zimmerberg, & Kozlov, 2004; Tamm, 2003).

A second possible mechanism of biological fusion machines may entail the use of force produced by protein conformational changes to stress the bilayer. SNAREpins and viral fusion protein hairpins have been proposed to use such a mechanism (Harrison, 2008; F. Li et al., 2007; Sudhof & Rothman, 2009). Few models have addressed this important issue. Kuzmin et al. (2001) predicted that protein forces resulting from conformational changes may locally bend apposing membranes into abutting nipples. The small diameter of the nipples (~20 nm) would facilitate membrane contact since only small membrane regions suffer significant repulsive hydration forces (Kuzmin et al., 2001). Additionally, the authors suggested that bending energies accumulated in the curved nipples promote hemifusion: as compared to flat, unperturbed membranes the nipple configuration has higher energy such that a hemifused state of yet higher energy could be more readily accessible. Similarly, Kozlov and Chernomordik (1998) proposed that a ring of influenza HAs can locally buckle the viral membrane into a highly curved dimple, which could facilitate membrane contact and hemifusion.

A related mechanism of fusion proteins may directly induce membrane curvature. It was proposed that synaptotagmin, the calcium sensor for synaptic transmission, can insert its calcium-binding C2 domains into the membrane to induce positive curvature and locally buckle the plasma membrane to facilitate contact and hemifusion with the synaptic vesicle (Martens, Kozlov, & McMahon, 2007). Another hypothesis described hemifusion induced by a coat of fusion proteins aggregated on the membrane surface (Kozlov & Chernomordik, 2002). The coat would have negative spontaneous curvature and be more rigid than the membrane such that the membrane-coat couple would buckle into an outwardly protruding dimple. The base of the dimple, coated by proteins, would have negative curvature while the uncoated tip of the dimple would have positive curvature and contact the apposing membrane with bending stresses that facilitate hemifusion.

Computer simulations have captured the nonlamellar structures of prehemifusion transition states lying on the fusion pathway between membrane contact and hemifusion. In a number of simulation studies, prehemifusion transition states consisted of one or more lipids straddling the intermembrane gap, with each hydrocarbon tail of the lipids inserted into opposite membranes (Kasson, Lindahl, & Pande, 2010; Marrink, de Vries, & Tieleman, 2009; Smeijers, Markvoort, Pieterse, & Hilbers, 2006; Stevens, Hoh, & Woolf, 2003). Smeijers et al. (2006) found these lipidic bridges were preceded by the formation of a small hydrophobic patch in which lipid head groups had parted to reveal the hydrocarbon tail groups. Although membrane tension was not controlled in this study, such hydrophobic patches are thought to be created by membrane tension (Leckband & Israelachvili, 2001). Interestingly, in the simulations of Shillcock and Lipowsky (2005) membrane tension induced by reduction of lipid packing density promoted hemifusion.

B. *Initial Hemifusion Connection*

The initial hemifused connection between membranes has been proposed to be a stalk, or minimally sized HD (see Fig. 1BI) (Kozlov, Leikin, Chernomordik, Markin, & Chizmadzhev, 1989; Kozlovsky & Kozlov, 2002; Markin & Albanesi, 2002). As compared to extended HD intermediates, stalk-like structures may be more readily accessible from membrane contact since fewer lipid rearrangements are presumably involved. Stalk-like structures have been observed in multilamellar membrane systems by X-ray diffraction (Yang &

Huang, 2002) and in computer simulations of membrane fusion (Kasson et al., 2010; Marrink et al., 2009; Noguchi & Takasu, 2001; Smeijers et al., 2006). However, stalks have not been experimentally observed in fusion systems. This could be because the stalk size is well below optical resolution. Alternatively, stalks may be bypassed altogether, or they may have short lifetimes, possibly able to revert to separate membranes or to rapidly mature into more advanced intermediates such as the extended HD or fusion pore (Arnold, 1995; Leikina & Chernomordik, 2000).

Several calculations of the energy of stalks have examined the feasibility of the proposed structure (Cohen & Melikyan, 2004; Katsov, Muller, & Schick, 2004; Kozlov et al., 1989; Kozlovsky & Kozlov, 2002; Kuzmin et al., 2001; Markin & Albanesi, 2002). The calculation of Kozlovsky and Kozlov (2002), which realistically allowed for lipid tilt in addition to monolayer bending, predicted stalk energies of 40 kT for planar DOPC membranes. These stalks were deemed to be energetically achievable on the basis of electroporation studies, which found energy barriers to bilayer pore formation of ~40 kT were accessible on experimental timescale-of-seconds (Chernomordik & Kozlov, 2003).

Continuum models have found that stalks have net negative curvature and predicted that stalk energies decrease with decreasing spontaneous curvature of the membrane monolayers (Katsov et al., 2004; Kozlovsky & Kozlov, 2002; Markin & Albanesi, 2002). Thus, it was predicted that hemifusion is more likely to be achieved for membranes containing negative curvature lipids such as DOPE, and less likely for membranes containing positive curvature lipids such as LPC (Chernomordik & Kozlov, 2003). Indeed, negative (positive) curvature lipids promoted (blocked) hemifusion in diverse systems mediated by fusion proteins or by divalent cations (Chernomordik et al., 1995, 1997; Liu et al., 2008; Melia, You, Tareste, & Rothman, 2006).

Stalk models provide information about the energy landscape near the transition from membrane contact to nucleation of hemifusion. Stalk energies may represent the rate-limiting energy barrier to achieving hemifusion. However, it is possible that this barrier is instead associated with a prehemifusion transition state involving discontinuous leaflets, which then relaxes into a hemifusion connection (Marrink et al., 2009). Nevertheless stalk energies may set the lower bound for the true energy barrier to hemifusion.

C. Expansion of the Hemifusion Diaphragm

Models have predicted that the stalk is a metastable state, which tends to yield an expanded HD on the pathway to fusion of SUVs (J. Y. Lee & Schick, 2008; Malinin & Lentz, 2004) and GUVs (Warner & O'Shaughnessy, 2011). Indeed, directly observed hemifused intermediates in diverse systems mediated by proteins or divalent cations have been extended HDs (Chernomordik et al., 1987; Kuhl et al., 1996; Leckband & Israelachvili, 2001; Melikyan et al., 1995; Nikolaus et al., 2010; Zampighi et al., 2006). These were often observed to grow radially outward and reach a final equilibrium size (Chernomordik et al., 1987; Kuhl et al., 1996; Leckband & Israelachvili, 2001; Melikyan et al., 1995; Nikolaus et al., 2010), consistent with the proposal that hemifusion originates from narrow HD connections such as the stalk (Kozlovsky & Kozlov, 2002). Extended HDs have been documented in several computer simulations (Kasson et al., 2006; Marrink et al., 2009; Smeijers et al., 2006). Continuum models have attempted to determine the forces that determine the equilibrium hemifused state and the growth kinetics by which equilibrium is reached.

It is apparent from Fig. 1B that membrane leaflets are sharply bent along the HD perimeter. Models have analyzed how the stresses associated with such sharp bends affect the equilibrium HDs formed between tension-free membranes. An early model assumed that due to the sharp bending, the HD rim contains a void space, which cannot be filled by lipid tails (Siegel, 1993). This resulted in calculated energies of extended HD structures, which were prohibitively large (Siegel, 1993). Kozlovsky and Kozlov later predicted that lipids tilt relative to the monolayer plane to fill such void space (Kozlovsky et al., 2002). For infinite membranes under zero tension, they found that curvature stresses along the highly bent HD perimeter determined the equilibrium hemifused state as follows. For membranes of moderately negative spontaneous curvature such as those composed predominantly of DOPC, HD extension is disfavored by curvature stresses, which create an effective line tension along the HD perimeter; thus the stalk is predicted to be the equilibrium hemifused state. By contrast, strongly negative spontaneous curvature, such as that of a predominantly DOPE bilayer, favored indefinite HD extension, whereas membranes of intermediate spontaneous curvature, such as a roughly equal DOPC–DOPE mixture, were predicted to have equilibrium HDs a few nanometers in diameter (Kozlovsky et al., 2002).

Warner and O'Shaughnessy (2011) developed a general model for hemifusion of two vesicles. The authors found that membrane tension drives HD growth since the HD reduces total bilayer area, being a shared vesicle boundary. Divalent cations were predicted to provide additional forces driving HD growth by augmenting membrane tension and by inducing electrostatic forces, which selectively shrink outer leaflets. Since HD growth compresses (expands) the outer (inner) leaflets it is resisted by another form of tension, termed interleaflet tension, which disfavors a real difference between leaflets of a bilayer. Balance of the tensile and cationic forces determined an equilibrium HD of area a few percent of the mean vesicle area for typical conditions in divalent cation-mediated fusion studies. The predictions were in good quantitative agreement with the findings of Nikolaus et al. summarized above (see Section IV.C) where measured HD sizes were ~9% of the mean vesicle area. Warner and O'Shaughnessy predicted that an energy barrier to HD extension from the stalk state is determined by the interplay of the tensile and cationic forces driving growth and the effective line tension of the HD perimeter resisting growth. The calculated energy barrier depended strongly on membrane tension, divalent cation concentration, and lipid composition (Warner & O'Shaughnessy, 2011). This energy barrier, estimated to be ~23 kT for the conditions of Nikolaus et al., was presumably surmounted in those experiments to achieve the equilibrium extended HDs reported.

J. Y. Lee and Schick (2008) found that curvature stresses inherent to small ~20-nm vesicles induce an effective surface tension with drives HD extension. Similarly, Malinin and Lentz (2004) found that decreasing vesicle diameter promotes extension of PEG-mediated HDs between SUVs. These results suggest that HD extension in biological fusion could be driven in part by curvature induced by fusion proteins. It was also found that HD expansion between a vesicle and a planar bilayer could be driven by membrane tension in the planar bilayer (J. Y. Lee & Schick, 2008).

A general model for HD growth kinetics between protein-free vesicles was proposed in which HD growth is resisted by interleaflet friction emerging as outer leaflets slide over inner leaflets to make room for the HD (Warner & O'Shaughnessy, 2011). The authors found that HD growth rates increase with membrane tension and divalent cation concentration, the same forces that determined the equilibrium hemifused state. Predictions for growth of HD area agreed quantitatively with measurements of Nikolaus et al. summarized above (cf. Section IV.C). The authors also predicted that once the energy barrier to HD growth is overcome (see above) the time required to reach hemifusion equilibrium increases in proportion to vesicle area, from milliseconds for ~100 nm vesicles to seconds for micron-scale GUVs. Hed and Safran (2003) modeled the kinetics of HD growth between supported bilayers driven by PEG polymer and resisted by interleaflet friction. Their predictions were in quantitative agreement with the experiments of Kuhl et al. (1996) where ~100 μm HDs grew over ~600 s.

D. The Transition from Hemifusion to the Fusion Pore

It is thought that the HD is ruptured to yield a fusion pore (Fig. 2BIII). Depending on the system, rupture might be driven by one or more of several factors, including the membrane tension of the HD, curvature stresses along the HD perimeter or fusion protein-mediated stresses.

While experiments in several systems have shown that hemifusion precedes fusion, Nikolaus et al. (2010) were able to visualize rupture of the HD, which completed GUV merger. Warner and O'Shaughnessy (2011) calculated HD tensions in this system and found that they were above the lysis threshold; thus, it was proposed that HD rupture was driven by tension. Membrane tension may be important in other systems. Tension generated by osmotic swelling was required for fusion of vesicles with suspended bilayers (Akabas, Cohen, & Finkelstein, 1984; Cohen et al., 1980; Cohen, Akabas, Zimmerberg, & Finkelstein, 1984; Zimmerberg et al., 1980), suggesting that tension drove HD rupture. Osmotically generated tension also drove complete fusion in the cell–RBC system mediated by GPI-HA (Melikyan et al., 1995). Additionally, membrane tension promoted the transition from hemifusion to fusion in simulations (Shillcock & Lipowsky, 2005).

Kozlovsky and Kozlov found that curvature stresses of magnitude larger than the lysis threshold of membrane tension are generated in a narrow region along the perimeter of the HD and proposed that these stresses may drive pore formation (Kozlovsky et al., 2002). Consistent with this, HD rupture along the perimeter has been observed in simulations (Smeijers et al., 2006). However, this mechanism may not be universal since stable end-state HDs with perimeters of several micrometers have been documented in many systems (Chernomordik et al., 1987; Leckband & Israelachvili, 2001; Melikyan et al., 1995; Nikolaus et al., 2010). Addition of positive curvature agents to the distal monolayers of hemifused membranes is known to drive the transition to complete fusion of otherwise end-state hemifusion mediated by Ca^{2+} or HA (Chernomordik et al., 1995, 1997, 1998). This evidence is consistent with a mechanism in which these positive curvature agents destabilized the HDs to drive formation of a bilayer pore of positive curvature within the HD (Chernomordik & Kozlov, 2003).

Membrane tensions measured thus far in biological systems are well below lysis levels, suggesting that one role of fusion proteins may be to locally generate membrane tension or other stresses to drive HD rupture and expansion of the fusion pore. Protein coats on membranes have been proposed to accomplish this: by imposing negative curvature on the contacting membrane leaflets, the coat would favor a larger HD perimeter and thus create an outward pulling force on the HD (Kozlov & Chernomordik, 2002). If HD tension is induced by this pulling force it may drive HD rupture (Kozlov & Chernomordik, 2002). It was proposed that Ca^{2+} drives strong binding of synaptotagmin to the contacting proximal leaflets of an HD, and that the leaflets acquire negative curvature as they bend around the synaptotagmins, helping to drive HD expansion and fusion pore formation (Zimmerberg, Akimov, & Frolov, 2006).

Finally, though not the focus of this chapter, expansion of the fusion pore is thought to be the final step along the fusion pathway. The energetic requirements of this step have been addressed by several continuum models in protein-free or protein-mediated systems (Chizmadzhev, Kuzmin, Kumenko, Zimmerberg, & Cohen, 2000; Cohen & Melikyan, 2004; Kozlov et al., 1989; Kuzmin et al., 2001; Markin & Albanesi, 2002). In general, these models find that pore expansion is resisted by bending stresses along the edge of the pore, and that expansion may be the most energetically demanding stage of the fusion pathway. Membrane tension was predicted to drive pore expansion (Chizmadzhev et al., 2000). The interplay between bending stresses and membrane tension have been predicted to cause opening and closing of flickering pores (Nanavati, Markin, Oberhauser, & Fernandez, 1992), which have been documented experimentally in protein-free (Chanturiya et al., 1997) and biological systems (Monck & Fernandez, 1996).

In conclusion, modeling studies of hemifusion and fusion have made considerable progress toward understanding biophysical mechanisms. However, many basic questions remain unanswered, such as why lipid anchored SNAREs and GPI-HA lead to stable end-state hemifusion, while the full length proteins lead to complete fusion (Grote et al., 2000; Kemble et al., 1994). Understanding such issues will require a more complete and detailed elucidation of the mechanisms driving progress along the pathway to hemifusion and fusion.

References

Abdulreda, M. H., Bhalla, A., Chapman, E. R., & Moy, V. T. (2008). Atomic force microscope spectroscopy reveals a hemifusion intermediate during soluble N-ethylmaleimide-sensitive factor-attachment protein receptors-mediated membrane fusion. *Biophysical Journal, 94*(2), 648–655.

Akabas, M. H., Cohen, F. S., & Finkelstein, A. (1984). Separation of the osmotically driven fusion event from vesicle-planar membrane attachment in a model system for exocytosis. *Journal of Cell Biology, 98*(3), 1063–1071.

Alvarez de Toledo, G., Fernandez-Chacon, R., & Fernandez, J. M. (1993). Release of secretory products during transient vesicle fusion. *Nature, 363*(6429), 554–558.

Amatore, C., Arbault, S., Bouret, Y., Guille, M., Lemaitre, F., & Verchier, Y. (2006). Regulation of exocytosis in chromaffin cells by trans-insertion of lysophosphatidylcholine and arachidonic acid into the outer leaflet of the cell membrane. *ChemBioChem, 7*(12), 1998–2003.

Angelova, M. I., & Dimitrov, D. S. (1986). Liposome electroformation. *Faraday Discussions of the Chemical Society., 81*, 303–311.

Apel, C. L., Deamer, D. W., & Mautner, M. N. (2002). Self-assembled vesicles of monocarboxylic acids and alcohols: Conditions for stability and for the encapsulation of biopolymers. *Biochimica et Biophysica Acta, 1559*(1), 1–9.

Armstrong, R. T., Kushnir, A. S., & White, J. M. (2000). The transmembrane domain of influenza hemagglutinin exhibits a stringent length requirement to support the hemifusion to fusion transition. *Journal of Cell Biology, 151*(2), 425–437.

Arnold, K. (1995). Cation-induced vesicle fusion modulated by polymers and proteins. In R. Lipowsky & E. Sackmann (Eds.), *Handbook of Biological Physics* (Vol. 1, Part 2, Chap. 19, pp. 903–957). Amsterdam: North-Holland.

Borrego-Diaz, E., Peeples, M. E., Markosyan, R. M., Melikyan, G. B., & Cohen, F. S. (2003). Completion of trimeric hairpin formation of influenza virus hemagglutinin promotes fusion pore opening and enlargement. *Virology, 316*(2), 234–244.

Bowen, M., & Brunger, A. T. (2006). Conformation of the synaptobrevin transmembrane domain. *Proceedings of the National Academy of Sciences United States of America, 103*(22), 8378–8383.

Chanturiya, A., Chernomordik, L. V., & Zimmerberg, J. (1997). Flickering fusion pores comparable with initial exocytotic pores occur in protein-free phospholipid bilayers. *Proceedings of the National Academy of Sciences United States of America, 94*(26), 14423–14428.

Chanturiya, A., Scaria, P., & Woodle, M. C. (2000). The role of membrane lateral tension in calcium-induced membrane fusion. *Journal of Membrane Biology, 176*(1), 67–75.

Chen, J., Skehel, J. J., & Wiley, D. C. (1999). N- and C-terminal residues combine in the fusion-pH influenza hemagglutinin HA(2) subunit to form an N cap that terminates the triple-stranded coiled coil. *Proceedings of the National Academy of Sciences United States of America, 96*(16), 8967–8972.

Chernomordik, L. V. (1996). Non-bilayer lipids and biological fusion intermediates. *Chemistry and Physics of Lipids, 81*(2), 203–213.

Chernomordik, L. V., Chanturiya, A., Green, J., & Zimmerberg, J. (1995). The hemifusion intermediate and its conversion to complete fusion: Regulation by membrane composition. *Biophysical Journal, 69*(3), 922–929.

Chernomordik, L. V., Frolov, V. A., Leikina, E., Bronk, P., & Zimmerberg, J. (1998). The pathway of membrane fusion catalyzed by influenza hemagglutinin: Restriction of lipids, hemifusion, and lipidic fusion pore formation. *Journal of Cell Biology, 140*(6), 1369–1382.

Chernomordik, LV., & Kozlov, M. M. (2003). Protein-lipid interplay in fusion and fission of biological membranes. *Annual Review of Biochemistry, 72*, 175–207.

Chernomordik, L. V., & Kozlov, M. M. (2005). Membrane hemifusion: Crossing a chasm in two leaps. *Cell, 123*(3), 375–382.

Chernomordik, L. V., & Kozlov, M. M. (2008). Mechanics of membrane fusion. *Nature Structural & Molecular Biology, 15*(7), 675–683.

Chernomordik, L. V., Leikina, E., Frolov, V., Bronk, P., & Zimmerberg, J. (1997). An early stage of membrane fusion mediated by the low pH conformation of influenza hemagglutinin depends upon membrane lipids. *Journal of Cell Biology, 136*(1), 81–93.

Chernomordik, L. V., Melikyan, G. B., & Chizmadzhev, Y. A. (1987). Biomembrane fusion: A new concept derived from model studies using two interacting planar lipid bilayers. *Biochimica et Biophysica Acta, 906*(3), 309–352.

Chernomordik, L. V., Vogel, S. S., Sokoloff, A., Onaran, H. O., Leikina, E. A., & Zimmerberg, J. (1993). Lysolipids reversibly inhibit Ca(2+)-, GTP- and pH-dependent fusion of biological membranes. *FEBS Letters, 318*(1), 71–76.

Chernomordik, L. V., Zimmerberg, J., & Kozlov, M. M. (2006). Membranes of the world unite! *Journal of Cell Biology, 175*(2), 201–207.

Chizmadzhev, Y. A., Kuzmin, P. I., Kumenko, D. A., Zimmerberg, J., & Cohen, F. S. (2000). Dynamics of fusion pores connecting membranes of different tensions. *Biophysical Journal, 78*(5), 2241–2256.

Cleverley, D. Z., Geller, H. M., & Lenard, J. (1997). Characterization of cholesterol-free insect cells infectible by baculoviruses: Effects of cholesterol on VSV fusion and infectivity and on cytotoxicity induced by influenza M2 protein. *Experimental Cell Research, 233*(2), 288–296.

Cleverley, D. Z., & Lenard, J. (1998). The transmembrane domain in viral fusion: Essential role for a conserved glycine residue in vesicular stomatitis virus G protein. *Proceedings of the National Academy of Sciences United States of America, 95*(7), 3425–3430.

Cohen, F. S., Akabas, M. H., Zimmerberg, J., & Finkelstein, A. (1984). Parameters affecting the fusion of unilamellar phospholipidvesicles with planar bilayer membranes. *Journal of Cell Biology, 98*(3), 1054–1062.

Cohen, F. S., & Melikyan, G. B. (2004). The energetics of membrane fusion from binding, through hemifusion, pore formation, and pore enlargement. *Journal of Membrane Biology, 199*(1), 1–14.

Cohen, F S., Zimmerberg, J., & Finkelstein, A. (1980). Fusion of phospholipid vesicles with planar phospholipid bilayer membranes. II. Incorporation of a vesicular membrane marker into the planar membrane. *The Journal of General Physiology, 75*(3), 251–270.

Das, S., & Rand, R. P. (1984). Diacylglycerol causes major structural transitions in phospholipid bilayer membranes. *Biochemical and Biophysical Research Communications, 124*(2), 491–496.

Dennison, S. M., Bowen, M. E., Brunger, A. T., & Lentz, B. R. (2006). Neuronal SNAREs do not trigger fusion between synthetic membranes but do promote PEG-mediated membrane fusion. *Biophysical Journal, 90*(5), 1661–1675.

Diao, J., Su, Z., Ishitsuka, Y., Lu, B., Lee, K. S., & Lai, Y., et al. (2010). A single-vesicle content mixing assay for SNARE-mediated membrane fusion. *Nature Communications, 1*(5), 1–6.

Domanska, M. K., Kiessling, V., & Tamm, L. K. (2010). Docking and fast fusion of synaptobrevin vesicles depends on the lipid compositions of the vesicle and the acceptor SNARE complex-containing target membrane. *Biophys J, 99*(9), 2936–2946.

Earp, L. J., Delos, S. E., Park, H. E., & White, J. M. (2005). The many mechanisms of viral membrane fusion proteins. *Current Topics in Microbiology and Immunology, 285*, 25–66.

Efrat, A., Chernomordik, L. V., & Kozlov, M. M. (2007). Point-like protrusion as a prestalk intermediate in membrane fusion pathway. *Biophysical Journal, 92*(8), L61–63.

Feigenson, G. W. (1986). On the nature of calcium ion binding between phosphatidylserine lamellae. *Biochemistry, 25*(19), 5819–5825.

Floyd, D. L., Ragains, J. R., Skehel, J. J., Harrison, S. C., & van Oijen, A. M. (2008). Single-particle kinetics of influenza virus membrane fusion. *Proceedings of the National Academy of Sciences United States of America, 105*(40), 15382–15387.

Frolov, V. A., Cho, M. S., Bronk, P., Reese, T. S., & Zimmerberg, J. (2000). Multiple local contact sites are induced by GPI-linked influenza hemagglutinin during hemifusion and flickering pore formation. *Traffic, 1*(8), 622–630.

Gaudin, Y. (2000). Rabies virus-induced membrane fusion pathway. *Journal of Cell Biology, 150* (3), 601–612.

Gibbons, D. L., Vaney, M. C., Roussel, A., Vigouroux, A., Reilly, B., & Lepault, J., et al. (2004). Conformational change and protein-protein interactions of the fusion protein of Semliki Forest virus. *Nature, 427*(6972), 320–325.

Giraudo, C. G., Hu, C., You, D., Slovic, A. M., Mosharov, E. V., & Sulzer, D., et al. (2005). SNAREs can promote complete fusion and hemifusion as alternative outcomes. *Journal of Cell Biology, 170*(2), 249–260.

Graham, T. R., & Kozlov, M. M. (2010). Interplay of proteins and lipids in generating membrane curvature. *Current Opinion in Cell Biology, 22*(4), 430–436.

Grote, E., Baba, M., Ohsumi, Y., & Novick, P. J. (2000). Geranylgeranylated SNAREs are dominant inhibitors of membrane fusion. *Journal of Cell Biology, 151*(2), 453–466.

Gurezka, R., Laage, R., Brosig, B., & Langosch, D. (1999). A heptad motif of leucine residues found in membrane proteins can drive self-assembly of artificial transmembrane segments. *The Journal of Biological Chemistry, 274*(14), 9265–9270.

Haluska, C. K., Riske, K. A., Marchi-Artzner, V., Lehn, J. M., Lipowsky, R., & Dimova, R. (2006). Time scales of membrane fusion revealed by direct imaging of vesicle fusion with high temporal resolution. *Proceedings of the National Academy of Sciences United States of America, 103*(43), 15841–15846.

Han, X., Bushweller, J. H., Cafiso, D. S., & Tamm, L. K. (2001). Membrane structure and fusion-triggering conformational change of the fusion domain from influenza hemagglutinin. *Nature Structural Biology, 8*(8), 715–720.

Han, X., Steinhauer, D. A., Wharton, S. A., & Tamm, L. K. (1999). Interaction of mutant influenza virus hemagglutinin fusion peptides with lipid bilayers: Probing the role of hydrophobic residue size in the central region of the fusion peptide. *Biochemistry, 38*(45), 15052–15059.

Harrison, S. C. (2008). Viral membrane fusion. *Nature Structural & Molecular Biology, 15*(7), 690–698.

Hed, G., & Safran, S. A. (2003). Initiation and dynamics of hemifusion in lipid bilayers. *Biophysical Journal, 85*(1), 381–389.

Heldwein, E. E., Lou, H., Bender, F. C., Cohen, G. H., Eisenberg, R. J., & Harrison, S. C. (2006). Crystal structure of glycoprotein B from herpes simplex virus 1. *Science, 313*(5784), 217–220.

Helfrich, W. (1973). Elastic properties of lipid bilayers: Theory and possible experiments. *Zeitschrift für Naturforschung C, 28*(11), 693–703.

Helm, C. A., Israelachvili, J. N., & McGuiggan, P. M. (1989). Molecular mechanisms and forces involved in the adhesion and fusion of amphiphilic bilayers. *Science, 246*(4932), 919–922.

Heuvingh, J., Pincet, F., & Cribier, S. (2004). Hemifusion and fusion of giant vesicles induced by reduction of inter-membrane distance. *The European Physical Journal E - Soft Matter, 14*(3), 269–276.

Hofmann, M. W., Peplowska, K., Rohde, J., Poschner, B. C., Ungermann, C., & Langosch, D. (2006). Self-interaction of a SNARE transmembrane domain promotes the hemifusion-to-fusion transition. *Journal of Molecular Biology, 364*(5), 1048–1060.

Hofmann, M. W., Weise, K., Ollesch, J., Agrawal, P., Stalz, H., & Stelzer, W., et al. (2004). De novo design of conformationally flexible transmembrane peptides driving membrane fusion. *Proceedings of the National Academy of Sciences United States of America, 101*(41), 14776–14781.

Holthuis, J. C., & Levine, T. P. (2005). Lipid traffic: Floppy drives and a superhighway. *Nature Reviews of Molecular Cell Biology, 6*(3), 209–220.

Holthuis, J. C., van Meer, G., & Huitema, K. (2003). Lipid microdomains, lipid translocation and the organization of intracellular membrane transport (Review). *Molecular Membrane Biology, 20* (3), 231–241.

Huang, Q., Chen, C. L., & Herrmann, A. (2004). Bilayer conformation of fusion peptide of influenza virus hemagglutinin: A molecular dynamics simulation study. *Biophysical Journal, 87*(1), 14–22.

Huang, R. T. C., Rott, R., & Klenk, H. D. (1981). Influenza-viruses cause hemolysis and fusion of cells. *Virology, 110*(1), 243–247.

Imai, M., Mizuno, T., & Kawasaki, K. (2006). Membrane fusion by single influenza hemagglutinin trimers. Kinetic evidence from image analysis of hemagglutinin-reconstituted vesicles. *The Journal of Biological Chemistry, 281*(18), 12729–12735.

Ishiguro, R., Matsumoto, M., & Takahashi, S. (1996). Interaction of fusogenic synthetic peptide with phospholipid bilayers: Orientation of the peptide alpha-helix and binding isotherm. *Biochemistry, 35*(15), 4976–4983.

Jackson, M. B., & Chapman, E. R. (2006). Fusion pores and fusion machines in Ca^{2+}-triggered exocytosis. *Annual Review of Biophysics and Biomolecular Structure, 35*, 135–160.

Jahn, R., & Grubmuller, H. (2002). Membrane fusion. *Current Opinion in Cell Biology, 14*(4), 488–495.

Jahn, R., Lang, T., & Sudhof, T. C. (2003). Membrane fusion. *Cell, 112*(4), 519–533.

Jahn, R., & Scheller, R. H. (2006). SNAREs—Engines for membrane fusion. *Nature Reviews of Molecular Cell Biology, 7*(9), 631–643.

Joo, K. I., Tai, A., Lee, C. L., Wong, C., & Wang, P. (2010). Imaging multiple intermediates of single-virus membrane fusion mediated by distinct fusion proteins. *Microscopy Research and Technique, 73*(9), 886–900.

Jun, Y., & Wickner, W. (2007). Assays of vacuole fusion resolve the stages of docking, lipid mixing, and content mixing. *Proceedings of the National Academy of Sciences United States of America, 104*(32), 13010–13015.

Kasson, P. M., Kelley, N. W., Singhal, N., Vrljic, M., Brunger, A. T., & Pande, V. S. (2006). Ensemble molecular dynamics yields submillisecond kinetics and intermediates of membrane fusion. *Proceedings of the National Academy of Sciences United States of America, 103*(32), 11916–11921.

Kasson, P. M., Lindahl, E., & Pande, V. S. (2010). Atomic-resolution simulations predict a transition state for vesicle fusion defined by contact of a few lipid tails. *PLoS Computational Biology, 6*(6), e1000829.

Katsov, K., Muller, M., & Schick, M. (2004). Field theoretic study of bilayer membrane fusion. *I. Hemifusion mechanism. Biophysical Journal, 87*(5), 3277–3290.

Kemble, G. W., Danieli, T., & White, J. M. (1994). Lipid-anchored influenza hemagglutinin promotes hemifusion, not complete fusion. *Cell, 76*(2), 383–391.

Kesavan, J., Borisovska, M., & Bruns, D. (2007). v-SNARE actions during Ca(2+)-triggered exocytosis. *Cell, 131*(2), 351–363.

Kielian, M., & Rey, F. A. (2006). Virus membrane-fusion proteins: More than one way to make a hairpin. *Nature Reviews Microbiology, 4*(1), 67–76.

Klyachko, V. A., & Jackson, M. B. (2002). Capacitance steps and fusion pores of small and large-dense-core vesicles in nerve terminals. *Nature, 418*(6893), 89–92.

Knecht, V., & Grubmuller, H. (2003). Mechanical coupling via the membrane fusion SNARE protein syntaxin 1A: A molecular dynamics study. *Biophysical Journal, 84*(3), 1527–1547.

Korte, T., Ludwig, K., Krumbiegel, M., Zirwer, D., Damaschun, G., & Herrmann, A. (1997). Transient changes of the conformation of hemagglutinin of influenza virus at low pH detected by time-resolved circular dichroism spectroscopy. *The Journal of Biological Chemistry, 272* (15), 9764–9770.

Kozerski, C., Ponimaskin, E., Schroth-Diez, B., Schmidt, M. F., & Herrmann, A. (2000). Modification of the cytoplasmic domain of influenza virus hemagglutinin affects enlargement of the fusion pore. *Journal of Virology, 74*(16), 7529–7537.

Kozlov, M. M., & Chernomordik, L. V. (1998). A mechanism of protein-mediated fusion: Coupling between refolding of the influenza hemagglutinin and lipid rearrangements. *Biophysical Journal, 75*(3), 1384–1396.

Kozlov, M. M., & Chernomordik, L. V. (2002). The protein coat in membrane fusion: Lessons from fission. *Traffic, 3*(4), 256–267.

Kozlov, M. M., Leikin, S. L., Chernomordik, L. V., Markin, V. S., & Chizmadzhev, Y. A. (1989). Stalk mechanism of vesicle fusion—Intermixing of aqueous contents. *European Biophysics Journal with Biophysics Letters, 17*(3), 121–129.

Kozlov, M. M., & Markin, V. S. (1983). Possible mechanism of membrane fusion. *Biofizika, 28*(2), 242–247.

Kozlov, M. M., McMahon, H. T., & Chernomordik, L. V. (2010). Protein-driven membrane stresses in fusion and fission. *Trends Biochem Sci, 35*(12), 699–706.

Kozlovsky, Y., Chernomordik, L. V., & Kozlov, M. M. (2002). Lipid intermediates in membrane fusion: Formation, structure, and decay of hemifusion diaphragm. *Biophysical Journal, 83*(5), 2634–2651.

Kozlovsky, Y., Efrat, A., Siegel, D. P., & Kozlov, M. M. (2004). Stalk phase formation: Effects of dehydration and saddle splay modulus. *Biophysical Journal, 87*(4), 2508–2521.

Kozlovsky, Y., & Kozlov, M. M. (2002). Stalk model of membrane fusion: Solution of energy crisis. *Biophysical Journal, 82*(2), 882–895.

Kozlovsky, Y., Zimmerberg, J., & Kozlov, M. M. (2004). Orientation and interaction of oblique cylindrical inclusions embedded in a lipid monolayer: A theoretical model for viral fusion peptides. *Biophysical Journal, 87*(2), 999–1012.

Krumbiegel, M., Herrmann, A., & Blumenthal, R. (1994). Kinetics of the low pH-induced conformational changes and fusogenic activity of influenza hemagglutinin. *Biophysical Journal, 67*(6), 2355–2360.

Kuhl, T., Guo, Y., Alderfer, J. L., Berman, A. D., Leckband, D., & Israelachvili, J., et al. (1996). Direct measurement of polyethylene glycol induced depletion attraction between lipid bilayers. *Langmuir, 12*, 3003–3014.

Kuzmin, P. I., Zimmerberg, J., Chizmadzhev, Y. A., & Cohen, F. S. (2001). A quantitative model for membrane fusion based on low-energy intermediates. *Proceedings of the National Academy of Sciences United States of America, 98*(13), 7235–7240.

Lague, P., Roux, B., & Pastor, R. W. (2005). Molecular dynamics simulations of the influenza hemagglutinin fusion peptide in micelles and bilayers: Conformational analysis of peptide and lipids. *Journal of Molecular Biology, 354*(5), 1129–1141.

Langosch, D., Crane, J. M., Brosig, B., Hellwig, A., Tamm, L. K., & Reed, J. (2001). Peptide mimics of SNARE transmembrane segments drive membrane fusion depending on their conformational plasticity. *Journal of Molecular Biology, 311*(4), 709–721.

Leckband, D., & Israelachvili, J. (2001). Intermolecular forces in biology. *Quarterly Reviews of Biophysics, 34*(2), 105–267.

Lee, J., & Lentz, B. R. (1997). Evolution of lipidic structures during model membrane fusion and the relation of this process to cell membrane fusion. *Biochemistry, 36*(21), 6251–6259.

Lee, J. Y., & Schick, M. (2008). Calculation of free energy barriers to the fusion of small vesicles. *Biophysical Journal, 94*(5), 1699–1706.

Lei, G., & MacDonald, R. C. (2003). Lipid bilayer vesicle fusion: Intermediates captured by high-speed microfluorescence spectroscopy. *Biophysical Journal, 85*(3), 1585–1599.

Lei, G., & MacDonald, R. C. (2008). Effects on interactions of oppositely charged phospholipid vesicles of covalent attachment of polyethylene glycol oligomers to their surfaces: Adhesion, hemifusion, full fusion and "endocytosis". *Journal of Membrane Biology, 221*(2), 97–106.

Leikin, S L., Kozlov, M. M., Chernomordik, L. V., Markin, V. S., & Chizmadzhev, Y. A. (1987). Membrane fusion: Overcoming of the hydration barrier and local restructuring. *Journal of Theoretical Biology, 129*(4), 411–425.

Leikina, E., & Chernomordik, L. V. (2000). Reversible merger of membranes at the early stage of influenza hemagglutinin-mediated fusion. *Molecular Biology of the Cell, 11*(7), 2359–2371.

Lentz, B. R., Malinin, V., Haque, M. E., & Evans, K. (2000). Protein machines and lipid assemblies: Current views of cell membrane fusion. *Current Opinion in Structural Biology, 10*(5), 607–615.

Lentz, B. R., McIntyre, G. F., Parks, D. J., Yates, J. C., & Massenburg, D. (1992). Bilayer curvature and certain amphipaths promote poly(ethylene glycol)-induced fusion of dipalmitoylphosphatidylcholine unilamellar vesicles. *Biochemistry, 31*(10), 2643–2653.

Li, F., Pincet, F., Perez, E., Eng, W. S., Melia, T. J., & Rothman, J. E., et al. (2007). Energetics and dynamics of SNAREpin folding across lipid bilayers. *Nature Structural & Molecular Biology, 14*(10), 890–896.

Li, Y., Han, X., Lai, A. L., Bushweller, J. H., Cafiso, D. S., & Tamm, L. K. (2005). Membrane structures of the hemifusion-inducing fusion peptide mutant G1S and the fusion-blocking mutant G1V of influenza virus hemagglutinin suggest a mechanism for pore opening in membrane fusion. *Journal of Virology, 79*(18), 12065–12076.

Lindau, M., & Almers, W. (1995). Structure and function of fusion pores in exocytosis and ectoplasmic membrane fusion. *Current Opinion in Cell Biology, 7*(4), 509–517.

Liu, T., Wang, T., Chapman, E. R., & Weisshaar, J. C. (2008). Productive hemifusion intermediates in fast vesicle fusion driven by neuronal SNAREs. *Biophysical Journal, 94*(4), 1303–1314.

Lu, X., Zhang, F., McNew, J. A., & Shin, Y. K. (2005). Membrane fusion induced by neuronal SNAREs transits through hemifusion. *The Journal of Biological Chemistry, 280*(34), 30538–30541.

Macosko, J. C., Kim, C. H., & Shin, Y. K. (1997). The membrane topology of the fusion peptide region of influenza hemagglutinin determined by spin-labeling EPR. *Journal of Molecular Biology, 267*(5), 1139–1148.

Malinin, V. S., & Lentz, B. R. (2004). Energetics of vesicle fusion intermediates: Comparison of calculations with observed effects of osmotic and curvature stresses. *Biophysical Journal, 86*(5), 2951–2964.

Markin, V. S., & Albanesi, J. P. (2002). Membrane fusion: Stalk model revisited. *Biophysical Journal, 82*(2), 693–712.

Marrink, S. J., de Vries, A. H., & Tieleman, D. P. (2009). Lipids on the move: Simulations of membrane pores, domains, stalks and curves. *Biochimica et Biophysica Acta, 1788*(1), 149–168.

Marsh, M., & Helenius, A. (2006). Virus entry: Open sesame. *Cell, 124*(4), 729–740.

Martens, S., Kozlov, M. M., & McMahon, H. T. (2007). How synaptotagmin promotes membrane fusion. *Science, 316*(5828), 1205–1208.

Mattai, J., Hauser, H., Demel, R. A., & Shipley, G. G. (1989). Interactions of metal ions with phosphatidylserine bilayer membranes: Effect of hydrocarbon chain unsaturation. *Biochemistry, 28*(5), 2322–2330.

May, S. (2002). Structure and energy of fusion stalks: The role of membrane edges. *Biophysical Journal, 83*(6), 2969–2980.

McNew, J. A., Weber, T., Engelman, D. M., Sollner, T. H., & Rothman, J. E. (1999). The length of the flexible SNAREpin juxtamembrane region is a critical determinant of SNARE-dependent fusion. *Molecular Cell, 4*(3), 415–421.

McNew, J. A., Weber, T., Parlati, F., Johnston, R. J., Melia, T. J., & Sollner, T. H., et al. (2000). Close is not enough: SNARE-dependent membrane fusion requires an active mechanism that transduces force to membrane anchors. *Journal of Cell Biology, 150*(1), 105–117.

Melia, T. J., You, D. Q., Tareste, D. C., & Rothman, J. E. (2006). Lipidic antagonists to SNARE-mediated fusion. *Journal of Biological Chemistry, 281*(40), 29597–29605.

Melikyan, G. B., Barnard, R. J., Abrahamyan, L. G., Mothes, W., & Young, J. A. (2005). Imaging individual retroviral fusion events: From hemifusion to pore formation and growth. *Proceedings of the National Academy of Sciences United States of America, 102*(24), 8728–8733.

Melikyan, G. B., Brener, S. A., Ok, D. C., & Cohen, F. S. (1997). Inner but not outer membrane leaflets control the transition from glycosylphosphatidylinositol-anchored influenza hemagglutinin-induced hemifusion to full fusion. *Journal of Cell Biology, 136*(5), 995–1005.

Melikyan, G. B., Lin, S., Roth, M. G., & Cohen, F. S. (1999). Amino acid sequence requirements of the transmembrane and cytoplasmic domains of influenza virus hemagglutinin for viable membrane fusion. *Molecular Biology of the Cell, 10*(6), 1821–1836.

Melikyan, G. B., Markosyan, R. M., Hemmati, H., Delmedico, M. K., Lambert, D. M., & Cohen, F. S. (2000). Evidence that the transition of HIV-1 gp41 into a six-helix bundle, not the bundle configuration, induces membrane fusion. *Journal of Cell Biology, 151*(2), 413–423.

Melikyan, GB., Niles, W. D., & Cohen, F. S. (1993). Influenza virus hemagglutinin-induced cell-planar bilayer fusion: Quantitative dissection of fusion pore kinetics into stages. *The Journal of General Physiology, 102*(6), 1151–1170.

Melikyan, G. B., White, J. M., & Cohen, F. S. (1995). GPI-anchored influenza hemagglutinin induces hemifusion to both red blood cell and planar bilayer membranes. *Journal of Cell Biology, 131*(3), 679–691.

Mima, J., Hickey, C. M., Xu, H., Jun, Y., & Wickner, W. (2008). Reconstituted membrane fusion requires regulatory lipids, SNAREs and synergistic SNARE chaperones. *EMBO Journal, 27* (15), 2031–2042.

Mohler, W. A., Shemer, G., del Campo, J. J., Valansi, C., Opoku-Serebuoh, E., & Scranton, V., et al. (2002). The type I membrane protein EFF-1 is essential for developmental cell fusion. *Developmental Cell, 2*(3), 355–362.

Monck, J. R., & Fernandez, J. M. (1996). The fusion pore and mechanisms of biological membrane fusion. *Current Opinion in Cell Biology, 8*(4), 524–533.

Murata, M., Takahashi, S., Shirai, Y., Kagiwada, S., Hishida, R., & Ohnishi, S. (1993). Specificity of amphiphilic anionic peptides for fusion of phospholipid vesicles. *Biophysical Journal, 64*(3), 724–734.

Nanavati, C., Markin, V. S., Oberhauser, A. F., & Fernandez, J. M. (1992). The exocytotic fusion pore modeled as a lipidic pore. *Biophysical Journal, 63*(4), 1118–1132.

Neher, E., & Marty, A. (1982). Discrete changes of cell membrane capacitance observed under conditions of enhanced secretion in bovine adrenal chromaffin cells. *Proceedings of the National Academy of Sciences United States of America, 79*(21), 6712–6716.

Nikolaus, J., Stockl, M., Langosch, D., Volkmer, R., & Herrmann, A. (2010). Direct visualization of large and protein-free hemifusion diaphragms. *Biophysical Journal, 98*(7), 1192–1199.

Noguchi, H., & Takasu, M. (2001). Fusion pathways of vesicles: A Brownian dynamics simulation. *Journal of Chemical Physics, 115*(20), 9547–9551.

Nussler, F., Clague, M. J., & Herrmann, A. (1997). Meta-stability of the hemifusion intermediate induced by glycosylphosphatidylinositol-anchored influenza hemagglutinin. *Biophysical Journal, 73*(5), 2280–2291.

Ohki, S. (1982). A mechanism of divalent ion-induced phosphatidylserine membrane fusion. *Biochimica et Biophysica Acta, 689*(1), 1–11.

Ollesch, J., Poschner, B. C., Nikolaus, J., Hofmann, M. W., Herrmann, A., & Gerwert, K., et al. (2007). Secondary structure and distribution of fusogenic LV-peptides in lipid membranes. *European Biophysical Journal, 37*(4), 435–445.

Pantazatos, D. P., & MacDonald, R. C. (1999). Directly observed membrane fusion between oppositely charged phospholipid bilayers. *Journal of Membrane Biology, 170*(1), 27–38.

Papahadjopoulos, D., Nir, S., & Duzgunes, N. (1990). Molecular mechanisms of calcium-induced membrane fusion. *Journal of Bioenergetics and Biomembranes, 22*(2), 157–179.

Papahadjopoulos, D., Vail, W. J., Newton, C., Nir, S., Jacobson, K., & Poste, G., et al. (1977). Studies on membrane fusion. III. The role of calcium-induced phase changes. *Biochimica et Biophysica Acta, 465*(3), 579–598.

Park, H. E., Gruenke, J. A., & White, J. M. (2003). Leash in the groove mechanism of membrane fusion. *Nature Structural Biology, 10*(12), 1048–1053.

Parsegian, V. A., Rand, R. P., & Gingell, D. (1984). Lessons for the study of membrane fusion from membrane interactions in phospholipid systems. *Ciba Foundation Symposium, 103*, 9–27.

Podbilewicz, B., Leikina, E., Sapir, A., Valansi, C., Suissa, M., & Shemer, G., et al. (2006). The *C. elegans* developmental fusogen EFF-1 mediates homotypic fusion in heterologous cells and in vivo. *Developmental Cell, 11*(4), 471–481.

Qiao, H., Armstrong, R. T., Melikyan, G. B., Cohen, F. S., & White, J. M. (1999). A specific point mutant at position 1 of the influenza hemagglutinin fusion peptide displays a hemifusion phenotype. *Molecular Biology of the Cell, 10*(8), 2759–2769.

Rand, R. P., & Parsegian, V. A. (1984). Physical force considerations in model and biological membranes. *Canadian Journal of Biochemistry and Cell Biology, 62*(8), 752–759.

Rand, R. P., & Parsegian, V. A. (1989). Hydration forces between phospholipid bilayers. *Biochimica et Biophysica Acta, 988*(3), 351–376.

Razinkov, V. I., Melikyan, G. B., Epand, R. M., Epand, R. F., & Cohen, F. S. (1998). Effects of spontaneous bilayer curvature on influenza virus-mediated fusion pores. *The Journal of General Physiology, 112*(4), 409–422.

Reese, C., Heise, F., & Mayer, A. (2005). Trans-SNARE pairing can precede a hemifusion intermediate in intracellular membrane fusion. *Nature, 436*(7049), 410–414.

Rizo, J. (2006). Illuminating membrane fusion. *Proceedings of the National Academy of Sciences United States of America, 103*(52), 19611–19612.

Rizo, J., & Dai, H. (2007). How much can SNAREs flex their muscles? *Nature Structural & Molecular Biology, 14*(10), 880–882.

Robson Marsden, H., Elbers, N. A., Bomans, P. H., Sommerdijk, N. A., & Kros, A. (2009). A reduced SNARE model for membrane fusion. *Angewandte Chemie International Edition Engl, 48*(13), 2330–2333.

Roche, S., Bressanelli, S., Rey, F. A., & Gaudin, Y. (2006). Crystal structure of the low-pH form of the vesicular stomatitis virus glycoprotein G. *Science, 313*(5784), 187–191.

Russell, C. J., Jardetzky, T. S., & Lamb, R. A. (2001). Membrane fusion machines of paramyxoviruses: Capture of intermediates of fusion. *EMBO Journal, 20*(15), 4024–4034.

Salzwedel, K., Johnston, P. B., Roberts, S. J., Dubay, J. W., & Hunter, E. (1993). Expression and characterization of glycophospholipid-anchored human immunodeficiency virus type 1 envelope glycoproteins. *Journal of Virology, 67*(9), 5279–5288.

Sapir, A., Avinoam, O., Podbilewicz, B., & Chernomordik, L. V. (2008). Viral and developmental cell fusion mechanisms: Conservation and divergence. *Developmental Cell, 14*(1), 11–21.

Schroth-Diez, B., Ludwig, K., Baljinnyam, B., Kozerski, C., Huang, Q., & Herrmann, A. (2000). The role of the transmembrane and of the intraviral domain of glycoproteins in membrane fusion of enveloped viruses. *Bioscience Reports, 20*(6), 571–595.

Schroth-Diez, B., Ponimaskin, E., Reverey, H., Schmidt, M. F., & Herrmann, A. (1998). Fusion activity of transmembrane and cytoplasmic domain chimeras of the influenza virus glycoprotein hemagglutinin. *Journal of Virology, 72*(1), 133–141.

Shemer, G., & Podbilewicz, B. (2003). The story of cell fusion: Big lessons from little worms. *Bioessays, 25*(7), 672–682.

Shillcock, J. C., & Lipowsky, R. (2005). Tension-induced fusion of bilayer membranes and vesicles. *Nature Materials, 4*(3), 225–228.

Siddiqui, T. J., Vites, O., Stein, A., Heintzmann, R., Jahn, R., & Fasshauer, D. (2007). Determinants of synaptobrevin regulation in membranes. *Molecular Biology of the Cell, 18*(6), 2037–2046.

Siegel, D. P. (1993). Energetics of intermediates in membrane fusion: Comparison of stalk and inverted micellar intermediate mechanisms. *Biophysical Journal, 65*(5), 2124–2140.

Siegel, D. P., Banschbach, J., Alford, D., Ellens, H., Lis, L. J., & Quinn, P. J., et al. (1989). Physiological levels of diacylglycerols in phospholipid membranes induce membrane fusion and stabilize inverted phases. *Biochemistry, 28*(9), 3703–3709.

Siegel, D. P., Cherezov, V., Greathouse, D. V., Koeppe, R. E II., Killian, J. A., & Caffrey, M. (2006). Transmembrane peptides stabilize inverted cubic phases in a biphasic length-dependent manner: Implications for protein-induced membrane fusion. *Biophysical Journal, 90*(1), 200–211.

Sinn, C. G., Antonietti, M., & Dimova, R. (2006). Binding of calcium to phosphatidylcholine-phosphatidylserine membranes. *Colloids and Surfaces A, 282–283*, 410–419.

Smeijers, A. F., Markvoort, A. J., Pieterse, K., & Hilbers, P. A. (2006). A detailed look at vesicle fusion. *The Journal of Physical Chemistry B, 110*(26), 13212–13219.

Stegmann, T., Hoekstra, D., Scherphof, G., & Wilschut, J. (1985). Kinetics of pH-dependent fusion between influenza virus and liposomes. *Biochemistry, 24*(13), 3107–3113.

Stevens, M. J., Hoh, J. H., & Woolf, T. B. (2003). Insights into the molecular mechanism of membrane fusion from simulation: Evidence for the association of splayed tails. *Physical Review Letters, 91*(18), 188102.

Stockl, M., Nikolaus, J., & Herrmann, A. (2010). Visualization of lipid domain-specific protein sorting in giant unilamellar vesicles. *Methods in Molecular Biology, 606*, 115–126.

Struck, D. K., Hoekstra, D., & Pagano, R. E. (1981). Use of resonance energy transfer to monitor membrane fusion. *Biochemistry, 20*(14), 4093–4099.

Sudhof, T. C., & Rothman, J. E. (2009). Membrane fusion: Grappling with SNARE and SM proteins. *Science, 323*(5913), 474–477.

Sutton, R. B., Fasshauer, D., Jahn, R., & Brunger, A. T. (1998). Crystal structure of a SNARE complex involved in synaptic exocytosis at 2.4 A resolution. *Nature, 395*(6700), 347–353.

Tamm, L. K. (2003). Hypothesis: Spring-loaded boomerang mechanism of influenza hemagglutinin-mediated membrane fusion. *Biochimica et Biophysica Acta, 1614*(1), 14–23.

Tanford, C. (1980). *The hydrophobic effect. Formation of micelles and biological membranes. Wiley, New York.*

Tatulian, S. A., & Tamm, L. K. (2000). Secondary structure, orientation, oligomerization, and lipid interactions of the transmembrane domain of influenza hemagglutinin. *Biochemistry, 39*(3), 496–507.

Thomas, A., & Brasseur, R. (2006). Tilted peptides: The history. *Current Protein & Peptide Science, 7*(6), 523–527.

Tong, S., & Compans, R. W. (2000). Oligomerization, secretion, and biological function of an anchor-free parainfluenza virus type 2 (PI2) fusion protein. *Virology, 270*(2), 368–376.

Tse, F. W., Iwata, A., & Almers, W. (1993). Membrane flux through the pore formed by a fusogenic viral envelope protein during cell fusion. *Journal of Cell Biology, 121*(3), 543–552.

Ungermann, C., & Langosch, D. (2005). Functions of SNAREs in intracellular membrane fusion and lipid bilayer mixing. *Journal of Cell Science, 118*(Pt 17), 3819–3828.

Vaidya, N. K., Huang, H., & Takagi, S. (2010). Coarse grained molecular dynamics simulation of interaction between hemagglutinin fusion peptides and lipid bilayer membranes. *The Advances in Applied Mathematics and Mechanics, 2*(4), 430–450.

Veatch, S. L., & Keller, S. L. (2003). Separation of liquid phases in giant vesicles of ternary mixtures of phospholipids and cholesterol. *Biophysical Journal, 85*(5), 3074–3083.

Vicogne, J., Vollenweider, D., Smith, J. R., Huang, P., Frohman, M. A., & Pessin, J. E. (2006). Asymmetric phospholipid distribution drives in vitro reconstituted SNARE-dependent membrane fusion. *Proceedings of the National Academy of Sciences United States of America, 103* (40), 14761–14766.

Walde, P., Cosentino, K., Engel, H., & Stano, P. (2010). Giant vesicles: Preparations and applications. *Chembiochem, 11*(7), 848–865.

Wang, C. T., Lu, J. C., Bai, J., Chang, P. Y., Martin, T. F., & Chapman, E. R., et al. (2003). Different domains of synaptotagmin control the choice between kiss-and-run and full fusion. *Nature, 424* (6951), 943–947.

Wang, T., Smith, E. A., Chapman, E. R., & Weisshaar, J. C. (2009). Lipid mixing and content release in single-vesicle, SNARE-driven fusion assay with 1–5 ms resolution. *Biophysical Journal, 96* (10), 4122–4131.

Wang, Y., Dulubova, I., Rizo, J., & Sudhof, T. C. (2001). Functional analysis of conserved structural elements in yeast syntaxin Vam3p. *The Journal of Biological Chemistry, 276*(30), 28598–28605.

Warner, J.M., & O'Shaughnessy, B. (2011). Hemifusion on the pathway to membrane fusion. *preprint.*

Weiss, C. D., & White, J. M. (1993). Characterization of stable Chinese hamster ovary cells expressing wild-type, secreted, and glycosylphosphatidylinositol-anchored human immunodeficiency virus type 1 envelope glycoprotein. *Journal of Virology, 67*(12), 7060–7066.

Weissenhorn, W., Dessen, A., Harrison, S. C., Skehel, J. J., & Wiley, D. C. (1997). Atomic structure of the ectodomain from HIV-1 gp41. *Nature, 387*(6631), 426–430.

Weissenhorn, W., Hinz, A., & Gaudin, Y. (2007). Virus membrane fusion. *FEBS Letters, 581*(11), 2150–2155.

Wilschut, J., Duzgunes, N., Fraley, R., & Papahadjopoulos, D. (1980). Studies on the mechanism of membrane fusion: Kinetics of calcium ion induced fusion of phosphatidylserine vesicles followed by a new assay for mixing of aqueous vesicle contents. *Biochemistry, 19*(26), 6011–6021.

Wilson, I. A., Skehel, J. J., & Wiley, D. C. (1981). Structure of the haemagglutinin membrane glycoprotein of influenza virus at 3 A resolution. *Nature, 289*(5796), 366–373.

Xu, Y., Zhang, F., Su, Z., McNew, J. A., & Shin, Y. K. (2005). Hemifusion in SNARE-mediated membrane fusion. *Nature Structural & Molecular Biology, 12*(5), 417–422.

Yang, L., & Huang, H. W. (2002). Observation of a membrane fusion intermediate structure. *Science, 297*(5588), 1877–1879.

Yang, L., & Huang, H. W. (2003). A rhombohedral phase of lipid containing a membrane fusion intermediate structure. *Biophysical Journal, 84*(3), 1808–1817.

Yeagle, P. L., Smith, F. T., Young, J. E., & Flanagan, T. D. (1994). Inhibition of membrane fusion by lysophosphatidylcholine. *Biochemistry, 33*(7), 1820–1827.

Yoon, T. Y., Okumus, B., Zhang, F., Shin, Y. K., & Ha, T. (2006). Multiple intermediates in SNARE-induced membrane fusion. *Proceedings of the National Academy of Sciences United States of America, 103*(52), 19731–19736.

Zampighi, G. A., Zampighi, L. M., Fain, N., Lanzavecchia, S., Simon, S. A., & Wright, E. M. (2006). Conical electron tomography of a chemical synapse: Vesicles docked to the active zone are hemi-fused. *Biophysical Journal, 91*(8), 2910–2918.

Zimmerberg, J., Akimov, S. A., & Frolov, V. (2006). Synaptotagmin: Fusogenic role for calcium sensor? *Nature Structural & Molecular Biology, 13*(4), 301–303.

Zimmerberg, J., Cohen, F. S., & Finkelstein, A. (1980). Fusion of phospholipid vesicles with planar phospholipid bilayer membranes. I. Discharge of vesicular contents across the planar membrane. *The Journal of General Physiology, 75*(3), 241–250.

Zimmerberg, J., & Kozlov, M. M. (2006). How proteins produce cellular membrane curvature. *Nature Reviews of Molecular Cell Biology, 7*(1), 9–19.

CHAPTER 2

Dynamic Remodeling of Membranes Catalyzed by Dynamin

Thomas J. Pucadyil
Indian Institute of Science Education and Research, Pashan, Pune, Maharashtra, India

I. OVERVIEW

Members of the dynamin family of large GTPases are unique among proteins involved in remodeling cell membranes. They readily polymerize on favorable membrane templates and display reversible modes of membrane association that is coupled to their GTPase cycle. Recent studies with dynamin-1, a prototypical member of the dynamin superfamily, have highlighted their capacity to catalyze membrane fission under conditions of constant GTP turnover. Combined with novel structural information on the mechanism of GTP hydrolysis, these recent developments constitute a general paradigm to understand the mechanochemical basis of cell membrane remodeling processes.

Current Topics in Membranes, Volume 68

0065-230X/10 $35.00
DOI: 10.1016/B978-0-12-385891-7.00002-7

II. INTRODUCTION

Eukaryotic cells are characterized by an elaborate endomembrane system encapsulated by the plasma membrane. Comprised of the endoplasmic reticulum, Golgi apparatus, endosomes and lysosomes, the endomembrane system, and the plasma membrane constitute a dynamic membrane network. Physical and functional compartmentalization of this network is achieved by the production and consumption of pools of compositionally distinct transport vesicles. Vesicles are produced from a donor compartment by membrane budding and fission and consumed at the acceptor compartment by membrane fusion. Owing to the development of assays that accurately report lipid and content mixing in membrane vesicles in the late 1970s and early 1980s (Struck, Hoekstra, & Pagano, 1981; Wilschut & Papahadjopoulos, 1979), reconstitution of membrane fusion using viral and SNARE protein have significantly advanced our understanding of how transport vesicles are consumed at the acceptor compartment (Jahn & Sudhof, 1999). In contrast, mechanisms underlying membrane budding and fission remain largely unclear due to the lack of sensitive assays that can distinguish membrane fission from physical shearing of and protein desorption from membranes.

Of the numerous transport pathways involved in generation of vesicular intermediates in cells, the fission machinery involved in coated vesicular transport has been relatively well characterized. Coat complexes perform the dual role of concentrating proteins and budding membranes in the process of generating nascent transport vesicles (Brodsky, Chen, Knuehl, Towler, & Wakeham, 2001; Traub, 2009). Clathrin-coated vesicles mediate transport between the plasma membrane, endosome and *trans* Golgi compartments. Release of coated vesicles requires the membrane necks of coated buds to undergo scission, a process requiring GTP hydrolysis, and is catalyzed by members of the dynamin superfamily of large GTPases (Praefcke & McMahon, 2004). Early indication of the involvement of dynamin in scission of clathrin-coated endocytic buds came from analysis of a temperature sensitive *Drosophila* mutant *shibire* that displayed rapid paralysis at nonpermissive temperatures (Koenig & Ikeda, 1983). Thin-section electron micrographs of synaptic termini of paralyzed flies revealed a dramatic depletion of synaptic vesicles and accumulation of coated pits with electron dense material at the necks of coated pits resembling a collar. The *shibire* locus was subsequently identified as the *Drosophila* homologue of dynamin (van der Bliek & Meyerowitz, 1991).

In an attempt to understand the mechanistic basis of membrane fission, several reconstitution efforts have since been directed toward understanding dynamin behavior on artificial membrane templates. Much like the viral and SNARE proteins in membrane fusion research, these studies constitute a framework to understand membrane fission mechanisms.

III. CHARACTERISTICS OF DYNAMIN

Classical dynamins are characterized by the presence of an N-terminal G domain that catalyzes nucleotide hydrolysis, a middle domain that represents an oligomerization interface, a pleckstrin homology (PH) domain that provides docking sites for negatively charged lipid headgroups, a GTPase effector domain (GED) that confers structural stability to and modulates nucleotide hydrolysis at the G domain, and a C-terminal proline–arginine rich domain (PRD) that interacts with Src-homology 3 (SH3) domain-containing endocytic accessory proteins. Dynamin-1, the prototypical member of the dynamin superfamily, is involved in scission of clathrin-coated buds from the presynaptic membrane. Despite its PH domain-mediated ability to bind directly to membranes *in vitro*, PRD–SH3 domain interactions are required to recruit dynamin to clathrin-coated pits (Bethoney, King, Hinshaw, Ostap, & Lemmon, 2009; Okamoto, Herskovits, & Vallee, 1997).

Dynamin-1 displays a low affinity for GTP (K_m = 14–50 μM) and high rates of GDP dissociation (k_{off} = 60–90 s^{-1}) such that nucleotide exchange can occur spontaneously without external guanine nucleotide exchange factors (Song & Schmid, 2003). These attributes allow facile analysis of its nucleotide-dependent membrane remodeling events. Dynamin-1 has a high propensity to self-assemble under low ionic strength conditions or under physiological conditions on membrane templates (Hinshaw & Schmid, 1995; Sweitzer & Hinshaw, 1998). Self-assembly stimulates basal GTP hydrolysis, in part mediated by favorable positioning of the intramolecular GED (Chappie et al., 2009; Warnock, Hinshaw, & Schmid, 1996). GTP hydrolysis provides a reversible switch for membrane association (Ramachandran & Schmid, 2008). Thus, dynamin-1 shows higher binding to membranes in the apo- or GTP-bound state and lower binding in the GDP-bound state. In presence of GTP, dynamin-1 binds and dissociates from the membrane reflecting a cycle of association, spontaneous self-assembly, GTP hydrolysis, and dissociation.

IV. DYNAMIN-INDUCED MEMBRANE REMODELING

A. *Dynamin Behavior Under Conditions of Acute GTP Turnover*

Morphological features of electron-dense collars at the necks of arrested coated pits in the *shibire* mutant have been recreated upon treating synaptosomes with GTP-γ-S (Takei, McPherson, Schmid, & De Camilli, 1995). In addition, purified dynamin-1 can spontaneously self-assemble into collar-like structures under low ionic strength conditions (Hinshaw & Schmid, 1995) and into long helical spirals on negatively charged liposomes (Sweitzer & Hinshaw, 1998).

Together, these results for the first time conveyed the impression that self-assembly exerts forces strong enough to distort membranes. Exposure of helical spirals adsorbed on EM grids to GTP led to narrowing of membrane tubules (Sweitzer & Hinshaw, 1998). Subsequent analysis showed that such constriction was caused by GTP binding (Zhang & Hinshaw, 2001). Constriction reflects both an axial and radial condensation of the helical spiral with the outer diameter reducing from ~50 to ~40 nm and interrung spacing reducing of 13.2 to 9.4 nm (Zhang & Hinshaw, 2001), leading to a change from 14.2 to 13.8 subunits per turn of the spiral (Chen, Zhang, Egelman, & Hinshaw, 2004). Notably, constriction was apparent for helical spirals of dynamin-1 lacking the C-terminal PRD (dynamin-1 δPRD) and not so apparent with full-length protein. The unstructured PRD in full-length protein was found to sterically hinder condensation of the helical spiral upon GTP binding and SH3 domain proteins that interact with the PRD were proposed to act to remove this hindrance in order for the constriction to effect membrane fission *in vivo*. Similar observations have been made on the *Dictostellium* dynamin A that lacks the C-terminal PRD (Klockow et al., 2002).

Orientation of individual domains in the helical spiral was recently determined by docking crystal structures of the G and PH domains on to cryo-EM reconstructions of helical spirals (Mears, Ray, & Hinshaw, 2007). Based on such analysis, the authors proposed a corkscrew model for membrane fission where GTP binding-induced conformational changes are transmitted through the molecule leading to sliding and twisting of individual rungs of the spiral. Consequently, the strain induced on the membrane results in fission. Interestingly, cryo-EM analysis with helical spirals of dynamin-1 δPRD showed that GTP hydrolysis does not cause fission but instead causes disassembly of the spiral (Danino, Moon, & Hinshaw, 2004).

Similar observations of GTP hydrolysis-dependent dynamin disassembly were reported on helical spirals formed on lipid nanotubes—tubular lipid templates formed by the incorporation of nonhydroxylated fatty acid-containing galactosylcerebroside in the membrane (Marks et al., 2001; Stowell, Marks, Wigge, & McMahon, 1999). Exposure to GDP resulted in the helical spiral patterns to expand and disassembly on the nanotube, much like unwinding of a spring. Since these patterns remained unaltered in the presence of GTP-γ-S and the transition state mimic GDP–AlF_4, it was proposed that similar GTP hydrolysis-dependent helical expansion at the necks of clathrin-coated pits would result in membrane fission and release of a clathrin-coated vesicle.

GTP-dependent concerted conformational changes in helical spirals are perhaps best captured in a light microscopy aided visualization of dynamin behavior on a membrane template of lipid bilayer sheets (Roux, Uyhazi, Frost, & De Camilli, 2006). Dynamin addition to such templates led to rapid elongation of membrane tubules that attained lengths of several microns, eventually forming an apparent network. Growth occurred by incorporation of dynamin molecules at the base of membrane tubules. GTP addition led to a transient stiffening of the network followed by rapid retraction of each membrane tubule toward nodes of the network, a possible consequence of disassembly of the helical spiral and release of the underlying membrane. Tubules that were free at one end retracted back to the membrane reservoir but those tethered at both ends underwent rupture during the course of retraction. GTP addition revealed a rotatory movement of polystyrene beads attached to the spiral about the tubule axis, interpreted to reflect twisting of the helical spiral. The torque exerted on the membrane by such twisting activity was assumed to cause fission. A necessary requirement for fission therefore was tethering of the membrane, presumably provided by the actin cytoskeleton in cells.

B. Dynamin Behavior Under Conditions of Constant GTP Turnover

Roux et al. extended their studies on membrane tubules drawn out of giant unilamellar vesicles by the action of kinesin motor proteins. An assay system of this kind with a preformed tubular membrane template combined with the temporal resolution offered by light microscopy for the first time allowed analysis of dynamin behavior under constant GTP turnover. Remarkably, addition of dynamin to membrane tubules in the presence of GTP led to fragmentation of the tubule and release of numerous small vesicles. This observation for the first time conveyed the impression that dynamin could act on its own to mediated membrane fission.

Recent analyses of dynamin behavior under constant GTP turnover corroborate the observations made on membrane tethers pulled out of GUVs using kinesin motor proteins. Dynamin-catalyzed membrane fission and vesicle release was demonstrated in a novel system of supported bilayers with excess reservoir or SUPER templates formed on plain silica beads (Pucadyil & Schmid, 2008). These unilamellar templates retain more membrane than is necessary to cover the surface and the excess reservoir is likely organized as membrane folds. SUPER templates are formed by a salt-mediated disproportionate increase in liposome binding rates compared to fusion rates during formation of supported bilayers (Pucadyil & Schmid, 2010). These templates offer the unique advantage of simultaneous microscopy- and sedimentation-based analysis of reactions leading to budding and fission of membranes.

Dynamin-1 addition to SUPER templates led to the growth and elongation of membrane tubules decorated with helical dynamin spirals, as seen earlier on liposome-based templates (Roux et al., 2006; Sweitzer & Hinshaw, 1998). GTP addition led to disassembly of spirals and retraction of membrane tubules, again similar to effects seen earlier (Danino et al., 2004; Roux et al., 2006; Sweitzer & Hinshaw, 1998). However, dynamin addition in presence of GTP leads to release of membrane vesicles from the templates. Dual channel recordings of fluorescently labeled dynamin added to templates containing a fluorescent lipid probe reveal short dynamin assemblies (seen as fluorescent punctae as opposed to long tubes) undergoing scission and released into solution. This assay accurately recapitulates dynamin function since membrane fission activity on SUPER templates correlates remarkably well to dynamin-catalyzed scission of clathrin-coated pits in cells (Pucadyil & Schmid, 2010). Fig. 1 shows a

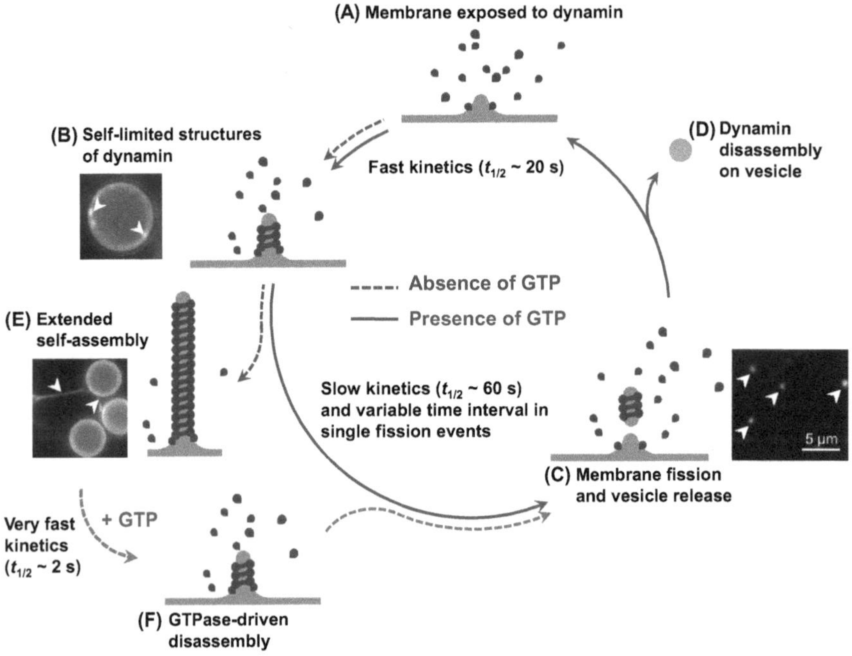

FIGURE 1 Schematic representation of cycles of dynamin-catalyzed membrane fission. (A) Dynamin (red dots) binds the membrane (blue) leading to formation of a bud, a process that requires both self-assembly and shallow insertion of hydrophobic residues in the PH domain. (B) Presence of GTP (red, solid lines) regulates dynamin assembly to a self-limited structure (arrowheads in image), reflecting a steady state where rates of assembly and stimulated GTP hydrolysis catalyzed disassembly are balanced. (C) Self-limited structures persist for variable time periods before membrane fission resulting in release of vesicles in solution. Partial assemblies of dynamin persist on the released vesicle (arrowheads in image) and complete disassembly occurs postfission, shown in (D). (E) In the absence of GTP (blue, dashed lines), dynamin assembly leads to formation of a stable scaffold or helical spirals on the membrane leading to tubule extension. (F) GTP addition at this step results in scaffold disassembly causing tubule retraction without fission (green, dashed lines). Membrane fission and vesicle release is observed only after complete retraction of the scaffold. Kinetic parameters shown in the figure are derived from dual-channel recordings of fluorescent dynamin (green) monitored on fluorescent (red) SUPER templates. Fluorescence images are snap shots from such recordings and are reproduced with permission from Pucadyil and Schmid (2008). (See Color Insert.)

schematic representation of the events taking place on such templates during membrane fission.

Behavior of dynamin prior to fission was analyzed on flow-induced tubular extensions from SUPER templates. Tubes ranging from 0.5 to 1 μm in diameter can be generated from these templates by the viscous drag caused by flow of glycerol-containing buffer across these templates (see Fig. 2A). The presence of

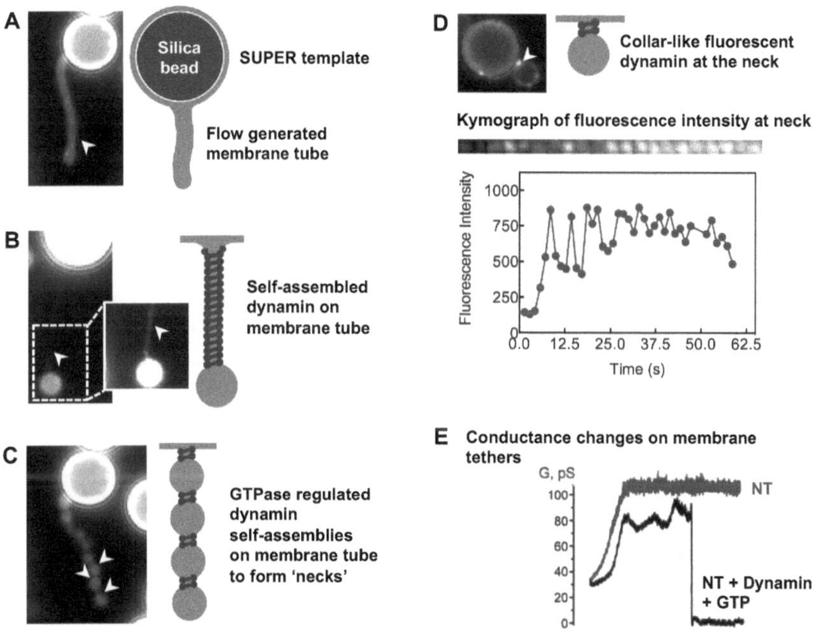

FIGURE 2 Monitoring dynamin behavior during individual fission events. (A) Snapshot from a time-lapse movie showing glycerol flow-induced extension of a membrane tube (marked by arrowhead) from a SUPER template. (B) Flowing dynamin suspended in glycerol over the template results in rapid buckling and capture of the tube to form a long neck (marked by arrowhead) with a large bud. Inset is the boxed field adjusted for contrast. (C) Flowing dynamin suspended in glycerol over templates bathed in buffer containing GTP results in buckling and capture of tube with multiple short necks (marked by arrowheads), an activity that requires both self-assembly as well as shallow insertion of hydrophobic residues in the PH domain. (D) Similar experiment as described in (C) but with fluorescent dynamin indicates that the neck is defined by a collar-like structure of dynamin (marked by arrowhead). Kymograph and plot show fluctuations in intensity of the collar-like structure at the neck prior to fission of the large bud. (E) Plot showing wave-like fluctuations in conductance in narrow tubes prior to fission in presence of dynamin and GTP (black trace, NT + dynamin + GTP). The instantaneous drop in conductance to baseline values marks fission. Conductance of narrow tubes (blue, NT) in the absence of dynamin and GTP is shown for comparison. Images in (A, B, and D) and plots in (E) are reproduced with permission from Pucadyil & Schmid (2008) and Bashkirov et al. (2008), respectively. (See Color Insert.)

dynamin in the glycerol containing buffer resulted in rapid buckling of the membrane to form a long neck with a large bud at the end (Fig. 2B), a telling example of the mechanical force exerted on the membrane and the high degree of cooperativity exhibited by dynamin self-assembly. In addition to self-assembly, shallow insertion of hydrophobic residues in the variable loop 1 of the PH domain appears to be necessary for driving such topological rearrangements in the membrane (Ramachandran et al., 2009), much like what has been described for the N-BAR domain containing proteins (Campelo, McMahon, & Kozlov, 2008; McMahon & Gallop, 2005). In presence of GTP, dynamin self-assembly and GTPase driven disassembly ensured that the tube buckled but never extended in length resulting in the formation of multiple necks (Fig. 2C). The large buds subsequently undergo fission at the necks to release large vesicles. Similar experiments carried out with fluorescently labeled dynamin showed that the necks were defined by collar-like structures (Fig. 2D). Fluorescence intensity at these collar-like structures displayed fluctuations even as the neck remained well defined, likely reflecting multiple rounds of dynamin self-assembly and GTPase driven partial disassembly before fission (Fig. 2D). Persistence of collar-like structures showed heterogeneity with lifetimes ranging from 25 to 75 s, which could reflect stochasticity in the fission reaction.

Conductance measurements to monitor topological changes occurring in membrane tethers pulled out of GUVs provide the first quantitative estimates of the mechanical forces involved in dynamin-induced membrane remodeling reactions (Bashkirov et al., 2008). When assayed on tethers composed of a lipid composition that produced narrow tubes (NT; luminal radius of $\sim$6 nm), assembly of dynamin to form helical spirals squeezed the tube to a final radius of $\sim$2.6 nm. Scaled geometries of outer diameters of dynamin helical spirals correspond to a change from $\sim$60 to 40 nm. The final outer diameter match earlier reported values for the constricted state of helical spirals (Zhang & Hinshaw, 2001). Thus, assembly by itself can impose severe curvature stress without the necessity of additional nucleotide-dependent conformational changes in the spiral. When GTP or GDP was added postassembly, conductance increased to levels seen before dynamin addition implying disassembly of helical spirals. In the constant presence of GTP, minor changes in conductance were followed by an instantaneous drop in conductance indicative of a non-leaky membrane fission event likely involving a hemifusion membrane intermediate. Similar analysis performed on short tubes, which reduces averaging of conductance across the tube and improves signal to noise, revealed slow, wave-like fluctuations in conductance prior to fission (Fig. 2E). The lifetime of fission events displayed remarkable heterogeneity in distribution with wave-like fluctuations in conductance persisting for intervals ranging from 20 to 80 s before fission, again indicative of stochasticity in the fission reaction. The wave-like changes in conductance were interpreted to reflect formation of short (three rungs) and transient assemblies of dynamin. Furthermore, modeling studies hypothesized that disassembly of short (two rungs of the helical spiral) greatly favors transformation of the stressed membrane into a hemifusion intermediate. For longer spirals, the membrane shows a greater tendency to bulge out upon spiral disassembly thus reducing the probability for generation of the hemifusion intermediate.

V. EVOLUTION OF MODELS FOR DYNAMIN-CATALYZED MEMBRANE FISSION

The striking observation that dynamin can spontaneously form highly ordered helical spirals to a large extent contributed to defining mechanochemical models for dynamin function. Preassembled helical spirals of dynamin were an obvious starting point since nucleotide-based conformational changes have traditionally compared the apo- and nucleotide-bound states. The obvious next step was to analyze the effect of GTP addition to such spirals. That dynamin-mediated fission required a power-stroke generated by a concerted and global conformational change was an intuitively simple interpretation of the remarkable effects seen in preassembled dynamin upon GTP addition. However, disassembly of the helical spiral and absence of membrane fission was a consistent outcome of these experiments. As a result, essential involvement of SH3 domain-containing partner proteins, coats, and contribution from the actin cytoskeleton as a provider of extrinsic membrane tension were considered necessary for membrane fission.

Results from increasingly sophisticated assays that monitor dynamics of lipid–protein interactions and the mechanical forces exerted on the lipid bilayer due to such interactions have led to an evolution in models describing dynamin-catalyzed membrane fission. It now appears that the proposed inability for dynamin to catalyze fission on its own may be a reflection of not having appropriately reconstituted a minimal membrane fission reaction. Indeed, results from three independent experiments monitoring dynamin behavior on widely different membrane templates but in the constant presence of GTP reveal that dynamin is indeed capable of catalyzing membrane fission without the necessity of accessory factors (Bashkirov et al., 2008; Pucadyil & Schmid, 2008; Roux et al., 2006). The model that now emerges is that cycling of dynamin between membrane-bound assembled and disassembled states due to constant turnover of GTP is necessary for membrane fission (Bashkirov et al., 2008; Pucadyil & Schmid, 2008). During these cycles, the membrane undergoes squeezing and relaxation leading to the stochastic generation of a hemifusion intermediate thereby causing fission (Bashkirov et al., 2008; Pucadyil & Schmid, 2008). This is in essence quite contrary to the precision and purpose previously associated with a motor-like function for dynamin (Sever, Damke, & Schmid, 2000; Song & Schmid, 2003). This model does not rule out global and concerted conformational changes occurring in dynamin. It only predicts that such changes may be associated with dynamin disassembly and not directly to membrane fission. Similar models have been proposed for membrane fission involving the Arf and Sar GTPases (Pucadyil & Schmid, 2009), the ESCRT complexes (Fabrikant et al., 2009), and the bacterial-dynamin like proteins (Low & Lowe, 2010). These models together put membrane fission mechanistically closer to membrane fusion where the energy for bringing lipid bilayers close enough for fusion is provided by structural rearrangements of the fusion machinery while energy from nucleotide hydrolysis is required for subsequent disassembly of the fusion machinery.

VI. CURRENT INSIGHTS FROM STRUCTURAL STUDIES

A. Structure of Dynamin-1 G Domain-GED Fusion

Structure of a minimal G domain-GED (GG) fusion protein crystallized in presence of the transition state mimic GDP–ALF_4 indicates a dimeric arrangement, assumed to represent the interface between two rungs of the helical spiral in assembled dynamin (Chappie, Acharya, Leonard, Schmid, & Dyda, 2010). Compared to the structure of the G domain in the apo-state (Reubold et al., 2005), the recent structure suggests that GTP binding and hydrolysis induces extensive conformational changes in the helical spiral. Interestingly, the structure of dynamin-3 G domain crystallized in the GDP bound form also reveals a dimeric arrangement suggesting that disassembly may require an additional step of nucleotide dissociation (Yang et al., 2009, PDB 3L43).

The GG fusion protein structure shows the two monomers in the GG dimer to be oriented at an angle of 120°. If one assumes a rigid body extension of this orientation along the full length of the molecule, this configuration renders only one of the two PH domains capable of interacting with the membrane. Thus, transition state-induced conformational uncoupling of either rung of the helical spiral with the membrane could represent a mechanism explaining how GTP hydrolysis leads to disassembly of the helical spiral. Since PH domain-membrane interactions are sensitive to the oligomeric state of dynamin-1 (Bethoney et al., 2009), uncoupling a few PH domains along the rung could lead to disassembly of the entire rung and subsequent dissociation from the membrane. Such a scheme of events is consistent with the earlier observation that dynamin disassembly is slower than its membrane dissociation (Ramachandran & Schmid, 2008).

The PH domains of individual dynamin molecules appear to be docked in a uniform orientation in cryo-EM reconstruction of the helical spiral (Mears et al.,

2007). Packing the GG dimer into the preassembled helical spiral may therefore be inconceivable implying that formation of the transition state may be possible only for dynamin molecules present at the outermost rungs of the helical spiral. Thus, GTP hydrolysis could result in sequential disassembly starting from the outermost rungs of the spiral. Even under conditions where long spirals are favored, such as on highly curved membrane tether, constant GTP turnover could ensure assembly and disassembly cycles at the ends of the spiral resulting in fission of the tether.

B. Structure of the MxA Stalk

MxA is an intereferon-inducible dynamin-like myxovirus resistance GTPase and is an important component of the innate immune system. Belonging to the dynamin-like protein family, Mx proteins lack the PH and C-terminal PRD domains. Like the classical dynamins, MxA assembles under low salt conditions, and binds and tubulates membranes (Accola, Huang, Al Masri, & McNiven, 2002). The recently reported crystal structure of the stalk region (composed of the middle and GED domains) of MxA suggests that formation of higher order oligomers and helical spirals depends on interactions between the stalk regions (Gao et al., 2010). G domains appear not to be involved in oligomerization but facilitate interrung contacts in the helical spiral. These results highlight the possibility that stimulated GTP hydrolysis occurs only upon formation of a complete rung of the helical spiral.

C. Nucleotide-Dependent Conformational Changes in BDLP

Recent crystal structures of the bacterial dynamin-like protein (BDLP) from the cyanobacterium *Nostoc punctiforme* in the apo- and GDP-bound state and cryo-EM reconstructions in the GTP-bound state assembled on membrane tubules reveal a dramatic intermolecular rearrangement upon GTP binding (Low & Lowe, 2006; Low, Sachse, Amos, & Lowe, 2009). The G domains homodimerize upon GTP binding and consequently display higher affinity for the membrane that favors self-assembly. Similar effects of GTP binding-induced membrane association have been observed for MxA (Gao et al., 2010) and dynamin A (Klockow et al., 2002).

Helical spirals formed by the assembly of BDLP in the GTP-bound state constrict the underlying membrane to form a water filled channel of $\sim$10 nm in diameter, again indicating that self-assembly imposes strong curvature stress on the membrane (Low et al., 2009). Membrane curvature appears to be generated by shallow insertion of residues present in the membrane-interacting paddle domain, similar to that seen for dynamin-1 (Ramachandran et al., 2009). GTP hydrolysis results in subsequent membrane dissociation. Fusion of membrane bilayers was proposed to be a passive outcome of the polymerization-induced curvature stress imposed on the membrane and the disassembly of the helical spiral (Low & Lowe, 2010), a mechanism similar to that proposed for dynamin-1 (Bashkirov et al., 2008; Pucadyil & Schmid, 2008).

VII. FUTURE PERSPECTIVES

The helical spiral structure of dynamin forms the basis of our current understanding of its assembly stimulated GTP hydrolysis, disassembly and membrane fission. This structure however represents a state of dynamin in the apo- and/or GTP-bound form poised for rapid disassembly in response to GTP hydrolysis. Gaining high-resolution structural information of the collar-like assembly seen at the necks of budding vesicles in constant presence of GTP will be crucial to elucidate the functional coupling between nucleotide-dependent conformational changes and membrane fission.

Models proposed for dynamin-catalyzed membrane fission are based on analysis of one member of the dynamin superfamily. Members of the dynamin superfamily of proteins display remarkable structural differences such as substitution of the PH domain for a lipid-binding loop or a transmembrane domain (Low & Lowe, 2010; Praefcke & McMahon, 2004). In addition, they display differences in kinetic parameters of nucleotide hydrolysis. Analyzing their role in membrane fusion or fission would offer significant insights into understanding the evolutionary significance of functional diversity associated with members of the dynamin family.

Dynamin has lately been referred to as a minimal membrane fission apparatus to suggest that the machinery necessary for membrane fission is self-contained in the molecule (Pucadyil & Schmid, 2008). Nevertheless, this apparatus works in a stochastic manner and additional factors in the cellular milieu may be necessary to improve fission efficiency. Dynamin-catalyzed membrane fission seen *in vitro* appears not to require accessory SH3 domain containing proteins or contributions from the actin cytoskeleton but these are essential for fission of coated vesicles (Itoh et al., 2005; Okamoto et al., 1997; Yarar, Waterman-Storer, & Schmid, 2005). Indeed, results from monitoring dynamics of budding and fission of clathrin-coated vesicles indicate an exquisite control both at the stage at which dynamin is recruited to and the duration for which it persists at the necks of clathrin-coated pits (Doyon et al., 2011; Merrifield, Perrais, & Zenisek, 2005). The challenge that lies ahead is to determining how the numerous cellular factors together manage membrane fission reactions with such precision and fidelity in cells.

Acknowledgments

I thank members of the Schmid lab and Vadim Frolov for many stimulating discussions that have shaped the ideas put forth in this manuscript. This work was supported by institutional grants and an intermediate fellowship from the Wellcome Trust-DBT India Alliance.

References

Accola, M. A., Huang, B., Al Masri, A., & McNiven, M. A. (2002). The antiviral dynamin family member, MxA, tubulates lipids and localizes to the smooth endoplasmic reticulum. *Journal of Biological Chemistry, 277*(24), 21829–21835.

Bashkirov, P V., Akimov, S. A., Evseev, A. I., Schmid, S. L., Zimmerberg, J., & Frolov, V. A. (2008). GTPase cycle of dynamin is coupled to membrane squeeze and release, leading to spontaneous fission. *Cell, 135*(7), 1276–1286.

Bethoney, K. A., King, M. C., Hinshaw, J. E., Ostap, E. M., & Lemmon, M. A. (2009). A possible effector role for the pleckstrin homology (PH) domain of dynamin. *Proceedings of the National Academy of Sciences of the United States of America, 106*(32), 13359–13364.

Brodsky, F. M., Chen, C. Y., Knuehl, C., Towler, M. C., & Wakeham, D. E. (2001). Biological basket weaving: Formation and function of clathrin-coated vesicles. *Annual Review of Cell and Developmental Biology, 17*, 517–568.

Campelo, F., McMahon, H. T., & Kozlov, M. M. (2008). The hydrophobic insertion mechanism of membrane curvature generation by proteins. *Biophysical Journal, 95*(5), 2325–2339.

Chappie, J. S., Acharya, S., Leonard, M., Schmid, S. L., & Dyda, F. (2010). G domain dimerization controls dynamin's assembly-stimulated GTPase activity. *Nature, 465*(7297), 435–440.

Chappie, J. S., Acharya, S., Liu, Y. W., Leonard, M., Pucadyil, T. J., & Schmid, S. L. (2009). An intramolecular signaling element that modulates dynamin function in vitro and in vivo. *Molecular Biology of the Cell, 20*(15), 3561–3571.

Chen, Y. J., Zhang, P., Egelman, E. H., & Hinshaw, J. E. (2004). The stalk region of dynamin drives the constriction of dynamin tubes. *Nature Structural & Molecular Biology, 11*(6), 574–575.

Danino, D., Moon, K. H., & Hinshaw, J. E. (2004). Rapid constriction of lipid bilayers by the mechanochemical enzyme dynamin. *Journal of Structural Biology, 147*(3), 259–267.

Doyon, J. B., Zeitler, B., Cheng, J., Cheng, A. T., Cherone, J. M., & Santiago, Y., et al. (2011). Rapid and efficient clathrin-mediated endocytosis revealed in genome-edited mammalian cells. *Nature Cell Biology, 13*(3), 331–337.

Fabrikant, G., Lata, S., Riches, J. D., Briggs, J. A., Weissenhorn, W., & Kozlov, M. M. (2009). Computational model of membrane fission catalyzed by ESCRT-III. *PLoS Computational Biology, 5*(11), e1000575.

Gao, S., von der Malsburg, A., Paeschke, S., Behlke, J., Haller, O., & Kochs, G., et al. (2010). Structural basis of oligomerization in the stalk region of dynamin-like MxA. *Nature, 465*(7297), 502–506.

Hinshaw, J. E., & Schmid, S. L. (1995). Dynamin self-assembles into rings suggesting a mechanism for coated vesicle budding. *Nature, 374*(6518), 190–192.

Itoh, T., Erdmann, K. S., Roux, A., Habermann, B., Werner, H., & De Camilli, P. (2005). Dynamin and the actin cytoskeleton cooperatively regulate plasma membrane invagination by BAR and F-BAR proteins. *Developmental Cell, 9*(6), 791–804.

Jahn, R., & Sudhof, T. C. (1999). Membrane fusion and exocytosis. *Annual Review of Biochemistry, 68*, 863–911.

Klockow, B., Tichelaar, W., Madden, D. R., Niemann, H. H., Akiba, T., & Hirose, K., et al. (2002). The dynamin A ring complex: Molecular organization and nucleotide-dependent conformational changes. *EMBO Journal, 21*(3), 240–250.

Koenig, J. H., & Ikeda, K. (1983). Evidence for a presynaptic blockage of transmission in a temperature-sensitive mutant of Drosophila. *Journal of Neurobiology, 14*(6), 411–419.

Low, H. H., & Lowe, J. (2006). A bacterial dynamin-like protein. *Nature, 444*(7120), 766–769.

Low, H. H., & Lowe, J. (2010). Dynamin architecture—From monomer to polymer. *Current Opinion in Structural Biology, 20*(6), 791–798.

Low, H. H., Sachse, C., Amos, L. A., & Lowe, J. (2009). Structure of a bacterial dynamin-like protein lipid tube provides a mechanism for assembly and membrane curving. *Cell, 139*(7), 1342–1352.

Marks, B., Stowell, M. H., Vallis, Y., Mills, I. G., Gibson, A., & Hopkins, C. R., et al. (2001). GTPase activity of dynamin and resulting conformation change are essential for endocytosis. *Nature, 410*(6825), 231–235.

McMahon, H. T., & Gallop, J. L. (2005). Membrane curvature and mechanisms of dynamic cell membrane remodelling. *Nature, 438*(7068), 590–596.

Mears, J. A., Ray, P., & Hinshaw, J. E. (2007). A corkscrew model for dynamin constriction. *Structure, 15*(10), 1190–1202.

Merrifield, C. J., Perrais, D., & Zenisek, D. (2005). Coupling between clathrin-coated-pit invagination, cortactin recruitment, and membrane scission observed in live cells. *Cell, 121*(4), 593–606.

Okamoto, P. M., Herskovits, J. S., & Vallee, R. B. (1997). Role of the basic, proline-rich region of dynamin in Src homology 3 domain binding and endocytosis. *Journal of Biological Chemistry, 272*(17), 11629–11635.

Praefcke, G J., & McMahon, H. T. (2004). The dynamin superfamily: Universal membrane tubulation and fission molecules? *Nature Reviews Molecular Cell Biology, 5*(2), 133–147.

Pucadyil, T. J., & Schmid, S. L. (2008). Real-time visualization of dynamin-catalyzed membrane fission and vesicle release. *Cell, 135*(7), 1263–1275.

Pucadyil, T. J., & Schmid, S. L. (2009). Conserved functions of membrane active GTPases in coated vesicle formation. *Science, 325*(5945), 1217–1220.

Pucadyil, T. J., & Schmid, S. L. (2010). Supported bilayers with excess membrane reservoir: A template for reconstituting membrane budding and fission. *Biophysical Journal, 99*(2), 517–525.

Ramachandran, R., Pucadyil, T. J., Liu, Y. W., Acharya, S., Leonard, M., & Lukiyanchuk, V., et al. (2009). Membrane insertion of the pleckstrin homology domain variable loop 1 is critical for dynamin-catalyzed vesicle scission. *Molecular Biology of the Cell, 20*(22), 4630–4639.

Ramachandran, R., & Schmid, S. L. (2008). Real-time detection reveals that effectors couple dynamin's GTP-dependent conformational changes to the membrane. *EMBO Journal, 27*(1), 27–37.

Reubold, T. F., Eschenburg, S., Becker, A., Leonard, M., Schmid, S. L., & Vallee, R. B., et al. (2005). Crystal structure of the GTPase domain of rat dynamin 1. *Proceedings of the National Academy of Sciences of the United States of America, 102*(37), 13093–13098.

Roux, A., Uyhazi, K., Frost, A., & De Camilli, P. (2006). GTP-dependent twisting of dynamin implicates constriction and tension in membrane fission. *Nature, 441*(7092), 528–531.

Sever, S., Damke, H., & Schmid, S. L. (2000). Garrotes, springs, ratchets, and whips: Putting dynamin models to the test. *Traffic, 1*(5), 385–392.

Song, B. D., & Schmid, S. L. (2003). A molecular motor or a regulator? Dynamin's in a class of its own. *Biochemistry, 42*(6), 1369–1376.

Stowell, M. H., Marks, B., Wigge, P., & McMahon, H. T. (1999). Nucleotide-dependent conformational changes in dynamin: Evidence for a mechanochemical molecular spring. *Nature Cell Biology, 1*(1), 27–32.

Struck, D. K., Hoekstra, D., & Pagano, R. E. (1981). Use of resonance energy transfer to monitor membrane fusion. *Biochemistry, 20*(14), 4093–4099.

Sweitzer, S. M., & Hinshaw, J. E. (1998). Dynamin undergoes a GTP-dependent conformational change causing vesiculation. *Cell, 93*(6), 1021–1029.

Takei, K., McPherson, P. S., Schmid, S. L., & De Camilli, P. (1995). Tubular membrane invaginations coated by dynamin rings are induced by GTP-gamma S in nerve terminals. *Nature, 374*(6518), 186–190.

Traub, L. M. (2009). Tickets to ride: Selecting cargo for clathrin-regulated internalization. *Nature Reviews Molecular Cell Biology, 10*(9), 583–596.

van der Bliek, A. M., & Meyerowitz, E. M. (1991). Dynamin-like protein encoded by the Drosophila shibire gene associated with vesicular traffic. *Nature, 351*(6325), 411–414.

Warnock, D. E., Hinshaw, J. E., & Schmid, S. L. (1996). Dynamin self-assembly stimulates its GTPase activity. *Journal of Biological Chemistry, 271*(37), 22310–22314.

Wilschut, J., & Papahadjopoulos, D. (1979). Ca^{2+}-induced fusion of phospholipid vesicles monitored by mixing of aqueous contents. *Nature, 281*(5733), 690–692.

Yang, S., Tempel, W., Tong, Y., Nedyalkova, L., Guan, X., Crombet, L., et al. (2009). Crystal structure of the dynamin 3 GTPase domain bound with GDP. PDB3L43.

Yarar, D., Waterman-Storer, C. M., & Schmid, S. L. (2005). A dynamic actin cytoskeleton functions at multiple stages of clathrin-mediated endocytosis. *Molecular Biology of the Cell, 16*(2), 964–975.

Zhang, P., & Hinshaw, J. E. (2001). Three-dimensional reconstruction of dynamin in the constricted state. *Nature Cell Biology, 3*(10), 922–926.

PART 2

VIRAL FUSION

CHAPTER 3

Structure and Working of Viral Fusion Machinery

Aurélie Albertini, Stéphane Bressanelli, Jean Lepault and Yves Gaudin
Centre de Recherche de Gif, Laboratoire de Virologie Moléculaire et Structurale, CNRS (UPR 3296), Gif sur Yvette, France

I. OVERVIEW

The entry of enveloped viruses into cells requires the fusion of viral and cellular membranes, driven by conformational changes in viral glycoproteins. Structural studies have defined three classes of viral membrane fusion proteins. Despite their different structural organizations, all seem to have a common mechanism of action that generates the same lipid organizations during the fusion pathway. In this chapter, the structures of the best characterized viral fusion glycoproteins of the three classes will be presented. We will focus on the interactions between fusion glycoproteins and membranes, presenting

Current Topics in Membranes, Volume 68 0065-230X/10 $35.00

DOI: 10.1016/B978-0-12-385891-7.00003-9

evidences for the existence of an extended intermediate conformation exposing the fusion peptides at the top of the molecule. We will show that the fusion peptides cannot penetrate deeply inside the membrane and that their rather interfacial interaction with the lipid bilayer is probably sufficient to initiate the fusion mechanism by facilitating the formation of stalk precursors. We will also review the regulatory mechanisms that allow fusion to be triggered at the right time and right place. Finally, many studies have demonstrated that fusion involves the cooperative action of a large number of fusion glycoproteins. We will discuss new data that have shown that the glycoproteins located outside the contact zone between virions and liposomes reorganize into regular arrays that have been proposed to drive the late stages of the fusion reaction.

II. INTRODUCTION

Entry of enveloped viruses into host cells requires binding of the virus to one or more receptors present at the cell surface, followed by fusion of the viral envelope with a cellular membrane. These steps are mediated by virally encoded glycoproteins that promote both receptor recognition and membrane fusion. The membrane fusion process involves large structural rearrangements of the fusogenic glycoproteins upon interaction with specific triggers (e.g., low pH environment and cellular receptors). These conformational changes result in the exposure of hydrophobic motifs (the so-called fusion peptides or fusion loops) that then interact with one or both of the participating membranes, resulting in their destabilization and merger. Triggering of the conformational change in the absence of a target membrane leads to inactivation of the fusion properties of the fusogenic glycoprotein.

Three different classes of viral fusion proteins have been hitherto identified based on their common postfusion structural motifs (see below). However, experimental data suggest that the membrane fusion pathway is very similar for all the enveloped viruses studied so far whatever the organization of their fusion machinery (Chernomordik, Frolov, Leikina, Bronk, & Zimmerberg, 1998; Chernomordik, Leikina, Frolov, Bronk, & Zimmerberg, J., 1997; Gaudin, 2000; Zaitseva, Mittal, Griffin, & Chernomordik, 2005). It is generally assumed that fusion proceeds via the formation of an intermediate stalk that is a local lipidic connection between the outer leaflets of the fusing membranes. Radial expansion of the stalk would induce the formation of a transient hemifusion diaphragm (i.e., a local bilayer made by the two initial inner leaflets) in which the formation of a pore and its enlargement would lead to complete fusion (Chernomordik & Kozlov, 2005, 2008; Chernomordik, Kozlov, & Zimmerberg, 1995). Even if fusion of two lipid bilayers is thermodynamically favorable, there are energetic barriers that have to be overcome at different steps of the process, the highest one being apparently encountered during the expansion of the initial fusion pore (Cohen & Melikyan, 2004). It has been proposed that the energy released during the favorable fusogenic structural transition is used to overcome these different barriers (Carr, Chaudhry, & Kim, 1997; Harrison, 2008).

In this chapter, the structures of the best characterized viral fusion glycoproteins of the three classes will be described, putting the emphasis on their similarities and their common mechanisms of action. Data that indicate the existence of an extended intermediate conformation, exposing the fusion peptides at the head of the molecule, will be presented. We will also discuss data from recent studies showing that fusion peptides cannot penetrate deeply inside the membrane and that their interfacial interaction with the lipid bilayer is probably sufficient to initiate the fusion mechanism by facilitating the formation of stalk precursors.

We will also review the regulatory mechanisms that allow fusion to be triggered at the right time and right place. We will particularly review the literature showing that viruses for which fusion is triggered by low pH have the ability to protect their fusion protein from undergoing irreversible conformational change and to avoid undesirable fusion reactions during their transport through the acidic compartments of the Golgi apparatus.

Finally, many studies have indicated that fusion involves the cooperative action of a large number of fusion glycoproteins. We will discuss new data showing that the glycoproteins located outside the contact zone between virion and liposomes reorganize into regular arrays that have been proposed to drive the late stages of the fusion reaction.

III. STRUCTURES OF FUSION PROTEINS

A. Class I Fusion Proteins

1. Influenza Hemagglutinin

The first crystal structure of a viral fusion protein ectodomain that has been determined is that of influenza virus hemagglutinin (HA) in its prefusion conformation (Wilson, Skehel, & Wiley, 1981) (Figure 1A, left). HA is synthesized as HA0, a precursor molecule cleaved by a cellular protease to produce HA1, a sialic acid binding domain, and HA2, a membrane-anchored fusion domain (Stieneke-Grober et al., 1992). HA0 cleavage primes the protein for the low pH-induced fusogenic conformational change.

HA is a homotrimer, each protomer consisting of an HA1 and an HA2 chain connected through a single disulfide bridge. Most of HA2 forms a fibrous triple coiled coil and HA1 is tethered to the top of this coiled coil and contains the major antigenic sites. The amino-terminal extremity of HA2, which has been generated by cleavage of HA0, is highly hydrophobic and constitutes the fusion

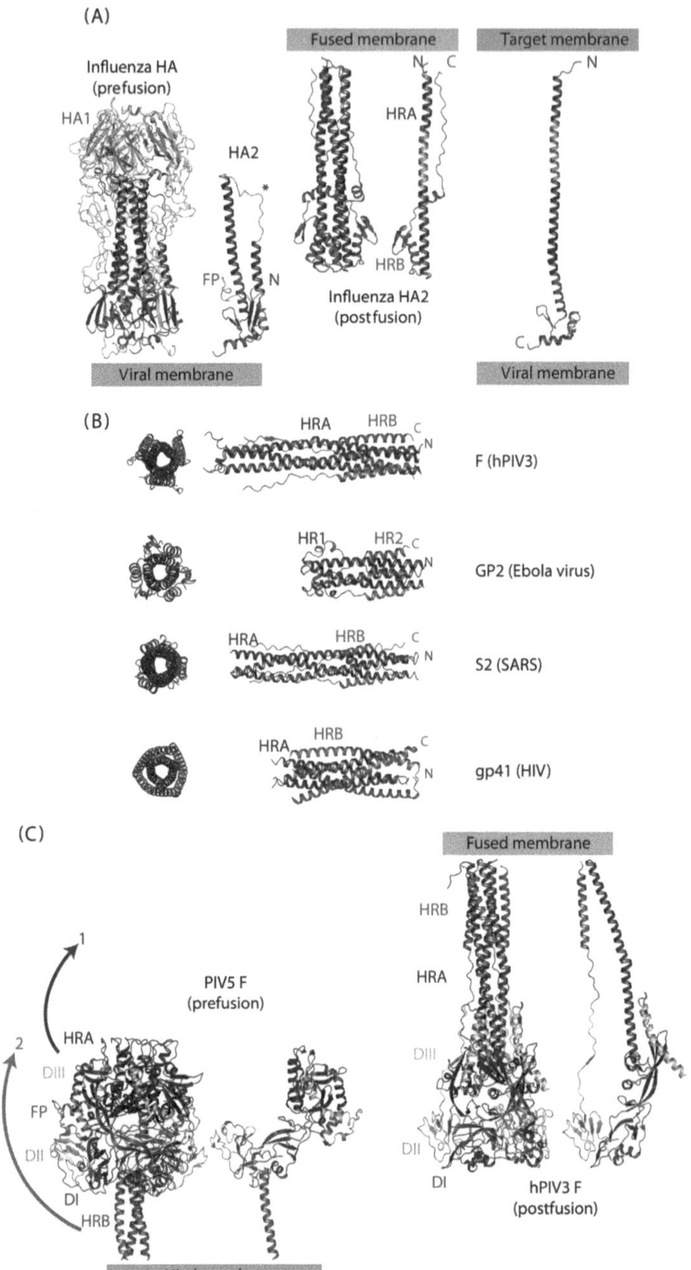

FIGURE 1 Class I fusion proteins. (A) Influenza virus hemagglutinin (HA) structures. *Left part:* Ribbon diagram of hemagglutinin trimer and HA2 protomer at neutral pH (pre-fusion

peptide. In the prefusion structure, it is buried along the threefold axis (Figure 1A, left).

When exposed to low pH (between pH 5 and 6, depending on the viral strain), HA undergoes a large-scale conformational change, the extent of which was visualized only in 1994 when the postfusion conformation of HA2 was crystallized (Bullough, Hughson, Skehel, & Wiley, 1994) (Figure 1A, middle). Low pH treatment results in the loss of HA1–HA2 interactions and dissociation of the HA1–HA1 interface stabilizing the prefusion structure. After the transition, the structure of HA1 is not significantly modified as it is still able to bind the viral receptor (Sauter et al., 1992) and is still recognized by monoclonal antibodies (Bizebard et al., 1995). In fact, HA low-pH-induced conformational change is essentially a dramatic refolding of HA2 (Figure 1A, middle). During this refolding, a long loop in native HA2 converts to a helix, resulting in length increase of the central coiled coil at its N-terminal end. This movement translocates the fusion peptide toward the head of the molecule, that is, toward the target membrane. At this stage, HA is in an extended intermediate conformation (Figure 1A, right). The structural transition is completed by a relocation of the carboxyterminal membrane anchor through a foldback mechanism to the amino-terminal end of the rod-shaped molecule. This results in the so-called

conformation) (Wilson et al., 1981). The receptor binding domain (HA1) subunits are colored in grey and the HA2 fusion proteins are colored in blue and red, the fusion peptide (FP) is depicted in green. *Middle part:* Ribbon diagram of HA2 trimer and protomer in the post-fusion conformation (Chen, Skehel, & Wiley, 1999). The central structure formed by HA2 folds into a long trimeric coiled-coil positioning the fusion peptides (absent in the crystalline structure) near the C-terminus of the molecule linked to the transmembrane anchor. *Right part:* Putative extended intermediate state adopted by HA2 during the structural transition. The cyan loop marked by an asterisk on the neutral HA2 protomer refolds into a helix resulting in a length increase of the central trimeric coiled-coil. This loop-helix transition translocates the fusion peptide toward the target membrane. (B) Trimeric hairpins structures of several class I viral fusion proteins. Side and top views of the trimeric core of the post-fusion structures from class I fusion proteins of four different viruses: hPIV3 F (Yin et al., 2006), Ebola virus GP2 (Weissenhorn et al., 1998), SRAS S2 (Supekar et al., 2004) and HIV gp41 (Weissenhorn et al., 1997). The N-terminal coiled-coil core (HRA or HR1 regions) is colored in blue and the C-terminal domain (HRB and HR2 regions) is in red. In all structures, the trimer-of-hairpins conformation brings the N- and C- termini (corresponding to the position of the fusion peptides and transmembrane domains) into close proximity. (C) Paramyxovirus F structures. Pre-fusion structure of parainfluenza virus 5 (PIV5) F (left part) and post-fusion structure of human parainfluenza virus 3 (hPIV3) F (right part) (Yin et al., 2005; Yin et al., 2006). For each conformation, trimer and protomer are presented. Domains I (DI), II (DII) and III (DIII), heptad repeats A (HRA) and B (HRB) are shown, hydrophobic fusion peptides (FP) are in green. The two conformations are related by flipping the fusion peptide and the C-terminal domain around a block made of DI and DII that do not change during the structural transition. HRA segments undergo a major reorganization that projects the hydrophobic fusion peptide toward the target membrane (movement illustrated by the blue arrow 1), then the HRB segments dissociate and turn around the globular head to form the post fusion structure (movement illustrated by red arrow 2). (See Color Insert.)

hairpin conformation in which the transmembrane segment and the fusion peptide are in close proximity at the same end of the elongated molecule (Figure 1A, middle).

2. Class I Structural Motif

The structural motif, found in the postfusion structure of HA2, is encountered in the postfusion structure of many other viral fusion proteins including those of retroviruses (Buzon et al., 2010; Weissenhorn, Dessen, Harrison, Skehel, & Wiley, 1997), paramyxoviruses (Baker, Dutch, Lamb, & Jardetzky, 1999; Yin, Paterson, Wen, Lamb, & Jardetzky, 2005), filoviruses (Weissenhorn, Carfi, Lee, Skehel, & Wiley, 1998), and coronaviruses (Duquerroy, Vigouroux, Rottier, Rey, & Bosch, 2005; Supekar et al., 2004; Xu et al., 2004) (Figure 1B). All these glycoproteins show a trimeric coiled coil at the N-termini of which are displayed the fusion peptides and against which are packed, in an antiparallel manner, the segments abutting the transmembrane (TM) domains. This motif has thus served to define the class I of fusion proteins. Most of the class I fusion proteins have also an internal, frequently furin-like, cleavage site located upstream a stretch of about 20 hydrophobic amino acids that will constitute the fusion peptide. Class I signature is the presence of a first heptad repeat region (named HRA or HR1) that is immediately downstream the fusion peptide and forms the central trimeric coiled coil. It is often (but not always) accompanied by a second one (named HRB or HR2) located in the C-terminal part of the ectodomain that constitutes the helices positioned, in an antiparallel manner, in the groove of the central coiled coil (Figure 1B).

Except for this common structural motif, the structure of the proteins are very different, as illustrated by the recent determination of the structure of the prefusion state of parainfluenza virus type 5 fusion protein (PIV5 F) (Yin, Wen, Paterson, Lamb, & Jardetzky, 2006) (Figure 1C) and that of GP of Ebola virus (Lee et al., 2008). The issue concerning whether there is a common origin to all these proteins or if they represent convergent solutions to developing a membrane fusion machinery is still an open question.

Furthermore, the mechanisms for activation of different class I viral fusion proteins are distinct. For many of them, activation is not dependent on low pH environment. Most of retroviruses env proteins and paramyxovirus F are activated by receptor binding events. This activation can be by direct interactions (HIV env) (Melikyan, 2008) or indirectly through a viral attachment protein (paramyxovirus F) (Lamb & Jardetzky, 2007). Finally, although Ebola virus enters cell by the endocytic pathway, this probably reflects a need for glycoprotein GP processing by lysosomal proteases, such as cathepsins, to trigger GP-mediated fusion, rather than simply low pH itself (Chandran, Sullivan, Felbor, Whelan, & Cunningham, 2005; Schornberg et al., 2006).

3. Structure of Paramyxovirus Fusion Protein

Structural studies of the paramyxovirus fusion proteins provided new insight into the molecular organization of a class I fusion machinery for which the folding and the mode of activation are very different from that of influenza virus. For paramyxoviruses, two viral glycoproteins are involved in cell entry. The first one, called HN, H, or G, depending on the virus, is an attachment protein. The second is the fusion protein F.

The structures of prefusion PIV5 F (Yin et al., 2006) and postfusion Newcastle disease virus (Swanson et al., 2010) and human PIV3 F proteins (Yin et al., 2005) have been solved (Figure 1C). F is a trimer that is proteolytically cleaved to liberate the N terminus of the fusion peptide (Lamb & Jardetzky, 2007). It is worth noting that in absence of its transmembrane segment, soluble F ectodomain can spontaneously convert into its postfusion conformation even in absence of the cleavage that primes the full-length glycoprotein for subsequent conformational change (Yin et al., 2005). Thus, to replace the missing transmembrane domain and the cytoplasmic tail and to stabilize the prefusion form, the trimeric GCN4 derivative was placed in register with C-terminal HRB of the secreted PIV5 F protein ectodomain (Yin et al., 2006).

After triggering, region HRA, adjacent to the fusion peptide and initially collapsed in a compact set of 11 small segments, undergoes a major conformational change that extends the fusion peptide outward for insertion into the target membrane (Figure 1C) (Yin et al., 2006). After this process, involving the refolding of the 11 segments into a single, extended α-helix, F forms an extended intermediate with the fusion peptide exposed at the top of the molecule. Region HRB, adjacent to the transmembrane domain, then translocates to bind to HRA, forming a six-helix bundle and the postfusion hairpin conformation.

How interaction of the attachment protein to the viral receptor promotes F conformational change is still a matter of debate. Two models have been proposed. On the one hand, the attachment protein may clamp and stabilize F in its prefusion state until the attachment protein binds the target cell (the "clamp model") (Aguilar et al., 2006; Bishop et al., 2007; Corey & Iorio, 2009; Plemper, Hammond, Gerlier, Fielding, & Cattaneo, 2002). On the other hand, the attachment protein itself may undergo a conformational change after receptor binding that destabilizes F and triggers its conformational change (the "provocateur model") (Connolly, Leser, Jardetzky, & Lamb, 2009).

The structures of several paramyxovirus receptor binding proteins have been resolved (Bowden, Aricescu, et al., 2008; Bowden, Crispin, et al., 2008; Bowden, Crispin, Harvey, Jones, & Stuart, 2010; Colf, Juo, & Garcia, 2007; Crennell, Takimoto, Portner, & Taylor, 2000; Lawrence et al., 2004; Hashiguchi et al., 2007; Xu et al., 2008; Yuan et al., 2005). Their structures are very similar,

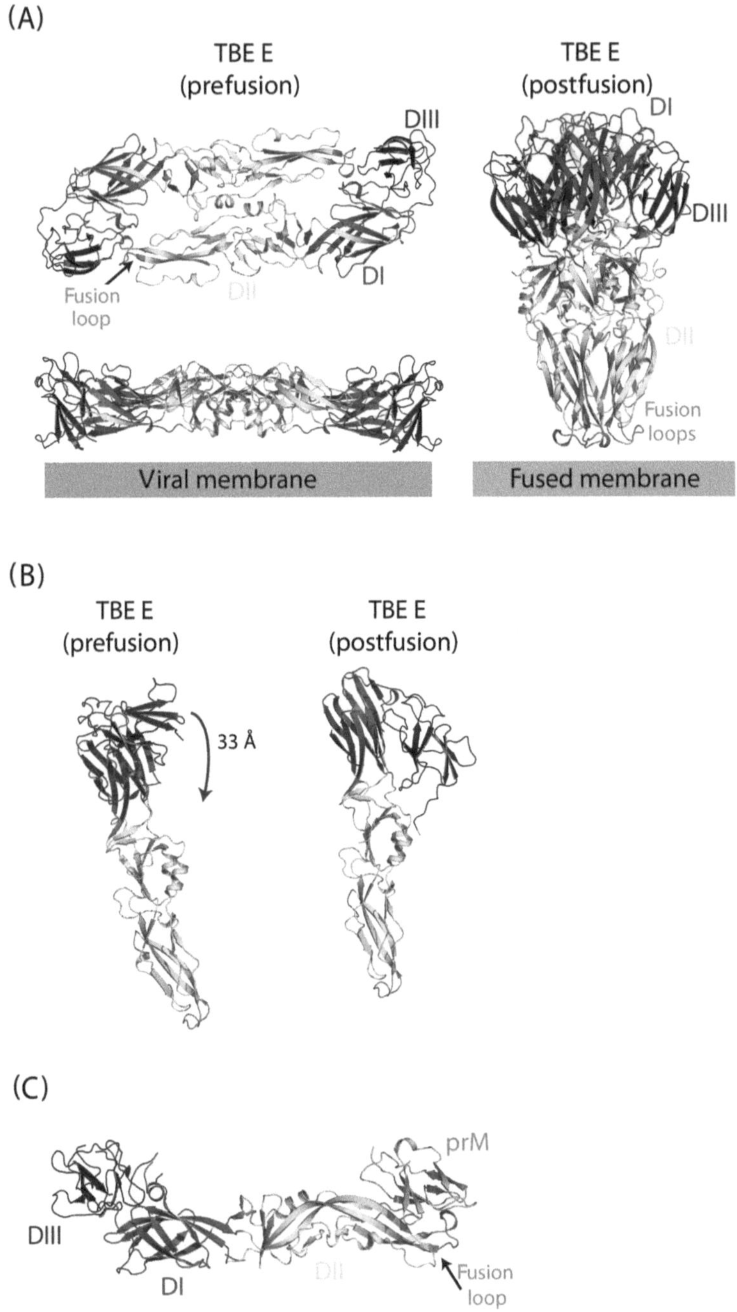

FIGURE 2 Class II fusion proteins. (A) TBEV glycoprotein E structures. Ribbon diagram of the pre-fusion E dimer (Rey et al., 1995) (left part) and post-fusion E trimer (Bressanelli et al., 2004)

suggesting that they trigger F conformational change through similar mechanisms. For some of these proteins, the structure of both the native free form and the receptor bound form has been solved without revealing any major change in the structure of the protein (Bowden, Aricescu, et al., 2008; Bowden, Crispin, et al., 2008; Xu et al., 2008; Yuan et al., 2005). It has to be noted that the stalk, connecting the head of the receptor binding protein to the transmembrane domain, was not resolved in any of the crystal structures. Multiple studies, however, have mapped the F-interacting region to the stalk of the receptor binding protein (Bishop et al., 2008; Bousse, Takimoto, Gorman, Takahashi, & Portner, 1994; Corey & Iorio, 2007; Porotto, Murrell, Greengard, & Moscona, 2003; Stone-Hulslander & Morrison, 1999).

B. Class II Fusion Proteins

The second class of fusion proteins was identified when the X-ray prefusion structure of the ectodomain of the fusion protein E1 of Semliki Forest virus (SFV) (Lescar et al., 2001), an alphavirus, was shown to have a remarkably similar secondary and tertiary structure to that of the fusion protein E of tick-borne encephalitis virus (TBEV) (Rey, Heinz, Mandl, Kunz, & Harrison, 1995), a flavivirus (Figure 2A). This indicated that the genes encoding these proteins were derived from a common ancestor. The alphaviruses and flaviviruses are members of the Togaviridae and Flaviviridae families, respectively, and other fusion proteins of viruses from these families, including West Nile virus E (Kanai et al., 2006; Nybakken, Nelson, Chen, Diamond, & Fremont, 2006), dengue virus E (Modis, Ogata, Clements, & Harrison, 2003), and hepatitis C virus E2 (Krey et al., 2010), have been demonstrated to belong to class II. It has also been proposed that the fusion glycoprotein of members of the Bunyaviridae family also belong to class II (Garry & Garry, 2004).

All class II fusion proteins identified so far fold cotranslationally with a chaperone protein, named p62 (or pE2) for alphaviruses (Wahlberg, Boere, & Garoff, 1989) and prM (promembrane protein) for flaviviruses (Heinz et al., 1994). This heterodimeric interaction is important for the correct folding and transport of the fusion protein (Sanchez-San Martin, Liu, & Kielian, 2009). The

(right part). Red yellow and blue part of the protein correspond to respectively the central domain (DI), the dimerization domain (DII) and the C-terminal domain (DIII); fusion loop is depicted in green and located at the tip of DII. (B) TBEV E protomer conforlmational change. Ribbon diagram of the pre- and post-fusion E protomer of TBEV. The arrow indicates the movement of domain III (connected to the transmembrane domain of the protein) toward the fusion loop. (C) Dengue virus precursor membrane-envelope protein complex (prM–E) structure. Ribbon diagram of the prM-E heterodimer (Li et al., 2008). The pr peptide, in light blue, protects the fusion loop in green. Domains of E are indicated and colored according to A. (See Color Insert.)

chaperone protein is cleaved by the cellular protease furin late in the secretory pathway (Stadler, Allison, Schalich, & Heinz, 1997; Zhang, Fugere, Day, & Kielian, 2003). The p62 is cleaved into E2 and E3 (Salminen, Wahlberg, Lobigs, Liljestrom, & Garoff, 1992; Wahlberg et al., 1989) and prM is cleaved into pr peptide and M protein (Heinz et al., 1994). This cleavage is crucial as it primes the protein, which is then in a metastable conformation, for subsequent low-pH-induced conformational change (Sanchez-San Martin, Liu, & Kielian, 2009). In their prefusion state anchored in the viral membrane, class II fusion proteins are organized following the icosahedral symmetry (Kuhn et al., 2002; Lescar et al., 2001).

The polypeptide chain of the class II proteins is folded into three domains essentially constituted by β-sheet (Lescar et al., 2001; Modis et al., 2003; Rey et al., 1995). The C terminus part of the ectodomain, which is linked to the transmembrane segment, and the internal fusion loop are found at the two ends of an extended molecule. The fusion loop is located at the tip of an elongated beta sheet. In the prefusion state, it is buried at the E1–E2 interface for alphaviruses and at the E homodimer interface for flaviviruses (Figure 2A).

Low pH exposure, in presence of a target membrane, triggers a complete rearrangement of the proteins (Bressanelli et al., 2004; Gibbons et al., 2004; Modis, Ogata, Clements, & Harrison, 2004). First, the alphavirus E2–E1 heterodimers or the flavivirus E–E homodimers dissociate and then reassociate to form E1 or E homotrimers (Figure 2A). The conformational change is less impressive than that observed for HA or for paramyxovirus F protein as there is essentially no change in secondary structure. Nevertheless, the interactions between the domains constituting the protomer are markedly different in the pre- and postfusion conformations. Particularly, domain III, which is connected to the C-terminal part of the molecule, moves by about 35 Å toward the fusion loop (Figure 2B). The movement of domain III redirects the polypeptide chain, so that the transmembrane segment is located near the fusion loop region in the postfusion trimeric form. Despite the different architecture of the proteins, the postfusion structure is thus reminiscent of the postfusion hairpin structure of class I fusion proteins.

A particularity of class II fusion proteins is that the presence of a target membrane catalyzes the formation of the postfusion trimer. In the absence of a membrane, soluble E ectodomain dimers dissociate at low pH but redimerize if the pH is returned to neutral (Stiasny, Allison, Marchler-Bauer, Kunz, & Heinz, 1996). Furthermore, several class II fusion proteins require the presence of specific lipids in the target membrane to achieve membrane fusion. SFV and Sindbis virus-induced fusion is strictly dependent on the presence of both cholesterol (Kielian & Helenius, 1984) and sphingolipids (Nieva, Bron, Corver, & Wilschut, 1994; Smit, Bittman, & Wilschut, 1999). Cholesterol does not seem to be absolutely required for flavivirus-induced membrane fusion (Umashankar et al., 2008) although its presence may facilitate the process for the hepatitis C virus glycoprotein (Lavillette et al., 2006) and strongly promotes West Nile virus fusion activity (Moesker, Rodenhuis-Zybert, Meijerhof, Wilschut, & Smit, 2010). Finally, effective dengue virus fusion to plasma and intracellular membranes, as well as to protein-free liposomes, requires the target membrane to contain anionic lipids such as bis(monoacylglycero)phosphate and phosphatidylserine (Zaitseva, Yang, Melikov, Pourmal, & Chernomordik, 2010).

C. Class III Fusion Proteins

The third class of fusion proteins was identified when the structures of the ectodomains of the fusogenic proteins G (G) of vesicular stomatitis virus (VSV) (Roche, Bressanelli, Rey, & Gaudin, 2006) and glycoprotein B (gB) of herpes simplex virus 1 were determined (Heldwein et al., 2006) (Figure 3). This revealed an unanticipated homology between the two proteins that belong to a negative-strand RNA and a DNA virus, respectively, and for which no sequence similarity was previously identified. Epstein Barr virus (EBV) gB (Backovic, Longnecker, & Jardetzky, 2009) and baculovirus gp64 (Kadlec, Loureiro, Abrescia, Stuart, & Jones, 2008), of which the structures were further determined, have been also demonstrated to belong to this new class (Figure 3B). Alignment of the amino acid sequences of G proteins from animal rhabdoviruses belonging to different genera also reveals that all of them share the same fold (Roche et al., 2006). The glycoprotein of Borna disease virus, another member of the Mononegavirales order, has been also proposed to belong to class III (Garry & Garry, 2009).

VSV G is the only class III fusion protein for which the structures of both the prefusion (Roche, Rey, Gaudin, & Bressanelli, 2007) and postfusion states (Roche et al., 2006) have been determined. The structures determined for other class III members are presumptive postfusion conformations, based on their structural similarity with the postfusion conformation of G.

The soluble ectodomain (G_{th}) of VSV, which has been crystallized, has been generated by limited proteolysis with thermolysin directly on purified virus, whereas the soluble form of HSV1 gB, EBV gB, and baculovirus gp64 were obtained through heterologous expression system.

The polypeptide chain of G_{th} folds into four distinct domains: a β-sheet-rich lateral domain (domain I), a central domain that is involved in the trimerization of the molecule (domain II), a pleckstrin homology domain (domain III), and the fusion domain (domain IV) (Figure 3A). The organization of class III fusion domain is similar (although not homologous) to that of class II fusion proteins. It contains an extended β-sheet structure at the tip of which are located two loops that constitute the membrane-interacting motif of the G ectodomain.

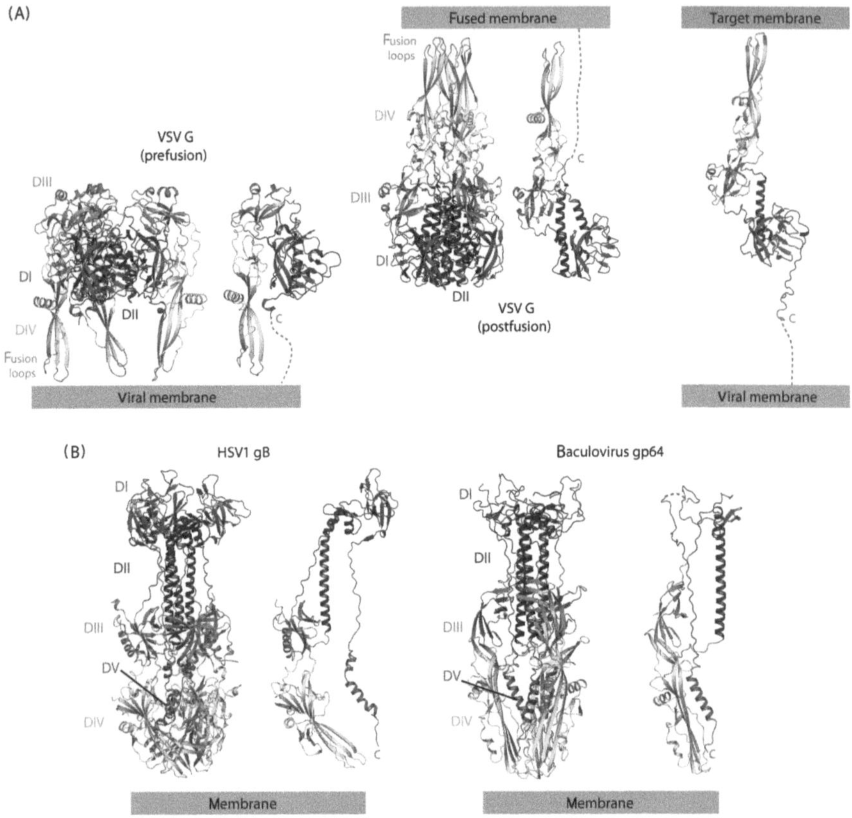

FIGURE 3 Class III fusion proteins. (A) Overall structures of the pre- and post-fusion forms of VSV glycoprotein. *Left part:* Ribbon diagram of the G pre-fusion trimer and protomer (Roche et al., 2007). *Middle part:* Ribbon diagrams of the G post-fusion trimer and protomer. During the structural transition from pre- to post-fusion conformation fusion domain and C-terminal domain turn around a rigid block made of DI and DII. The two conformations are thus related by flipping the fusion peptide and the C-terminal domain around this rigid block (Roche et al., 2007). *Right part:* Putative elongated monomeric conformation adopted by G during the structural transition. G is colored by domains: DI is the lateral domain (red), DII is the trimerization domain (blue), DIII is the pleckstrin homology domain (orange) and DIV is the fusion domain (yellow) fusion loops are in green. The C-terminus segment reaching the trans-membrane domain (not observed in the x-ray structures) is represented as a magenta dashed line. B. Overall structures of presumptive post-fusion states of HSV1-gb and baculovirus gp64. Left part: ribbon diagram of HSV1 trimer and protomer (Heldwein et al., 2006). Right part: ribbon diagram of baculovirus gp64 trimer and protomer (Kadlec et al., 2008). Domains on both proteins are colored according to their homologous counterparts of VSV G. The domain DV (not visible in the crystal structures of VSV G) is in magenta. (See Color Insert.)

In the prefusion structure, in striking contrast to class I and class II viral fusion proteins, the fusion loops are not buried at an oligomeric interface, but point toward the viral membrane (Figure 3A, left). It is therefore possible that in the prefusion conformation, the fusion loops also interact with the viral membrane.

After the end of the trimerization domain, there remain 40 amino acids for the polypeptide chain to reach G transmembrane domain, but the structure of this segment is unknown for both the pre- and the postfusion conformation. This missing part corresponds to the C-terminal segment of domain V of gB and gp64 that forms a α-helical structure that inserts between the other two protomers within the trimer (Backovic et al., 2009; Heldwein et al., 2006; Kadlec et al., 2008).

G_{th} structures revealed that the conformational change from pre- to postfusion state involves a dramatic reorganization of the glycoprotein (Figure 3A, middle). During the structural transition, domains I, III, and IV retain their tertiary structure but undergo large rearrangements in their relative orientation. This is due to both secondary structure modifications occurring in the hinge regions between the pleckstrin homology domain (domain III) and the fusion domain (domain IV) and complete refolding of the trimerization domain.

Global refolding of G exhibits striking similarities to that of class I proteins, such as paramyxovirus fusion protein F and influenza HA. The pre- and postfusion states of a protomer are related by flipping both the fusion domain and a C-terminal segment relative to a rigid block made by domain I and part of domain II. During this movement, both the fusion loops and the TM domain move $\sim$160 Å from one end of the molecule to the other. The fusion domain is projected toward the membrane through the combination of two movements: a rotation around the hinge between the pleckstrin homology domain and the fusion domain and the lengthening of the central helix (involved in the formation of the trimeric central core of the postfusion conformation). The movement of the transmembrane domain is due to the refolding of the C-terminal segment into an α-helix. In the trimeric postfusion state, the three C-terminal helices position themselves into the groove of the trimeric central core in an antiparallel manner to form a six-helix bundle.

The resulting overall organization of the postfusion state is reminiscent of that of class I protein. As in class I proteins, the fusion domains are N-terminal to the central helices and the transmembrane domains are located at the C-terminus of the antiparallel outer helices.

D. Other Viral Fusion Proteins

There are viral fusion proteins that cannot be classified in one of the three previously described classes. As an example, the fusion machinery of poxviruses (dsDNA viruses, of which the prototype is vaccinia virus) is poorly characterized. Genetic and biochemical studies have provided evidence for an entry/fusion complex, made of at least eight viral proteins, that takes part in entry of vaccinia virus into cells (Senkevich, Ojeda, Townsley, Nelson, & Moss, 2005).

1. Fusion Machinery of Herpesviruses

The fusion machinery of herpesviruses is more complex than that of other enveloped viruses. Although gB has the fold of a class III viral fusion protein (Figure 3B), it does not function alone. The conserved core fusion machinery consists of glycoproteins gB and the gH–gL heterodimer. Furthermore, binding of additional glycoproteins, such as gD in HSV-1 and gp42 in EBV, to the cellular receptors is essential for fusion to occur (Backovic & Jardetzky, 2009; Heldwein & Krummenacher, 2008).

In HSV-1, gD interacts with both gB and gH/gL and triggers formation of the core fusion machinery made of gB and gH/gL (Atanasiu et al., 2007). Recent data have suggested that gB may interact with target membranes prior to an interaction with gH/gL (Atanasiu, Saw, Cohen, & Eisenberg, 2010). The gB–gH/gL interaction seems to be required for the progression of fusion.

There are some conflicting accounts regarding whether gH/gL is a fusion protein. The gH/gL has been reported to be sufficient for induction of hemifusion in the presence of gD, consistent with the idea that gH/gL is a fusion protein (Subramanian & Geraghty, 2007). Two recent reports suggest that it is not the case (Atanasiu et al., 2010; Jackson & Longnecker, 2010) and that binding of gH/gL to gB rather triggers the conformational change of gB from prefusion to postfusion conformation. Consistent with this view, the crystal structure of the gH ectodomain bound to gL from HSV-2 has recently been determined. This complex has a unique architecture that does not resemble any known viral fusogen (Chowdary et al., 2010).

2. Fusion Proteins of Reoviruses

The nonenveloped avian, reptilian, and baboon reoviruses encode small fusion proteins (molecular weight ranging from 10 to 15 kDa) called FAST proteins (fusion-associated small transmembrane proteins) (Shmulevitz & Duncan, 2000). They are nonstructural proteins and thus are not involved in viral entry. They promote only syncytia formation. The small size and the lack of any homologues in previously characterized membrane fusion proteins indicate that the FAST proteins of reovirus represent a new class of membrane fusion proteins. Within the ectodomain of FAST proteins, an 11-residue sequence of apolar residues, termed the hydrophobic patch (HP), flanked by two highly conserved cysteines that form a disulfide bridge and create a cystine loop, constitutes a new class of fusion peptide (Barry, Key, Haddad, & Duncan, 2010). Remarkably, membrane curvature agents that inhibit hemifusion or promote pore formation mediated by other fusion proteins have no effect on FAST-induced cell–cell fusion, indicating that FAST-mediated fusion reaction may somehow differ from other virus-induced fusion reactions (Clancy, Barry, Ciechonska, & Duncan, 2010).

IV. REGULATION OF THE CONFORMATIONAL CHANGE

A. pH Sensitive Molecular Switches

The structures of viral fusion glycoproteins, of which the conformational change is triggered at low pH, has allowed the identification of amino acid residues that play the role of pH-sensitive molecular switches.

For TBEV, the initiation of fusion is crucially dependent on the protonation of one conserved histidine located at the interface between domains I and III of E. Its protonation results into the dissolution of domain interactions and to the exposure of the fusion peptide (Fritz, Stiasny, & Heinz, 2008). A similar study performed on SFV has also identified a histidine residue that acts to regulate the low-pH-dependent refolding of E1 (Qin, Zheng, & Kielian, 2009).

In the case of VSV, it was known that modification of G histidines with diethylpyrocarbonate inhibits its fusion properties (Carneiro et al., 2003). In the prefusion state, three conserved histidines (H60 and H162, both located in the fusion domain, and H407, located in the C-terminal segment of the protein) cluster together (Roche et al., 2007). Protonation of these residues at low pH is likely to destabilize the interaction between the C-terminal segment of G_{th} and the fusion domain in the prefusion conformation. This might trigger the movement of the fusion domain toward the target membrane.

Conversely, in the postfusion form, a large number of acidic amino acids are brought close together in the six-helix bundle (Roche et al., 2006). In the postfusion state, these residues are protonated and form hydrogen bonds. The deprotonation of these residues at higher pH induces strong repulsive forces that destabilize the trimer and trigger the conformational change back to the prefusion state. These residues are buried in the postfusion state, but are solvent-exposed in the prefusion state (Roche et al., 2006, 2007). As a consequence, they have a pK_a that is much higher in the postfusion than in the prefusion conformation. This stronger affinity of the postfusion state for the protons explains the cooperativity of rhabdoviral structural transition as a function of pH (Roche & Gaudin, 2002).

In the case of influenza HA, mutations at distinct positions affect the stability of trimer interfaces, that break during the rearrangement, and therefore alter the threshold pH for fusion (Daniels et al., 1985), and no specific amino acid residue has been identified, of which the protonation would trigger HA conformational change.

B. Dealing with Low pH During Transport Through the Golgi Apparatus

During their maturation and transport through the Golgi apparatus, fusion proteins encounter acidic compartments. It appears that viruses for which fusion is triggered by low pH have the ability to protect their fusion protein from undergoing irreversible conformational change and to avoid undesirable fusion reactions.

For influenza virus, the pH of the transport vesicles is regulated by the M2 protein (Sugrue et al., 1990), a proton channel (Pinto, Holsinger, & Lamb, 1992). Furthermore, the activating cleavage of HA0 into HA1 and HA2 occurs at a late stage of transport (Stieneke-Grober et al., 1992) or after viral budding in the extracellular medium (Skehel & Wiley, 2000).

As mentioned above, class II fusion proteins are transported in association with a chaperone protein. The cleavage of this accompanying protein is required to promote viral fusion under biological conditions. The recent structure of the immature dengue virus at low pH (Yu et al., 2008) has explained how this virus, which buds into the ER, avoids undesirable fusion in the acidic compartments of the Golgi. Immature particles bud into ER as spiky virions with 60 icosahedrally arranged protrusions, each consisting of a trimer of prM–E heterodimers (Y. Zhang et al., 2003b). The atomic structure of dengue virus prM–E heterodimer has been determined (Figure 2C) and fits well into the cryo-electron microscopy density of immature virus (Li et al., 2008). The immature particles are transported into the trans-Golgi network (TGN), where the low pH environment induces a conformational change in the virion, resulting in particles having a much smoother surface with E proteins arranged in a herringbone pattern. This structural transition exposes the furin cleavage site, and proteolysis of prM into pr peptide and M takes place in the TGN. After cleavage, pr peptide remains stably associated with the virion. The pr peptide stabilizes the metastable E dimer, preventing its dissociation and the formation of the postfusion homotrimers. It also covers the fusion loop in E, preventing interaction with host cell membranes. In fact, the dissociation of pr from the virion is pH dependent and pr dissociates only when the virion is released into the slightly alkaline extracellular medium. This definitively primes the metastable E dimer for subsequent low-pH-induced fusogenic conformational change.

For alphaviruses, the recently determined structures of E1 in complex with p62 (Voss et al., 2010), mature E1–E2–E3 glycoprotein complexes (Voss et al., 2010), and trimeric E1–E2 spikes crystallized at low pH (Li, Jose, Xiang, Kuhn, & Rossmann, 2010) have also provided a detailed picture of the maturation of the viral glycoproteins during transport and of the finely tuned regulation of their conformational changes.

For rhabdoviruses such as VSV or rabies virus, the native, prefusion conformation is not metastable, and thus the low-pH-induced structural transition is reversible. In fact, there is a pH-dependent equilibrium between the different conformations of G (Roche & Gaudin, 2002). This allows G to be transported through the acidic compartment of the Golgi apparatus and to recover its native prefusion state at the viral surface (Gaudin, Tuffereau, Durrer, Flamand, & Ruigrok, 1995).

V. WORKING OF THE FUSION MACHINERY

A. Interaction Between the Fusion Peptides and Membranes

Interactions between not only fusion peptides and target membranes but also with viral membranes have been demonstrated for both rhabdoviruses and influenza virus using hydrophobic photolabeling (Durrer, Gaudin, Ruigrok, Graf, & Brunner, 1995; Stegmann, Delfino, Richards, & Helenius, 1991; Tsurudome et al., 1992).

The low pH structures of TBE virus E, dengue fever virus E and SFV E1 (class II), and VSV G (class III) indicate that only hydrophobic side chains of the loops insert into the hydrocarbon chains of the outer leaflet of a target membrane. Any deep penetration of the fusion domain inside the membrane is precluded by the presence of charged residues and polar groups in the vicinity of the fusion loops (Gibbons et al., 2004; Modis et al., 2004; Roche et al., 2006). Residues located in the loops have been shown to be essential to the fusion process by directed mutagenesis experiments on both VSV G (Fredericksen & Whitt, 1995; Sun, Belouzard, & Whittaker, 2007) and herpes virus gB (Backovic, Jardetzky, & Longnecker, 2007; Hannah, Heldwein, Bender, Cohen, & Eisenberg, 2007).

Remarkably, all these fusion proteins have at least one polar aromatic residue in their fusion loops, a feature that is conserved across the rhabdovirus family. Tyrosines and tryptophans are residues that often position themselves at the interface between the fatty acid chains and the polar head group layers of lipids (Wimley & White, 1992). Such an interfacial interaction involving only a few residues cannot create a strong point of anchoring that can be used to pull the target membrane toward the viral one. Most probably, by perturbating the outer leaflet of the target bilayer, it facilitates the formation of point-like protrusions (Campelo, McMahon, & Kozlov, 2008) that have been proposed to be stalk precursors (Efrat, Chernomordik, & Kozlov, 2007). In addition, membrane deformation is probably facilitated by the large number of glycoproteins at the viral surface and their oligomeric status, allowing multiple fusion loops to interact with the external leaflet of the target bilayer.

As for class I protein, the structure of the amino-terminal fusion peptide of HA2 has been determined in a membrane mimicking environment by combining NMR and EPR spectroscopies (Han, Bushweller, Cafiso, & Tamm, 2001). At both pH 7.4 and 5, the domain is a kinked, predominantly helical amphipathic structure. Nevertheless, there are slight differences between both structures that may explain a deeper insertion of the fusion domain into the core of the lipid bilayer at low pH. Nevertheless, here again, in the proposed model of insertion, HA fusion peptides do not penetrate deep into the bilayer structure but rather induce local membrane curvature similar to class II and class III fusion loops. Finally, some class I proteins seem to have also internal fusion loops. This is the case for Ebola virus Gp2 whose fusion loop could expose a conserved WIPYF sequence at its tip (Weissenhorn et al., 1998).

Finally, as mentioned above, in G_{th} prefusion conformation, the fusion loops point toward the viral membrane. Whether they interact with it remains an intriguing question. Indeed, a direct destabilization of the viral bilayer could facilitate the initiation of the fusion process.

B. Other Putative Membrane Destabilizing Domains

Besides the fusion peptides or fusion loops, two other domains have been postulated to play a major role in the fusion process. The first one is the TM domain. Replacement of the TM domain of influenza HA or VSV G by a glycerophosphatidylinosotol (GPI) anchor results in the loss of the fusion properties (Kemble, Danieli, & White, 1994; Odell, Wanas, Yan, & Ghosh, 1997). In the case of GPI-anchored HA, fusion is blocked at an early stage (Kemble et al., 1994). In fact, it appears that GPI-anchored HA is able to open the initial fusion pore but is unable to expand it for complete membrane merger (Frolov, Cho, Bronk, Reese, & Zimmerberg 2000; Markosyan, Cohen, & Melikyan, 2000; Nüssler, Clague, Herrmann, 1997).

It has also been demonstrated that the transmembrane domain of influenza HA exhibits a stringent length requirement to support the hemifusion to fusion transition and that it must span the bilayer to promote full fusion (Armstrong, Kushnir, & White, 2000). Furthermore, mutations of some transmembrane residues of many fusion proteins have been shown to affect fusion (Cleverley & Lenard, 1998; Helseth et al., 1990; Melikyan, Markosyan, Roth, & Cohen, 2000). Mutagenesis studies performed on VSV revealed a crucial role for glycine residues in the TM domain. Mutations of these residues block fusion at a hemifusion stage (Cleverley & Lenard, 1998).

From all these observations, it has been proposed that the TM domain acts downstream of the formation of the initial fusion pore and contributes to its expansion. Nevertheless, the precise role played by the TM domain at a molecular level is still far from being understood.

The second domain is the membrane-proximal domain. For many fusion proteins, this rather hydrophobic and frequently tryptophan-rich domain has been shown to play a role in the fusion reaction (Howard et al., 2008; Vishwanathan & Hunter, 2008).

For VSV, a deletion of the 13 membrane-proximal amino acids dramatically reduces cell–cell fusion activity and reduces virus infectivity approximately 100-fold (Jeetendra et al., 2003). However, mutations of conserved aromatic residues (W441, F442, and W445) indicated that these residues are not important for fusion (Jeetendra et al., 2003). The structure of this segment, absent in the crystal structure, is not known but in the case of rabies virus G and in agreement with the structure of other class III proteins, it has been proposed to adopt an α-helical conformation having a strong amphipathic signature (Maillard et al., 2003).

Similar data have been obtained with HIV for which NMR experiments suggest that the membrane-proximal segment could form a bent amphipathic helix lying at the surface of the membrane bilayer (Schibli, Montelaro, & Vogel, 2001; Sun et al., 2008). Finally, the recent crystalline structure of a gp41 fragment including both fusion peptide and membrane-proximal external regions confirms that three aromatic side chains (Trp 678, Trp 680, and Tyr 681) are exposed toward the membrane and well positioned to insert their side chains into the bilayer (Buzon et al., 2010). Whether these residues are anchored in the viral membrane only in the postfusion conformation or in both the pre- and postfusion conformations remains an open question.

C. Structural Intermediates During the Conformational Changes

1. The Extended Intermediate Conformation

All the data obtained are consistent with the formation of an extended intermediate conformation (Figure 1A, right; Figure 3A, right). In this conformation, the fusion peptides or fusion loops are exposed at the top of the molecule and directed toward the target membrane. This intermediate is formed early during the fusion process as demonstrated by hydrophobic photolabeling for rhabdoviruses and influenza virus (Durrer et al., 1995; Stegmann et al., 1991; Tsurudome et al., 1992).

This extended intermediate has been particularly studied in the case of gp41. For this intermediate, it is proposed that the central coiled coil (made of HR1 segments) is already formed but that the relocation of HR2 segments against this central core has not occurred. Indeed, peptides corresponding to the HR2 segment can inhibit the fusion process (Furuta, Wild, Weng, & Weiss, 1998; Wild, Shugars, Greenwell, McDanal, & Matthews, 1994). A similar inhibition has been observed for paramyxoviruses with peptides derived from the carboxyterminal HRB region of F protein (Russell, Jardetzky, & Lamb, 2001). Such inhibitors do not function in the case of influenza HA. This is probably due to the shorter lifetime of HA extended conformation compared to that of gp41 (several minutes) (Melikyan et al., 2000; Munoz-Barroso, Durell, Sakaguchi, Appella, & Blumenthal, 1998).

For class II fusion proteins, it has been shown that domain III is also acting as a dominant-negative inhibitor of virus membrane fusion (Liao & Kielian, 2005). For SFV, it has been shown to bind to an intermediate trimeric E1 conformation and is thus supposed to inhibit the folding back of the molecule into the hairpin conformation.

For TBEV, exposure of virions to alkaline instead of acidic pH traps the glycoprotein in a monomeric intermediate conformation that is perpendicular to the viral membrane and interacts with target membranes via the fusion loop without proceeding to the merger of the membranes. Further treatment at low pH, however, leads to fusion, suggesting that these monomers correspond to an intermediate required to convert the prefusion dimer into the postfusion trimer (Stiasny, Kossl, Lepault, Rey, & Heinz, 2007).

2. Monomerization During the Refolding Mechanism

Refolding of class II fusion proteins generates trimers assembled from intermediate monomer conformations. Whether monomerization also occurs during class I and class III proteins refolding remains a matter of debate. Although the initial steps leading to fusion peptide exposure and its interaction with the target membrane may maintain strict trimeric symmetry, the folding back of the C-terminal part of the molecule requires to break the threefold symmetry of the molecule (Weissenhorn, Hinz, & Gaudin, 2007).

For both VSV G and paramyxovirus F proteins, there are large differences between the trimeric interfaces of the pre- and postfusion structures (Roche et al., 2006, 2007; Yin et al., 2005, 2006). A plausible scheme for the structural transition is that it goes through transient monomerization (Roche, Albertini, Lepault, Bressanelli, & Gaudin, 2008). Consistent with this hypothesis, it has been demonstrated that VSV G is a monomer at pH 7.4 (Doms, Keller, Helenius & Balch, 1987) after detergent solubilization. It is possible that this monomer is an intermediate state between the pre-and postfusion conformations.

D. Cooperativity

The activation energy of the fusion process has been estimated to be in the range of 40 kcal/mol (Lee & Lentz, 1998). For class I and class II fusion proteins, it has been proposed that the energy released during the irreversible transition from the metastable prefusion state to the stable postfusion state is used to overcome the high energetic barrier encountered during the fusion process. Indeed, energy recovery via refolding of a single env trimer is theoretically sufficient to overcome the free energy barrier of fusion (Kuzmin, Zimmerberg, Chizmadzhev, & Cohen, 2001) and experimental data have suggested that in the case of HIV, a single env glycoprotein trimer is sufficient for fusion (Yang, Kurteva, Ren, Lee, & Sodroski, 2005, 2006). However, work on many viruses (also including HIV) suggests that several fusion proteins are involved during the fusion process.

For HIV, another analysis of the data used by Yang and coworkers, with less restricted mathematical models, indicated that HIV fusion machinery is made of several env proteins (Klasse, 2007). This latter conclusion is consistent with other experimental data, indicating that both cellular receptor density and env density affect the kinetics of HIV-1 env-mediated fusion (Reeves et al., 2002) and that four to six coreceptor molecules are recruited into fusion complexes (Kuhmann, Platt, Kozak, & Kabat, 2000). Such interactions would require more than one Env trimer per virion.

Work on influenza virus hemagglutinin also indicates that HA surface density is important for fusion (Danieli, Pelletier, Henis, & White, 1996) and that its decrease arrests fusion at the hemifusion stage (Chernomordik et al., 1998). The restriction to lipid diffusion observed downstream of stalk formation has been attributed to the formation of a ring of HA surrounding the initial membrane contacts (Chernomordik et al., 1998). Such a ring structure of HA around a lipid neck joining the viral and the liposome membranes has been recently observed by cryo-electron microscopy (Lee, 2010). Finally, it has been demonstrated that HA molecules located outside the direct contact zone between fusing membranes are involved in the late stages of the fusion process (Leikina et al., 2004).

In the case of rhabdoviruses, the pH-dependent equilibrium between pre- and postfusion conformations of G implies that the energy released during the structural transition of one trimer cannot be sufficient to overcome the activation energy barrier and drive the fusion reaction. Indeed, the minimal number of spikes involved in the formation of a rabies virus fusion complex has been estimated to be about 15 (Roche & Gaudin, 2002).

How proteins cooperate to facilitate the formation of the different intermediates on the fusion pathway is still not known. However, some recent data provide new insights into the organization of the fusion machinery and suggest that distinct assemblies of fusion glycoproteins are involved at distinct stages of the fusion process.

For rabies virus (another member of the rhabdovirus family), a local hexagonal lattice of glycoproteins was observed at the surface of some rabies virus mutants affected in the kinetics of their structural transition when briefly incubated under suboptimal fusion conditions (Gaudin, Raux, Flamand, & Ruigrok, 1996). Each hexagon was made up of six G trimers. Interestingly, a similar organization of G_{th} was found in the crystals of the prefusion form (that had the P622 symmetry) (Roche et al., 2007).

Furthermore, an EM study performed on VSV revealed that the flat base of the virion (as other rhabdoviruses, VSV has a typical bullet shape with a flat base and a round tip) is a privileged site for fusion (Libersou et al., 2010). The plane nature of this part of the virion would help the protein to form the P6 lattice. Such a local organization could favor not only the formation of the initial lipid intermediates but also a concerted transition for the proteins making up the local network.

A second network, helical this time, of glycoproteins in their postfusion conformations can be observed on the cylindrical part of VSV particles incubated below pH 6 (Libersou et al., 2010). When formed in absence of a target membrane, this helical array disrupts the viral membrane, indicating that this network formation induces tension in the viral membranes. Furthermore, when G_{th} alone is incubated with target liposomes at low pH, this also leads to the formation of tubular structures at the surface of which G_{th} adopts a helical symmetry. This indicates that the postfusion state of G ectodomain is independently able to self-associate into a quasicrystalline array, thereby deforming membranes. It has been proposed that these interactions between glycoproteins G, in their postfusion conformation and located outside the contact zone, drive the enlargement of the initial fusion pore (Libersou et al., 2010).

A similar two-stage model can also be proposed for class II fusion glycoproteins. In their prefusion conformation, they adopt an icosahedral organization (Kuhn et al., 2002; Lescar et al., 2001). Here again, this organization could favor a concerted conformational change (at least locally) of the protein making the network. In their postfusion conformation, the spikes self-associate to form more or less regular networks, different from their initial organization observed before fusion (Gibbons et al., 2003; Sanchez-San Martin, Sosa, & Kielian, 2008; Stiasny, Bressanelli, Lepault, Rey, & Heinz, 2004).

Finally, for paramyxovirus PIV5, micrographs indicate that F proteins in their postfusion conformation tend to cluster at the viral surface (Ludwig et al., 2008). Thus, the tendency of fusion glycoproteins in their postfusion conformation to cluster and form closely packed arrays seems to be general. We propose that the glycoproteins located outside the contact zone between fusing membranes exert a role during the fusion process via the formation of these networks by inducing tension in the viral membrane.

An objection to this model can be raised that retroviruses, and particularly HIV, have very few spikes at their surface (Zhu et al., 2006). This is hardly reconcilable with network formation. For these viruses, it is possible that membrane tension needed at the late stages of the fusion process to drive the enlargement of the initial fusion pore is generated by cellular proteins in response to signalization induced by env binding to coreceptors. Interestingly, recent data revealed that dynamin is required for efficient HIV-1 fusion (Miyauchi, Kim, Latinovic, Morozov, & Melikyan, 2009). The authors have proposed that the ability of dynamin to regulate actin remodeling and associate with membrane-bending proteins (Kruchten & McNiven, 2006) could provide a driving force for initial pore expansion.

VI. CONCLUDING REMARKS

The determination of an increasing number of structures of viral fusion proteins and recent electron microscopy studies have shed light on the molecular mechanisms of fusion machinery. This has revealed similar principles of action for machinery made up of proteins having very different shapes. Nevertheless, some new hypotheses remain to be challenged and many questions remain unanswered.

As mentioned above, it is known that transmembrane domains of viral fusion proteins play an important role at the late stages of the fusion process, but no structure of full-length viral fusion glycoprotein (i.e., with its TM domain) has been determined. Thus, further structural studies of complete pre- and postfusion conformations of viral glycoproteins containing both membrane anchors (TM and fusion peptides) are needed. This should indicate how both regions cooperate to regulate the transition from hemifusion to fusion pore opening and pore enlargement.

The structures of intermediate conformations of fusion proteins, how they cooperate, and how they interact with and deform the fusing membranes are still very elusive. Understanding the structure of full-length intermediates in associations with the membranes is now a major challenge in the field.

Acknowledgments

This project was supported by the CNRS, INRA, and grants from the ANR, including postdoctoral fellowships to AA (ANR-08-BLAN-0256) and from Région Ile-de-France (SESAME 2006, # 1-07-579/R).

References

Aguilar, H. C., Matreyek, K. A., Filone, C. M., Hashimi, S. T., Levroney, E. L., Negrete, O. A., Bertolotti-Ciarlet, A., Choi, D. Y., McHardy, I., Fulcher, J. A., Su, S. V., Wolf, M. C., Kohatsu, L., Baum, L. G., & Lee, B. (2006). N-glycans on Nipah virus fusion protein protect against neutralization but reduce membrane fusion and viral entry. *Journal of Virology, 80*, 4878–4889.

Armstrong, R. T., Kushnir, A. S., & White, J. M. (2000). The transmembrane domain of influenza hemagglutinin exhibits a stringent length requirement to support the hemifusion to fusion transition. *Journal of Cell Biology, 151*, 425–437.

Atanasiu, D., Saw, W. T., Cohen, G. H., & Eisenberg, R. J. (2010). Cascade of events governing cell–cell fusion induced by herpes simplex virus glycoproteins gD, gH/gL, and gB. *Journal of Virology, 84*, 12292–12299.

Atanasiu, D., Whitbeck, J. C., Cairns, T. M., Reilly, B., Cohen, G. H., & Eisenberg, R. J. (2007). Bimolecular complementation reveals that glycoproteins gB and gH/gL of herpes simplex virus interact with each other during cell fusion. *Proceedings of the National Academy of Sciences of the United States of America, 104*, 18718–18723.

Backovic, M., & Jardetzky, T. S. (2009). Class III viral membrane fusion proteins. *Current Opinion in Structural Biology, 19*, 189–196.

Backovic, M., Jardetzky, T. S., & Longnecker, R. (2007). Hydrophobic residues that form putative fusion loops of Epstein-Barr virus glycoprotein B are critical for fusion activity. *Journal of Virology, 81*, 9596–9600.

Backovic, M., Longnecker, R., & Jardetzky, T. S. (2009). Structure of a trimeric variant of the Epstein-Barr virus glycoprotein B. *Proceedings of the National Academy of Sciences of the United States of America, 106*, 2880–2885.

Baker, K. A., Dutch, R. E., Lamb, R. A., & Jardetzky, T. S. (1999). Structural basis for paramyxovirus-mediated membrane fusion. *Molecular Cell, 3*, 309–319.

Barry, C., Key, T., Haddad, R., & Duncan, R. (2010). Features of a spatially constrained cystine loop in the p10 FAST protein ectodomain define a new class of viral fusion peptides. *Journal of Biological Chemistry, 285*, 16424–16433.

Bishop, K. A., Hickey, A. C., Khetawat, D., Patch, J. R., Bossart, K. N., Zhu, Z., Wang, L. F., Dimitrov, D. S., & Broder, C. C. (2008). Residues in the stalk domain of the hendra virus g glycoprotein modulate conformational changes associated with receptor binding. *Journal of Virology, 82*, 11398–11409.

Bishop, K. A., Stantchev, T. S., Hickey, A. C., Khetawat, D., Bossart, K. N., Krasnoperov, V., Gill, P., Feng, Y. R., Wang, L., Eaton, B. T., Wang, L. F., & Broder, C. C. (2007). Identification of Hendra virus G glycoprotein residues that are critical for receptor binding. *Journal of Virology, 81*, 5893–5901.

Bizebard, T., Gigant, B., Rigolet, P., Rasmussen, B., Diat, O., Bosecke, P., Wharton, S. A., Skehel, J. J., & Knossow, M. (1995). Structure of influenza virus haemagglutinin complexed with a neutralizing antibody. *Nature, 376*, 92–94.

Bousse, T., Takimoto, T., Gorman, W. L., Takahashi, T., & Portner, A. (1994). Regions on the hemagglutinin-neuraminidase proteins of human parainfluenza virus type-1 and Sendai virus important for membrane fusion. *Virology, 204*, 506–514.

Bowden, T. A., Aricescu, A. R., Gilbert, R. J., Grimes, J. M., Jones, E. Y., & Stuart, D. I. (2008a). Structural basis of Nipah and Hendra virus attachment to their cell-surface receptor ephrin-B2. *Nature Structural & Molecular Biology, 15*, 567–572.

Bowden, T. A., Crispin, M., Harvey, D. J., Aricescu, A. R., Grimes, J. M., Jones, E. Y., & Stuart, D. I. (2008b). Crystal structure and carbohydrate analysis of Nipah virus attachment glycoprotein: a template for antiviral and vaccine design. *Journal of Virology, 82*, 11628–11636.

Bowden, T. A., Crispin, M., Harvey, D. J., Jones, E. Y., & Stuart, D. I. (2010). Dimeric architecture of the Hendra virus attachment glycoprotein: evidence for a conserved mode of assembly. *Journal of Virology, 84*, 6208–6217.

Bressanelli, S., Stiasny, K., Allison, S. L., Stura, E. A., Duquerroy, S., Lescar, J., Heinz, F. X., & Rey, F. A. (2004). Structure of a flavivirus envelope glycoprotein in its low-pH-induced membrane fusion conformation. *EMBO Journal, 23*, 728–738.

Bullough, P. A., Hughson, F. M., Skehel, J. J., & Wiley, D. C. (1994). Structure of influenza haemagglutinin at the pH of membrane fusion. *Nature, 371*, 37–43.

Buzon, V., Natrajan, G., Schibli, D., Campelo, F., Kozlov, M. M., & Weissenhorn, W. (2010). Crystal structure of HIV-1 gp41 including both fusion peptide and membrane proximal external regions. *PLoS Pathogens, 6*, e1000880.

Campelo, F., McMahon, H. T., & Kozlov, M. M. (2008). The hydrophobic insertion mechanism of membrane curvature generation by proteins. *Biophysical Journal, 95*, 2325–2339.

Carneiro, F. A., Stauffer, F., Lima, C. S., Juliano, M. A., Juliano, L., & Da Poian, A. T. (2003). Membrane fusion induced by vesicular stomatitis virus depends on histidine protonation. *Journal of Biological Chemistry, 278*, 13789–13794.

Carr, C. M., Chaudhry, C., & Kim, P. S. (1997). Influenza hemagglutinin is spring-loaded by a metastable native conformation. *Proceedings of the National Academy of Sciences of the United States of America, 94*, 14306–14313.

Chandran, K., Sullivan, N. J., Felbor, U., Whelan, S. P., & Cunningham, J. M. (2005). Endosomal proteolysis of the Ebola virus glycoprotein is necessary for infection. *Science, 308*, 1643–1645.

Chen, J., Skehel, J. J., & Wiley, D. C. (1999). N- and C-terminal residues combine in the fusion-pH influenza hemagglutinin HA(2) subunit to form an N cap that terminates the triple-stranded coiled coil. *Proceedings of the National Academy of Sciences of the United States of America, 96*, 8967–8972.

Chernomordik, L. V., Frolov, V. A., Leikina, E., Bronk, P., & Zimmerberg, J. (1998). The pathway of membrane fusion catalyzed by influenza hemagglutinin: restriction of lipids, hemifusion, and lipidic fusion pore formation. *Journal of Cell Biology, 140*, 1369–1382.

Chernomordik, L., Kozlov, M. M., & Zimmerberg, J. (1995). Lipids in biological membrane fusion. *Journal of Membrane Biology, 146*, 1–14.

Chernomordik, L. V., & Kozlov, M. M. (2005). Membrane hemifusion: crossing a chasm in two leaps. *Cell, 123*, 375–382.

Chernomordik, L. V., & Kozlov, M. M. (2008). Mechanics of membrane fusion. *Nature Structural & Molecular Biology, 15*, 675–683.

Chernomordik, L. V., Leikina, E., Frolov, V., Bronk, P., & Zimmerberg, J. (1997). An early stage of membrane fusion mediated by the low pH conformation of influenza hemagglutinin depends upon membrane lipids. *Journal of Cell Biology, 136*, 81–93.

Chowdary, T. K., Cairns, T. M., Atanasiu, D., Cohen, G. H., Eisenberg, R. J., & Heldwein, E. E. (2010). Crystal structure of the conserved herpesvirus fusion regulator complex gH-gL. *Nature Structural & Molecular Biology, 17*, 882–888.

Clancy, E. K., Barry, C., Ciechonska, M., & Duncan, R. (2010). Different activities of the reovirus FAST proteins and influenza hemagglutinin in cell–cell fusion assays and in response to membrane curvature agents. *Virology, 397*, 119–129.

Cleverley, D. Z., & Lenard, J. (1998). The transmembrane domain in viral fusion: essential role for a conserved glycine residue in vesicular stomatitis virus G protein. *Proceedings of the National Academy of Sciences of the United States of America, 95*, 3425–3430.

Cohen, F. S., & Melikyan, G. B. (2004). The energetics of membrane fusion from binding, through hemifusion, pore formation, and pore enlargement. *Journal of Membrane Biology, 199*, 1–14.

Colf, LA., Juo, Z. S., & Garcia, K. C. (2007). Structure of the measles virus hemagglutinin. *Nature Structural & Molecular Biology, 14*, 1227–1228.

Connolly, S. A., Leser, G. P., Jardetzky, T. S., & Lamb, R. A. (2009). Bimolecular complementation of paramyxovirus fusion and hemagglutinin-neuraminidase proteins enhances fusion: implications for the mechanism of fusion triggering. *Journal of Virology, 83*, 10857–10868.

Corey, E. A., & Iorio, R. M. (2007). Mutations in the stalk of the measles virus hemagglutinin protein decrease fusion but do not interfere with virus-specific interaction with the homologous fusion protein. *Journal of Virology, 81*, 9900–9910.

Corey, E. A., & Iorio, R. M. (2009). Measles virus attachment proteins with impaired ability to bind CD46 interact more efficiently with the homologous fusion protein. *Virology, 383*, 1–5.

Crennell, S., Takimoto, T., Portner, A., & Taylor, G. (2000). Crystal structure of the multifunctional paramyxovirus hemagglutinin-neuraminidase. *Natural Structural Biology, 7*, 1068–1074.

Danieli, T., Pelletier, S. L., Henis, Y. I., & White, J. M. (1996). Membrane fusion mediated by the influenza virus hemagglutinin requires the concerted action of at least three hemagglutinin trimers. *Journal of Cell Biology, 133*, 559–569.

Daniels, R. S., Downie, J. C., Hay, A. J., Knossow, M., Skehel, J. J., Wang, M. L., & Wiley, D. C. (1985). Fusion mutants of the influenza virus hemagglutinin glycoprotein. *Cell, 40*, 431–439.

Doms, R. W., Keller, D. S., Helenius, A., & Balch, W. E. (1987). Role for adenosine triphosphate in regulating the assembly and transport of vesicular stomatitis virus G protein trimers. *J Cell Biol, 105*, 1957–1969.

Duquerroy, S., Vigouroux, A., Rottier, P. J., Rey, F. A., & Bosch, B. J. (2005). Central ions and lateral asparagine/glutamine zippers stabilize the post-fusion hairpin conformation of the SARS coronavirus spike glycoprotein. *Virology, 335*, 276–285.

Durrer, P., Gaudin, Y., Ruigrok, R. W., Graf, R., & Brunner, J. (1995). Photolabeling identifies a putative fusion domain in the envelope glycoprotein of rabies and vesicular stomatitis viruses. *Journal of Biological Chemistry, 270*, 17575–17581.

Efrat, A., Chernomordik, L. V., & Kozlov, M. M. (2007). Point-like protrusion as a prestalk intermediate in membrane fusion pathway. *Biophysical Journal, 92*, L61–L63.

Fredericksen, B. L., & Whitt, M. A. (1995). Vesicular stomatitis virus glycoprotein mutations that affect membrane fusion activity and abolish virus infectivity. *Journal of Virology, 69*, 1435–1443.

Fritz, R., Stiasny, K., & Heinz, F. X. (2008). Identification of specific histidines as pH sensors in flavivirus membrane fusion. *Journal of Cell Biology, 183*, 353–361.

Frolov, V. A., Cho, M. S., Bronk, P., Reese, T. S., & Zimmerberg, J. (2000). Multiple local contact sites are induced by GPI-linked influenza hemagglutinin during hemifusion and flickering pore formation. *Traffic, 1*, 622–630.

Furuta, R. A., Wild, C. T., Weng, Y., & Weiss, C. D. (1998). Capture of an early fusion-active conformation of HIV-1 gp41. *Natural Structural Biology, 5*, 276–279.

Garry, C. E., & Garry, R. F. (2004). Proteomics computational analyses suggest that the carboxyl terminal glycoproteins of Bunyaviruses are class II viral fusion protein (beta-penetrenes). *Theoretical Biology and Medical Modelling, 1*, 10.

Garry, C. E., & Garry, R. F. (2009). Proteomics computational analyses suggest that the bornavirus glycoprotein is a class III viral fusion protein (gamma penetrene). *Virology Journal, 6*, 145.

Gaudin, Y. (2000). Rabies virus-induced membrane fusion pathway. *Journal of Cell Biology, 150*, 601–612.

Gaudin, Y., Raux, H., Flamand, A., & Ruigrok, R. W. (1996). Identification of amino acids controlling the low-pH-induced conformational change of rabies virus glycoprotein. *Journal of Virology, 70*, 7371–7378.

Gaudin, Y., Tuffereau, C., Durrer, P., Flamand, A., & Ruigrok, R. W. (1995). Biological function of the low-pH, fusion-inactive conformation of rabies virus glycoprotein (G): G is transported in a fusion-inactive state-like conformation. *Journal of Virology, 69*, 5528–5534.

Gibbons, D. L., Erk, I., Reilly, B., Navaza, J., Kielian, M., Rey, F. A., & Lepault, J. (2003). Visualization of the target-membrane-inserted fusion protein of Semliki Forest virus by combined electron microscopy and crystallography. *Cell, 114*, 573–583.

Gibbons, D. L., Vaney, M. C., Roussel, A., Vigouroux, A., Reilly, B., Lepault, J., Kielian, M., & Rey, F. A. (2004). Conformational change and protein–protein interactions of the fusion protein of Semliki Forest virus. *Nature, 427*, 320–325.

Han, X., Bushweller, J. H., Cafiso, D. S., & Tamm, L. K. (2001). Membrane structure and fusion-triggering conformational change of the fusion domain from influenza hemagglutinin. *Natural Structural Biology, 8*, 715–720.

Hannah, B. P., Heldwein, E. E., Bender, F. C., Cohen, G. H., & Eisenberg, R. J. (2007). Mutational evidence of internal fusion loops in herpes simplex virus glycoprotein B. *Journal of Virology, 81*, 4858–4865.

Harrison, S. C. (2008). Viral membrane fusion. *Nature Structural & Molecular Biology, 15*, 690–698.

Hashiguchi, T., Kajikawa, M., Maita, N., Takeda, M., Kuroki, K., Sasaki, K., Kohda, D., Yanagi, Y., & Maenaka, K. (2007). Crystal structure of measles virus hemagglutinin provides insight into effective vaccines. *Proceedings of the National Academy of Sciences of the United States of America, 104*, 19535–19540.

Heinz, F. X., Stiasny, K., Puschner-Auer, G., Holzmann, H., Allison, S. L., Mandl, C. W., & Kunz, C. (1994). Structural changes and functional control of the tick-borne encephalitis virus glycoprotein E by the heterodimeric association with protein prM. *Virology, 198*, 109–117.

Heldwein, E. E., & Krummenacher, C. (2008). Entry of herpesviruses into mammalian cells. *Cellular and Molecular Life Sciences, 65*, 1653–1668.

Heldwein, E. E., Lou, H., Bender, F. C., Cohen, G. H., Eisenberg, R. J., & Harrison, S. C. (2006). Crystal structure of glycoprotein B from herpes simplex virus 1. *Science, 313*, 217–220.

Helseth, E., Olshevsky, U., Gabuzda, D., Ardman, B., Haseltine, W., & Sodroski, J. (1990). Changes in the transmembrane region of the human immunodeficiency virus type 1 gp41 envelope glycoprotein affect membrane fusion. *Journal of Virology, 64*, 6314–6318.

Howard, M. W., Travanty, E. A., Jeffers, S. A., Smith, M. K., Wennier, S. T., Thackray, L. B., & Holmes, K. V. (2008). Aromatic amino acids in the juxtamembrane domain of severe acute respiratory syndrome coronavirus spike glycoprotein are important for receptor-dependent virus entry and cell–cell fusion. *Journal of Virology, 82*, 2883–2894.

Jackson, J. O., & Longnecker, R. (2010). Reevaluating herpes simplex virus hemifusion. *Journal of Virology, 84*, 11814–11821.

Jeetendra, E., Ghosh, K., Odell, D., Li, J., Ghosh, H. P., & Whitt, M. A. (2003). The membrane-proximal region of vesicular stomatitis virus glycoprotein G ectodomain is critical for fusion and virus infectivity. *Journal of Virology, 77*, 12807–12818.

Kadlec, J., Loureiro, S., Abrescia, N. G., Stuart, D. I., & Jones, I. M. (2008). The postfusion structure of baculovirus gp64 supports a unified view of viral fusion machines. *Nature Structural & Molecular Biology, 15*, 1024–1030.

Kanai, R., Kar, K., Anthony, K., Gould, L. H., Ledizet, M., Fikrig, E., Marasco, W. A., Koski, R. A., & Modis, Y. (2006). Crystal structure of West Nile virus envelope glycoprotein reveals viral surface epitopes. *Journal of Virology, 80*, 11000–11008.

Kemble, G. W., Danieli, T., & White, J. M. (1994). Lipid-anchored influenza hemagglutinin promotes hemifusion, not complete fusion. *Cell, 76*, 383–391.

Kielian, M. C., & Helenius, A. (1984). Role of cholesterol in fusion of Semliki Forest virus with membranes. *Journal of Virology, 52*, 281–283.

Klasse, P. J. (2007). Modeling how many envelope glycoprotein trimers per virion participate in human immunodeficiency virus infectivity and its neutralization by antibody. *Virology, 369*, 245–262.

Krey, T., d'Alayer, J., Kikuti, C. M., Saulnier, A., Damier-Piolle, L., Petitpas, I., Johansson, D. X., Tawar, R. G., Baron, B., Robert, B., England, P., Persson, M. A., Martin, A., & Rey, F. A. (2010). The disulfide bonds in glycoprotein E2 of hepatitis C virus reveal the tertiary organization of the molecule. *PLoS Pathogens, 6*, e1000762.

Kruchten, A. E., & McNiven, M. A. (2006). Dynamin as a mover and pincher during cell migration and invasion. *Journal of Cell Science, 119*, 1683–1690.

Kuhmann, S. E., Platt, E. J., Kozak, S. L., & Kabat, D. (2000). Cooperation of multiple CCR5 coreceptors is required for infections by human immunodeficiency virus type 1. *Journal of Virology, 74*, 7005–7015.

Kuhn, R. J., Zhang, W., Rossmann, M. G., Pletnev, S. V., Corver, J., Lenches, E., Jones, C. T., Mukhopadhyay, S., Chipman, P. R., Strauss, E. G., Baker, T. S., & Strauss, J. H. (2002). Structure of dengue virus: implications for flavivirus organization, maturation, and fusion. *Cell, 108*, 717–725.

Kuzmin, P. I., Zimmerberg, J., Chizmadzhev, Y. A., & Cohen, F. S. (2001). A quantitative model for membrane fusion based on low-energy intermediates. *Proceedings of the National Academy of Sciences of the United States of America, 98*, 7235–7240.

Lamb, R. A., & Jardetzky, T. S. (2007). Structural basis of viral invasion: lessons from paramyxovirus F. *Current Opinion in Structural Biology, 17*, 427–436.

Lavillette, D., Bartosch, B., Nourrisson, D., Verney, G., Cosset, F. L., Penin, F., & Pecheur, E. I. (2006). Hepatitis C virus glycoproteins mediate low pH-dependent membrane fusion with liposomes. *Journal of Biological Chemistry, 281*, 3909–3917.

Lawrence, M. C., Borg, N. A., Streltsov, V. A., Pilling, P. A., Epa, V. C., Varghese, J. N., McKimm-Breschkin, J. L., & Colman, P. M. (2004). Structure of the haemagglutinin-neuraminidase from human parainfluenza virus type III. *Journal of Molecular Biology, 335*, 1343–1357.

Lee, J., & Lentz, B. R. (1998). Secretory and viral fusion may share mechanistic events with fusion between curved lipid bilayers. *Proceedings of the National Academy of Sciences of the United States of America, 95*, 9274–9279.

Lee, J. E., Fusco, M. L., Hessell, A. J., Oswald, W. B., Burton, D. R., & Saphire, E. O. (2008). Structure of the Ebola virus glycoprotein bound to an antibody from a human survivor. *Nature, 454*, 177–182.

Lee, K. K. (2010). Architecture of a nascent viral fusion pore. *EMBO Journal, 29*, 1299–1311.

Leikina, E., Mittal, A., Cho, M. S., Melikov, K., Kozlov, M. M., & Chernomordik, L. V. (2004). Influenza hemagglutinins outside of the contact zone are necessary for fusion pore expansion. *Journal of Biological Chemistry, 279*, 26526–26532.

Lescar, J., Roussel, A., Wien, M. W., Navaza, J., Fuller, S. D., Wengler, G., Wengler, G., & Rey, F. A. (2001). The fusion glycoprotein shell of Semliki Forest virus: an icosahedral assembly primed for fusogenic activation at endosomal pH. *Cell, 105*, 137–148.

Li, L., Jose, J., Xiang, Y., Kuhn, R. J., & Rossmann, M. G. (2010). Structural changes of envelope proteins during alphavirus fusion. *Nature, 468*, 705–708.

Li, L., Lok, S. M., Yu, I. M., Zhang, Y., Kuhn, R. J., Chen, J., & Rossmann, M. G. (2008). The flavivirus precursor membrane-envelope protein complex: structure and maturation. *Science, 319*, 1830–1834.

Liao, M., & Kielian, M. (2005). Domain III from class II fusion proteins functions as a dominant-negative inhibitor of virus membrane fusion. *Journal of Cell Biology, 171*, 111–120.

Libersou, S., Albertini, A. A., Ouldali, M., Maury, V., Maheu, C., Raux, H., de Haas, F., Roche, S., Gaudin, Y., & Lepault, J. (2010). Distinct structural rearrangements of the VSV glycoprotein drive membrane fusion. *Journal of Cell Biology, 191*, 199–210.

Ludwig, K., Schade, B., Bottcher, C., Korte, T., Ohlwein, N., Baljinnyam, B., Veit, M., & Herrmann, A. (2008). Electron cryomicroscopy reveals different F1 + F2 protein states in intact parainfluenza virions. *Journal of Virology, 82*, 3775–3781.

Maillard, A., Domanski, M., Brunet, P., Chaffotte, A., Guittet, E., & Gaudin, Y. (2003). Spectroscopic characterization of two peptides derived from the stem of rabies virus glycoprotein. *Virus Research, 93*, 151–158.

Markosyan, R. M., Cohen, F. S., & Melikyan, G. B. (2000). The lipid-anchored ectodomain of influenza virus hemagglutinin (GPI-HA) is capable of inducing nonenlarging fusion pores. *Molecular Biology of the Cell, 11*, 1143–1152.

Melikyan, G. B. (2008). Common principles and intermediates of viral protein-mediated fusion: the HIV-1 paradigm. *Retrovirology, 5*, 111.

Melikyan, G. B., Markosyan, R. M., Hemmati, H., Delmedico, M. K., Lambert, D. M., & Cohen, F. S. (2000a). Evidence that the transition of HIV-1 gp41 into a six-helix bundle, not the bundle configuration, induces membrane fusion. *Journal of Cell Biology, 151*, 413–423.

Melikyan, G. B., Markosyan, R. M., Roth, M. G., & Cohen, F. S. (2000b). A point mutation in the transmembrane domain of the hemagglutinin of influenza virus stabilizes a hemifusion intermediate that can transit to fusion. *Molecular Biology of the Cell, 11*, 3765–3775.

Miyauchi, K., Kim, Y., Latinovic, O., Morozov, V., & Melikyan, G. B. (2009). HIV enters cells via endocytosis and dynamin-dependent fusion with endosomes. *Cell, 137*, 433–444.

Modis, Y., Ogata, S., Clements, D., & Harrison, S. C. (2003). A ligand-binding pocket in the dengue virus envelope glycoprotein. *Proceedings of the National Academy of Sciences of the United States of America, 100*, 6986–6991.

Modis, Y., Ogata, S., Clements, D., & Harrison, S. C. (2004). Structure of the dengue virus envelope protein after membrane fusion. *Nature, 427*, 313–319.

Moesker, B., Rodenhuis-Zybert, I. A., Meijerhof, T., Wilschut, J., & Smit, J. M. (2010). Characterization of the functional requirements of West Nile virus membrane fusion. *Journal of General Virology, 91*, 389–393.

Munoz-Barroso, I., Durell, S., Sakaguchi, K., Appella, E., & Blumenthal, R. (1998). Dilation of the human immunodeficiency virus-1 envelope glycoprotein fusion pore revealed by the inhibitory action of a synthetic peptide from gp41. *Journal of Cell Biology, 140*, 315–323.

Nieva, J. L., Bron, R., Corver, J., & Wilschut, J. (1994). Membrane fusion of Semliki Forest virus requires sphingolipids in the target membrane. *EMBO Journal, 13*, 2797–2804.

Nüssler, F., Clague, M. J., & Herrmann, A. (1997). Meta-stability of the hemifusion intermediate induced by glycosylphosphatidylinositol-anchored influenza hemagglutinin. *Biophysical Journal, 73*, 2280–2291.

Nybakken, G. E., Nelson, C. A., Chen, B. R., Diamond, M. S., & Fremont, D. H. (2006). Crystal structure of the West Nile virus envelope glycoprotein. *Journal of Virology, 80*, 11467–11474.

Odell, D., Wanas, E., Yan, J., & Ghosh, H. P. (1997). Influence of membrane anchoring and cytoplasmic domains on the fusogenic activity of vesicular stomatitis virus glycoprotein G. *Journal of Virology, 71*, 7996–8000.

Pinto, L. H., Holsinger, L. J., & Lamb, R. A. (1992). Influenza virus M2 protein has ion channel activity. *Cell, 69*, 517–528.

Plemper, R. K., Hammond, A. L., Gerlier, D., Fielding, A. K., & Cattaneo, R. (2002). Strength of envelope protein interaction modulates cytopathicity of measles virus. *Journal of Virology, 76*, 5051–5061.

Porotto, M., Murrell, M., Greengard, O., & Moscona, A. (2003). Triggering of human parainfluenza virus 3 fusion protein (F) by the hemagglutinin-neuraminidase (HN) protein: an HN mutation diminishes the rate of F activation and fusion. *Journal of Virology, 77*, 3647–3654.

Qin, Z. L., Zheng, Y., & Kielian, M. (2009). Role of conserved histidine residues in the low-pH dependence of the Semliki Forest virus fusion protein. *Journal of Virology, 83*, 4670–4677.

Reeves, J. D., Gallo, S. A., Ahmad, N., Miamidian, J. L., Harvey, P. E., Sharron, M., Pohlmann, S., Sfakianos, J. N., Derdeyn, C. A., Blumenthal, R., Hunter, E., & Doms, R. W. (2002). Sensitivity of HIV-1 to entry inhibitors correlates with envelope/coreceptor affinity, receptor density, and fusion kinetics. *Proceedings of the National Academy of Sciences of the United States of America, 99*, 16249–16254.

Rey, F. A., Heinz, F. X., Mandl, C., Kunz, C., & Harrison, S. C. (1995). The envelope glycoprotein from tick-borne encephalitis virus at 2 A resolution. *Nature, 375*, 291–298.

Roche, S., Albertini, A. A., Lepault, J., Bressanelli, S., & Gaudin, Y. (2008). Structures of vesicular stomatitis virus glycoprotein: membrane fusion revisited. *Cellular and Molecular Life Sciences, 65*, 1716–1728.

Roche, S., Bressanelli, S., Rey, F. A., & Gaudin, Y. (2006). Crystal structure of the low-pH form of the vesicular stomatitis virus glycoprotein G. *Science, 313*, 187–191.

Roche, S., & Gaudin, Y. (2002). Characterization of the equilibrium between the native and fusion-inactive conformation of rabies virus glycoprotein indicates that the fusion complex is made of several trimers. *Virology, 297*, 128–135.

Roche, S., Rey, F. A., Gaudin, Y., & Bressanelli, S. (2007). Structure of the prefusion form of the vesicular stomatitis virus glycoprotein G. *Science, 315*, 843–848.

Russell, C. J., Jardetzky, T. S., & Lamb, R. A. (2001). Membrane fusion machines of paramyxoviruses: capture of intermediates of fusion. *EMBO Journal, 20*, 4024–4034.

Salminen, A., Wahlberg, J. M., Lobigs, M., Liljestrom, P., & Garoff, H. (1992). Membrane fusion process of Semliki Forest virus. II: Cleavage-dependent reorganization of the spike protein complex controls virus entry. *Journal of Cell Biology, 116*, 349–357.

Sanchez-San Martin, C., Liu, C. Y., & Kielian, M. (2009). Dealing with low pH: entry and exit of alphaviruses and flaviviruses. *Trends in Microbiology, 17*, 514–521.

Sanchez-San Martin, C., Sosa, H., & Kielian, M. (2008). A stable prefusion intermediate of the alphavirus fusion protein reveals critical features of class II membrane fusion. *Cell Host & Microbe, 4*, 600–608.

Sauter, N. K., Hanson, J. E., Glick, G. D., Brown, J. H., Crowther, R. L., Park, S. J., Skehel, J. J., & Wiley, D. C. (1992). Binding of influenza virus hemagglutinin to analogs of its cell-surface receptor, sialic acid: analysis by proton nuclear magnetic resonance spectroscopy and X-ray crystallography. *Biochemistry, 31*, 9609–9621.

Schibli, D. J., Montelaro, R. C., & Vogel, H. J. (2001). The membrane-proximal tryptophan-rich region of the HIV glycoprotein, gp41, forms a well-defined helix in dodecylphosphocholine micelles. *Biochemistry, 40*, 9570–9578.

Schornberg, K., Matsuyama, S., Kabsch, K., Delos, S., Bouton, A., & White, J. (2006). Role of endosomal cathepsins in entry mediated by the Ebola virus glycoprotein. *Journal of Virology, 80*, 4174–4178.

Senkevich, T. G., Ojeda, S., Townsley, A., Nelson, G. E., & Moss, B. (2005). Poxvirus multiprotein entry-fusion complex. *Proceedings of the National Academy of Sciences of the United States of America, 102*, 18572–18577.

Shmulevitz, M., & Duncan, R. (2000). A new class of fusion-associated small transmembrane (FAST) proteins encoded by the non-enveloped fusogenic reoviruses. *EMBO Journal, 19*, 902–912.

Skehel, J. J., & Wiley, D. C. (2000). Receptor binding and membrane fusion in virus entry: the influenza hemagglutinin. *Annual Review of Biochemistry, 69*, 531–569.

Smit, J. M., Bittman, R., & Wilschut, J. (1999). Low-pH-dependent fusion of Sindbis virus with receptor-free cholesterol- and sphingolipid-containing liposomes. *Journal of Virology, 73*, 8476–8484.

Stadler, K., Allison, S. L., Schalich, J., & Heinz, F. X. (1997). Proteolytic activation of tick-borne encephalitis virus by furin. *Journal of Virology, 71*, 8475–8481.

Stegmann, T., Delfino, J. M., Richards, F. M., & Helenius, A. (1991). The HA2 subunit of influenza hemagglutinin inserts into the target membrane prior to fusion. *Journal of Biological Chemistry, 266*, 18404–18410.

Stiasny, K., Allison, S. L., Marchler-Bauer, A., Kunz, C., & Heinz, F. X. (1996). Structural requirements for low-pH-induced rearrangements in the envelope glycoprotein of tick-borne encephalitis virus. *Journal of Virology, 70*, 8142–8147.

Stiasny, K., Bressanelli, S., Lepault, J., Rey, F. A., & Heinz, F. X. (2004). Characterization of a membrane-associated trimeric low-pH-induced form of the class II viral fusion protein E from tick-borne encephalitis virus and its crystallization. *Journal of Virology, 78*, 3178–3183.

Stiasny, K., Kossl, C., Lepault, J., Rey, F. A., & Heinz, F. X. (2007). Characterization of a structural intermediate of flavivirus membrane fusion. *PLoS Pathogens, 3*, e20.

Stieneke-Grober, A., Vey, M., Angliker, H., Shaw, E., Thomas, G., Roberts, C., Klenk, H. D., & Garten, W. (1992). Influenza virus hemagglutinin with multibasic cleavage site is activated by furin, a subtilisin-like endoprotease. *EMBO Journal, 11*, 2407–2414.

Stone-Hulslander, J., & Morrison, T. G. (1999). Mutational analysis of heptad repeats in the membrane-proximal region of Newcastle disease virus HN protein. *Journal of Virology, 73*, 3630–3637.

Subramanian, R. P., & Geraghty, R. J. (2007). Herpes simplex virus type 1 mediates fusion through a hemifusion intermediate by sequential activity of glycoproteins D, H, L, and B. *Proceedings of the National Academy of Sciences of the United States of America, 104*, 2903–2908.

Sugrue, R. J., Bahadur, G., Zambon, M. C., Hall-Smith, M., Douglas, A. R., & Hay, A. J. (1990). Specific structural alteration of the influenza haemagglutinin by amantadine. *EMBO Journal, 9*, 3469–3476.

Sun, X., Belouzard, S., & Whittaker, G. R. (2007). Molecular architecture of the bipartite fusion loops of vesicular stomatitis virus glycoprotein G, a class III viral fusion protein. *Journal of Biological Chemistry, 283*, 6418–6427.

Sun, Z. Y., Oh, K. J., Kim, M., Yu, J., Brusic, V., Song, L., Qiao, Z., Wang, J. H., Wagner, G., & Reinherz, E. L. (2008). HIV-1 broadly neutralizing antibody extracts its epitope from a kinked gp41 ectodomain region on the viral membrane. *Immunity, 28*, 52–63.

Supekar, V. M., Bruckmann, C., Ingallinella, P., Bianchi, E., Pessi, A., & Carfi, A. (2004). Structure of a proteolytically resistant core from the severe acute respiratory syndrome coronavirus S2 fusion protein. *Proceedings of the National Academy of Sciences of the United States of America, 101*, 17958–17963.

Swanson, K., Wen, X., Leser, G. P., Paterson, R. G., Lamb, R. A., & Jardetzky, T. S. (2010). Structure of the Newcastle disease virus F protein in the post-fusion conformation. *Virology, 402*, 372–379.

Tsurudome, M., Gluck, R., Graf, R., Falchetto, R., Schaller, U., & Brunner, J. (1992). Lipid interactions of the hemagglutinin HA2 NH2-terminal segment during influenza virus-induced membrane fusion. *Journal of Biological Chemistry, 267*, 20225–20232.

Umashankar, M., Sanchez-San Martin, C., Liao, M., Reilly, B., Guo, A., Taylor, G., & Kielian, M. (2008). Differential cholesterol binding by class II fusion proteins determines membrane fusion properties. *Journal of Virology, 82*, 9245–9253.

Vishwanathan, S. A., & Hunter, E. (2008). Importance of the membrane-perturbing properties of the membrane-proximal external region of human immunodeficiency virus type 1 gp41 to viral fusion. *Journal of Virology, 82*, 5118–5126.

Voss, J. E., Vaney, M. C., Duquerroy, S., Vonrhein, C., Girard-Blanc, C., Crublet, E., Thompson, A., Bricogne, G., & Rey, F. A. (2010). Glycoprotein organization of Chikungunya virus particles revealed by X-ray crystallography. *Nature, 468*, 709–712.

Wahlberg, J. M., Boere, W. A., & Garoff, H. (1989). The heterodimeric association between the membrane proteins of Semliki Forest virus changes its sensitivity to low pH during virus maturation. *Journal of Virology, 63*, 4991–4997.

Weissenhorn, W., Dessen, A., Harrison, S. C., Skehel, J. J., & Wiley, D. C. (1997). Atomic structure of the ectodomain from HIV-1 gp41. *Nature, 387*, 426–430.

Weissenhorn, W., Carfi, A., Lee, K. H., Skehel, J. J., & Wiley, D. C. (1998). Crystal structure of the Ebola virus membrane fusion subunit, GP2, from the envelope glycoprotein ectodomain. *Molecular Cell, 2*, 605–616.

Weissenhorn, W., Hinz, A., & Gaudin, Y. (2007). Virus membrane fusion. *FEBS Letters, 581*, 2150–2155.

Wild, C. T., Shugars, D. C., Greenwell, T. K., McDanal, C. B., & Matthews, T. J. (1994). Peptides corresponding to a predictive alpha-helical domain of human immunodeficiency virus type 1

gp41 are potent inhibitors of virus infection. *Proceedings of the National Academy of Sciences of the United States of America, 91*, 9770–9774.
Wilson, I. A., Skehel, J. J., & Wiley, D. C. (1981). Structure of the haemagglutinin membrane glycoprotein of influenza virus at 3 Å resolution. *Nature, 289*, 366–373.
Wimley, W. C., & White, S. H. (1992). Partitioning of tryptophan side-chain analogs between water and cyclohexane. *Biochemistry, 31*, 12813–12818.
Xu, Y., Liu, Y., Lou, Z., Qin, L., Li, X., Bai, Z., Pang, H., Tien, P., Gao, G. F., & Rao, Z. (2004). Structural basis for coronavirus-mediated membrane fusion: Crystal structure of mouse hepatitis virus spike protein fusion core. *Journal of Biological Chemistry, 279*, 30514–30522.
Xu, K., Rajashankar, K. R., Chan, Y. P., Himanen, J. P., Broder, C. C., & Nikolov, D. B. (2008). Host cell recognition by the henipaviruses: crystal structures of the Nipah G attachment glycoprotein and its complex with ephrin-B3. *Proceedings of the National Academy of Sciences of the United States of America, 105*, 9953–9958.
Yang, X., Kurteva, S., Ren, X., Lee, S., & Sodroski, J. (2005). Stoichiometry of envelope glycoprotein trimers in the entry of human immunodeficiency virus type 1. *Journal of Virology, 79*, 12132–12147.
Yang, X., Kurteva, S., Ren, X., Lee, S., & Sodroski, J. (2006). Subunit stoichiometry of human immunodeficiency virus type 1 envelope glycoprotein trimers during virus entry into host cells. *Journal of Virology, 80*, 4388–4395.
Yin, H. S., Paterson, R. G., Wen, X., Lamb, R. A., & Jardetzky, T. S. (2005). Structure of the uncleaved ectodomain of the paramyxovirus (hPIV3) fusion protein. *Proceedings of the National Academy of Sciences of the United States of America, 102*, 9288–9293.
Yin, H. S., Wen, X., Paterson, R. G., Lamb, R. A., & Jardetzky, T. S. (2006). Structure of the parainfluenza virus 5 F protein in its metastable, prefusion conformation. *Nature, 439*, 38–44.
Yu, I. M., Zhang, W., Holdaway, H. A., Li, L., Kostyuchenko, V. A., Chipman, P. R., Kuhn, R. J., Rossmann, M. G., & Chen, J. (2008). Structure of the immature dengue virus at low pH primes proteolytic maturation. *Science, 319*, 1834–1837.
Yuan, P., Thompson, T. B., Wurzburg, B. A., Paterson, R. G., Lamb, R. A., & Jardetzky, T. S. (2005). Structural studies of the parainfluenza virus 5 hemagglutinin-neuraminidase tetramer in complex with its receptor, sialyllactose. *Structure, 13*, 803–815.
Zaitseva, E., Mittal, A., Griffin, D. E., & Chernomordik, L. V. (2005). Class II fusion protein of alphaviruses drives membrane fusion through the same pathway as class I proteins. *Journal of Cell Biology, 169*, 167–177.
Zaitseva, E., Yang, S. T., Melikov, K., Pourmal, S., & Chernomordik, L. V. (2010). Dengue virus ensures its fusion in late endosomes using compartment-specific lipids. *PLoS Pathogens, 6*, .
Zhang, X., Fugere, M., Day, R., & Kielian, M. (2003a). Furin processing and proteolytic activation of Semliki Forest virus. *Journal of Virology, 77*, 2981–2989.
Zhang, Y., Corver, J., Chipman, P. R., Zhang, W., Pletnev, S. V., Sedlak, D., Baker, T. S., Strauss, J. H., Kuhn, R. J., & Rossmann, M. G. (2003b). Structures of immature flavivirus particles. *EMBO Journal, 22*, 2604–2613.
Zhu, P., Liu, J., Bess Jr., J., Chertova, E., Lifson, J. D., Grise, H., Ofek, G. A., Taylor, K. A., & Roux, K. H. (2006). Distribution and three-dimensional structure of AIDS virus envelope spikes. *Nature, 441*, 847–852.

CHAPTER 4

Membrane Fusion Mediated by Human Immunodeficiency Virus Envelope Glycoprotein

Gregory B. Melikyan
Department of Pediatrics, Infectious Diseases, Emory University, Atlanta, GA, USA

I. OVERVIEW

Human Immunodeficiency Virus (HIV) infects cells by engaging the cognate receptor (CD4) and coreceptors (CCR5 and/or CXCR4) at the cells surface and fusing the viral envelope to the cell membrane. Recent structural, biochemical and functional studies elucidated key steps of membrane fusion induced by HIV envelope glycoprotein and revealed the previously unappreciated complexity of this process. New data imply that, contrary to common perception, HIV enters cells through receptor/coreceptor-mediated endocytosis and fusion with endosomes. Moreover, there is evidence that HIV fusion with the plasma membrane does not proceed beyond the lipid mixing stage. This review discusses the molecular mechanism of envelope glycoprotein-mediated membrane merger and new challenges in understanding the HIV entry pathways and regulation of endosomal fusion.

Current Topics in Membranes, Volume 68 0065-230X/10 $35.00

DOI: 10.1016/B978-0-12-385891-7.00004-0

II. INTRODUCTION

The nucleoprotein complexes of enveloped viruses are encapsulated within a lipid membrane derived from a host cell. These viruses must therefore merge their envelopes with the target cell membrane to initiate infection, a step that is catalyzed by viral fusion proteins. Recent advances in the structural, biochemical, and functional characterization of viral fusogens revealed that, in spite of their striking structural diversity, these proteins are likely to mediate fusion through a common mechanism (Backovic & Jardetzky, 2009; Harrison, 2008; Kielian & Rey, 2006; Lamb & Jardetzky, 2007; White, Delos, Brecher, & Schornberg, 2008). Viral fusion is thought to progress in two major steps—the insertion of hydrophobic domains (referred to as fusion peptides or fusion loops) into a target membrane and the subsequent folding into a hairpin-like structure. This final structure brings the fusion peptides and transmembrane domains (TMDs) together, thus forcing a merger of the membranes to which these domains are anchored (Fig. 1A).

Human immunodeficiency virus (HIV) encodes a single envelope glycoprotein (Env) that mediates fusion with a host cell membrane. Env is activated by sequential interactions with its receptor (CD4) and coreceptors (CCR5 or CXCR4) (Berger, Murphy, & Farber, 1999). Although this protein is highly immunogenic, most antibodies recognize variable and/or non-neutralizing epitopes of this protein, while antibodies capable of neutralizing diverse HIV isolates are rare. This is in part due to shielding of conserved, functionally important domains by variable and/or highly glycosylated regions of Env. Although the engagement of CD4 and coreceptors (CRs) leads to the formation and exposure of cryptic Env epitopes (Eckert & Kim, 2001; Furuta, Wild, Weng, & Weiss, 1998; Kilgore, Salzwedel, Reddick, Allaway, & Wild, 2003; Sattentau & Moore, 1993; Wyatt & Sodroski, 1998), these sites are short lived and poorly accessible to antibodies. Thus, knowledge of the molecular mechanism of HIV Env-mediated fusion is essential for vaccine efforts and for developing new antiviral strategies. This review discusses the current views of the mechanism of Env-mediated fusion and of HIV entry pathways into host cells.

III. HIV ASSEMBLY AND STRUCTURE

HIV particles are believed to form within the lipid raft-like domains of the plasma membrane or, in some cell types, in specialized intracellular compartments (Benaroch, Billard, Gaudin, Schindler, & Jouve, 2010; Bieniasz, 2009; Chu, Wang, & Spearman, 2009; Freed & Mouland, 2006; Morita & Sundquist, 2004; Pelchen-Matthews, Kramer, & Marsh, 2003). Consistent with this notion, the HIV membrane is enriched in cholesterol and sphingomyelin (Aloia, Tian, & Jensen, 1993; R. Chan et al., 2008; Waheed & Freed, 2010). The main structural

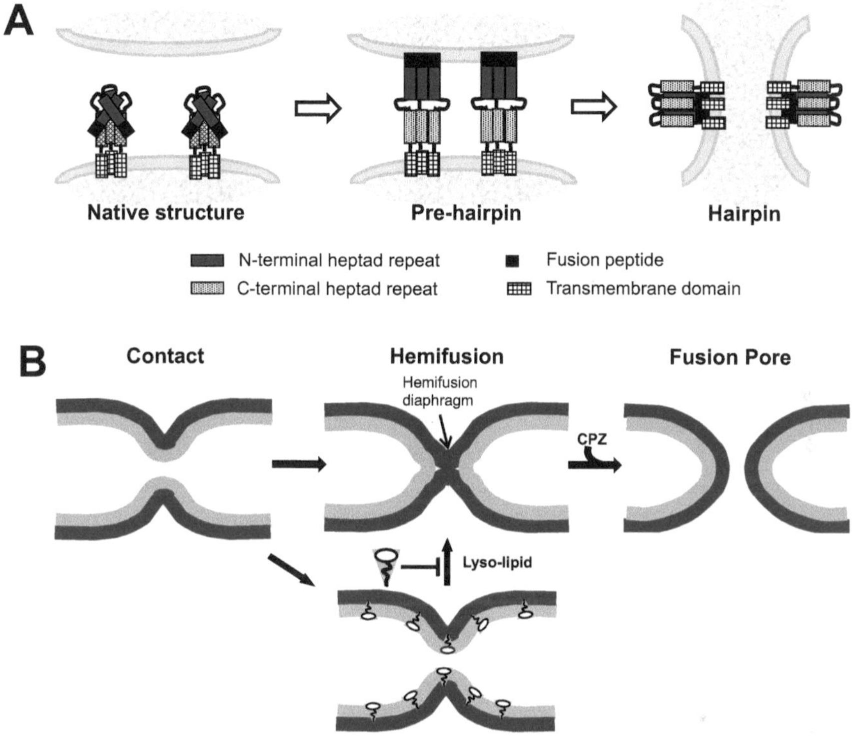

FIGURE 1 Consensus mechanism of viral protein-mediated membrane fusion (A) and lipid intermediates of bilayer fusion (B). (A) A diagram of a class I viral protein conformational changes progressing from a native conformation (left) to an extended prehairpin (middle) and, finally, to the postfusion hairpin structure. The proximity of the membrane-associated TMDs and fusion peptides in the hairpin structure implies that the transition from prehairpin to hairpin should force membranes into contact and promote their fusion. The receptor binding subunits of a fusion protein are not shown for visual clarity. (B) Lipid intermediates of bilayer fusion—hemifusion and small fusion pore. Lysolipids incorporated into contacting leaflets of two bilayers block hemifusion by disfavoring the bending of lipid monolayers required to form a lipid stalk. Chlorpromazine (CPZ) destabilizes a hemifusion diaphragm and thereby promotes the pore formation.

HIV protein, Gag, is both necessary and sufficient to form and release virus-like particles in the absence of other viral proteins (Bieniasz, 2009; Chu et al., 2009; Freed & Mouland, 2006; Morita & Sundquist, 2004). Each virion contains approximately 2000 Gag molecules, whereas only a small number of Env (around 14 molecules (Zhu et al., 2006)) is recruited into virions through interactions between Gag and the cytoplasmic tail (CT) of Env (Murakami & Freed, 2000; Wyma, Kotov, & Aiken, 2000). As a part of viral maturation occurring at the time or soon after budding, the HIV protease cleaves Gag, permitting the formation of a characteristic cone-shaped core by the Gag fragments (e.g., Benjamin, Ganser-Pornillos, Tivol, Sundquist, & Jensen, 2005; Briggs, Wilk, Welker, Krausslich, & Fuller, 2003). Immature HIV particles produced upon inhibiting the viral protease or eliminating the cleavage sites within the Gag polyprotein are not infectious due, in part, to their inability to fuse with target cells (Jiang & Aiken, 2006; Murakami, Ablan, Freed, & Tanaka, 2004). There is evidence that the HIV fusion activity is regulated "from within" through interactions between the immature Gag and the CT of Env (Jiang & Aiken, 2006, 2007; Wyma et al., 2004), which is thought to prevent premature activation of Env on virions being released from infected cells.

HIV Env is a trimer each monomer of which consists of two noncovalently associated subunits, the N-terminal receptor-binding subunit (gp120), and the transmembrane subunit (gp41) that drives membrane fusion (reviewed in Freed & Martin, 1995). These subunits are generated upon intracellular cleavage of the inactive gp160 precursor by furin-like proteases (Willey, Bonifacino, Potts, Martin, & Klausner, 1988). The heavily glycosylated gp120 consists of five conserved (C1–C5) domains interspersed with five highly variable (V1–V5) regions. The structure of monomeric gp120 and the proposed architecture of Env trimers revealed that variable and/or heavily glycosylated domains are positioned peripherally (Bennett et al., 2007; B. Chen et al., 2005; Wyatt & Sodroski, 1998; Wyatt et al., 1998; Yuan, Bazick, & Sodroski, 2006; Zanetti, Briggs, Grunewald, Sattentau, & Fuller, 2006), thereby shielding the conserved regions and restricting the antibody binding.

The gp41 ectodomain contains a conserved N-terminal stretch of hydrophobic residues (the fusion peptide, FP), followed by a polar region (PR) and the two heptad repeat domains, HR1 and HR2, separated by the disulfide-linked loop region (Fig. 2A). Downstream of the gp41 HR2 region is the pretransmembrane domain rich in aromatic residues. This region contains an important neutralization epitope (reviewed in Montero, van Houten, Wang, & Scott, 2008)) and has been reported to be essential for HIV fusion (Bellamy-McIntyre et al., 2007; Munoz-Barroso, Salzwedel, Hunter, & Blumenthal, 1999; Vishwanathan & Hunter, 2008). The unusually long CT of gp41 (ca. 150 residues) contains a consensus internalization signal, YXXΦ (where Y is tyrosine, X is any residue, and Φ is hydrophobic residue), which downregulates the Env expression on the cell surface (Ohno et al., 1997). Three amphipathic regions referred to as lentiviral lytic peptides (reviewed in Costin, 2007) have been identified within the CT and reported to be responsible for Env association with lipid rafts (P. Yang et al., 2010) and for regulation of the Env stability and fusion activity (Chernomordik, Chanturiya, Suss-Toby, Nora, & Zimmerberg, 1994; Lee, Ko, Wang, & Chen, 2002; Wyss et al., 2005).

Attempts to crystallize HIV Env were unsuccessful owing to its low stability and heavy glycosylation. The atomic structure has only been solved for a gp120 monomer, either alone (B. Chen et al., 2005) or in complex with the

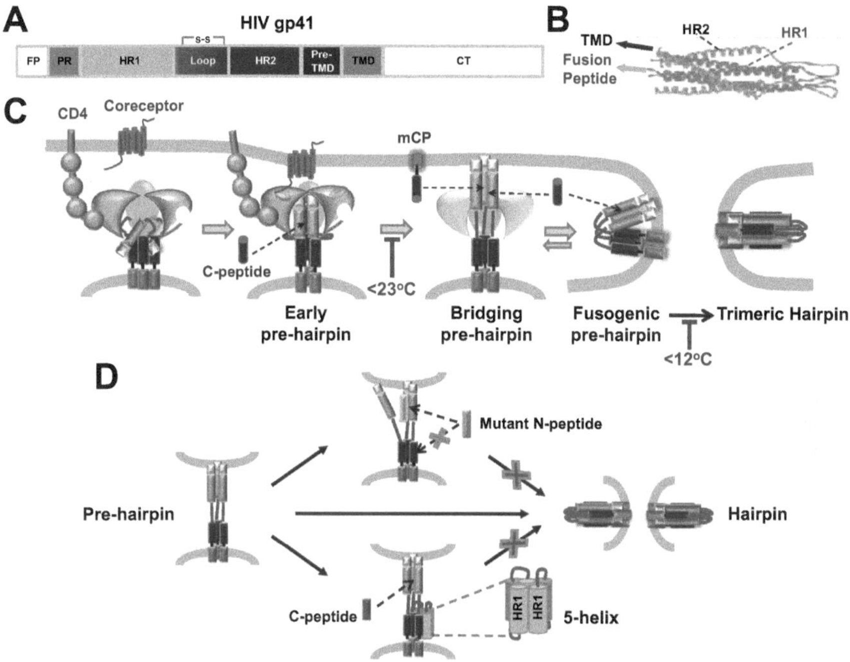

FIGURE 2 The structure of HIV gp41 and the proposed mechanism of Env-mediated membrane fusion. (A) Domains of the HIV gp41: FP, fusion peptide; PR, polar region; HR1 and HR2, N- and C-terminal heptad repeat domains, respectively; Loop, disulfide-bound loop region; pre-TMD, pretransmembrane domain segment; TMD, transmembrane domain; CT, cytoplasmic tail. (B) The NMR structure of the core fragment of SIV gp41 in solution (Caffrey et al., 1998). The HR1 domains are colored red and the HR2 domains are blue. Putative positions of the fusion peptide and TMD that are not parts of the hairpin structure are shown by colored block arrows. (C) Intermediates of HIV Env-induced membrane fusion. The gp41 HR domains are colored as in A, the fusion peptides are orange and the TMDs are gray. The purple globular subunit is gp120 (for visual clarity, not shown in late fusion intermediates). mCP is the target membrane-anchored C-peptide derived from the gp41 HR2 domain. Three subsets of gp41 prehairpins (early, bridging, and fusogenic) identified by functional experiments using inhibitory peptides are shown. Reversibility of fusogenic prehairpins observed at low temperature and in the presence of C-peptides is indicated by the small block arrow. (D) Trapping gp41 prehairpin intermediates with inhibitory peptides. The transition from an extended prehairpin to the final hairpin is blocked by the HR1-derived mutant peptide (yellow) that interacts with the gp41 HR1 domains forming a mixed coiled coil. The recombinant 5-helix protein, which mimics the late gp41 prehairpin structure, interacts with the HR2 domain of a gp41 prehairpin. The HR2-derived C-peptides bind to the complementary HR1 domain. These interactions block the hairpin formation.

soluble CD4 fragment (sCD4) and an antibody against the CR binding site (Kwong et al., 1998). The comparison of these structures referred to as "unliganded" and "liganded" gp120, respectively, revealed that the monomeric gp120 undergoes dramatic conformational changes upon engaging CD4, most notable of which is the formation and exposure of the CR binding site.

The crystal structure of a native gp41 ectodomain or its fragments has not been solved. Only the postfusion structure of a core gp41 fragment encompassing the HR1 and HR2 domains (reviewed in Eckert & Kim, 2001; Skehel & Wiley, 1998) and, most recently, the FP, HR1, HR2 and the pretransmembrane domains (Buzon et al., 2010) have been reported. The most complete postfusion structure was solved by NMR for the ectodomain of Simian Immunodeficiency Virus gp41 lacking the fusion peptide, but containing the HR1 and HR2 domains joined by the disulfide-linked loop (Caffrey et al., 1998) (Fig. 2B). Similar to other class I viral proteins, the gp41 core is a trimer of hairpins, in which the three C-terminal HR2 domains are packed antiparallel and peripheral to the central trimeric coiled coil formed by the N-terminal HR1 domains.

Although HIV Env proved to be difficult to crystallize, the low-resolution maps of HIV and SIV Env glycoproteins on the viral surface were published by several groups, using cryoelectron tomography (Liu, Bartesaghi, Borgnia, Sapiro, & Subramaniam, 2008; Zanetti et al., 2006; Zhu et al., 2006; Zhu, Winkler, Chertova, Taylor, & Roux, 2008) and cryoelectron microscopy (S. R. Wu et al., 2010). However, the architecture of the membrane-proximal fragment of gp41 remains controversial. According to some groups, this region has a tripod-like configuration (S. R. Wu et al., 2010; Zhu et al., 2006, 2008), while the others propose this to be a single stem extending into the viral membrane (Liu et al., 2008; Zanetti et al., 2006). On the other hand, it is universally accepted that CD4 binding induces major restructuring of Env, although the observed changes in the overall shape of viral spikes appear to differ (Liu et al., 2008; S. R. Wu et al., 2010).

In the absence of the atomic structure of Env in native and final conformations, the mechanism of HIV fusion has been modeled after better characterized class I viral proteins, such as influenza hemagglutinin (Eckert & Kim, 2001). On the other hand, a vast number of antibodies and inhibitors targeting distinct steps of HIV fusion facilitated the functional dissection of key intermediate steps of this process (reviewed in Gallo et al., 2003; Melikyan, 2008). These studies (discussed below) provided new insights into the mechanism of HIV entry.

IV. STRUCTURAL CHANGES IN HIV ENV LEADING TO MEMBRANE FUSION

A. *Conformational Changes Upstream of Membrane Merger*

1. CD4-Induced Conformational Changes

Antigenic changes caused by Env-receptor interactions are well documented. These changes involve rearrangement of the gp120 hypervariable loops and, most importantly, the formation and exposure of the CR binding site (B. Chen et al., 2005; Kwong et al., 1998; Sattentau, Moore, Vignaux, Traincard, &

Poignard, 1993). Of note, Env glycoproteins from CD4-independent HIV variants, which are capable of infecting receptor-negative cells expressing coreceptors, have been shown to exist in a "pretriggered" state reminiscent of the CD4-bound conformation (Edwards et al., 2001; L. Wu et al., 2009). Following the receptor engagement, the CR binding sites appear to form at the distal tip of Env facing the target membrane (Fig. 2C). The proximity to a target membrane is believed to limit the antibody access to this conserved region (Labrijn et al., 2003). In contrast, interactions between Env and sCD4 allow unrestricted antibody access to the CR binding site, as evidenced by their enhanced neutralizing potency against viruses pretreated with the soluble receptor fragment (Decker et al., 2005; Salzwedel, Smith, Dey, & Berger, 2000; Thali et al., 1993). It has been shown that CD4 binding facilitates the dissociation of gp120 from the transmembrane gp41 subunit, a phenomenon often referred to as gp120 shedding (Fu, Hart, Jonak, & Bugelski, 1993; Willey, Martin, & Peden, 1994). It remains controversial whether or not gp120 must come off gp41 in order to trigger fusion (Chien, Jiang, & Chang, 2008; Fischer, Karlsson, Dwek, & Platt, 1996; Thali, Furman, Helseth, Repke, & Sodroski, 1992), but there is evidence that interactions between the two subunits must weaken to permit full-scale conformational changes in gp41 (Abrahamyan, Markosyan, Moore, Cohen, & Melikyan, 2003).

CD4-induced conformational changes in HIV Env are not limited to gp120. The gp41 subunit also alters its conformation by forming/exposing the HR1 and HR2 domains, which are not detectable on a native trimer (Fig. 2C). Intermediate gp41 structures exposing the HR1/HR2 domains are operationally defined as prehairpin conformations (Figs. 2 and 3). The formation of prehairpins is evident from the increased reactivity of the CD4-bound Env with antibodies against the gp41 heptad repeat domains (Dimitrov, Louis, Bewley, Clore, & Blumenthal, 2005; Dimitrov et al., 2007; Finnegan, Berg, Lewis, & DeVico, 2002; Mische et al., 2005). In addition, the receptor-primed conformation of gp41, but not its native structure, interacts with the HR1- and HR2-derived synthetic peptides, referred to as N- and C-peptides, respectively (Furuta et al., 1998; Kilgore et al., 2003). These peptides selectively recognize the complementary gp41 HR domains involved in the hairpin formation. Similarly, the recombinant 5-helix protein mimicking an intermediate conformation of gp41 (Fig. 3) acquires the ability to bind to the HR2 domain of the CD4-bound Env (Dimitrov et al., 2005; Kahle, Steger, & Root, 2009).

In addition to biochemical and immunological evidence for the existence of gp41 prehairpins, the exposure of the HR1 and HR2 domains was detected through the functional "prime-wash" experiments employing the inhibitory N- and C-peptides (Gallo, Puri, & Blumenthal, 2001; Melikyan et al., 2000). These peptides block fusion by binding to the complementary heptad repeat regions on gp41 prehairpins and competitively inhibiting the transition to the final hairpin (reviewed in Eckert & Kim, 2001). Thus, the peptide–prehairpin interaction can

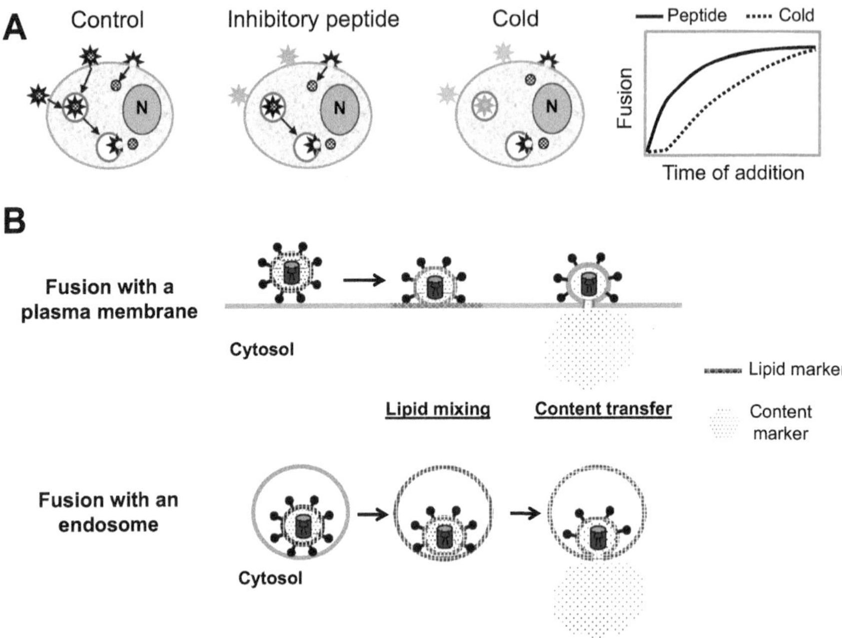

FIGURE 3 Strategies to determine the sites of HIV entry into cells. (A) The kinetics of the HIV escape from C-peptides that inhibit fusion of surface-accessible virions and from low temperature that blocks all fusion events. HIV-cell fusion is stopped at desired times of incubation at permissive temperature by introducing the peptide or by lowering the temperature. The alternative entry pathways (left) are shown by arrows. The middle and right panels show a scenario when C-peptides (middle) or low temperature (right) are introduced at some point after the virus fusion and uptake are commenced. N denoted the cell nucleus. Unfused viruses that are rendered inactive by either of the two inhibitors are shown in light gray. Intracellular fusion occurring in the presence of C-peptides is shown by an arrow in the middle panel. By contrast, low temperature blocks all subsequent fusion events irrespective of the virus location. The graph shows the relative escape kinetics from the peptide and from the temperature block, which suggest that most viruses enter endocytic compartments prior to fusion. (B) Single virus imaging-based strategy to distinguish between the HIV fusion at the cell surface (top) and endosomal fusion (bottom). A lipophilic marker is shown by a patterned line and a diffusible viral content marker is shown by black dots. The diagram illustrates the transfer of viral lipid and content markers into the cell membrane and cytosol.

be indirectly assessed by the peptide's ability to inhibit fusion when added after the formation of Env-sCD4 complexes and washed before coincubating effector and target cells. These experiments revealed that transient exposure of cells primed with sCD4, but not control cells, to inhibitory peptides reduced the efficiency of fusion (Gallo et al., 2001; Melikyan et al., 2000), thereby confirming that the respective heptad repeat domains were exposed as a result of receptor binding. The fact that these domains are no longer accessible to C- or N-peptides on the final hairpin structure (Kliger & Shai, 2000) allowed pinpointing the fusion step at which gp41 completes its refolding into the hairpin (Markosyan, Cohen, & Melikyan, 2003; Melikyan et al., 2000 and see below).

It is worth pointing out that the extent of CD4-induced conformational changes in Env depends on experimental conditions and varies for different HIV isolates. The higher CD4 binding affinity of Env from the laboratory-adapted strains (Kozak et al., 1997; Sattentau & Moore, 1993; Sullivan et al., 1998) appears to result in a greater exposure of the HR1/HR2 domains compared to primary isolates, which may need to engage CR to fully expose these regions (Furuta et al., 1998). Pretreatment with high concentrations of sCD4 at physiological temperature renders the virus noninfectious (Fu et al., 1993; Orloff, Kennedy, Belperron, Maddon, & McDougal, 1993). This inactivation has been shown to proceed through a transient activation of Env followed by an irreversible loss of function (Haim et al., 2009). Thus, whereas the fusion-related Env refolding requires both CD4 and coreceptors, CD4 alone can drive off-path conformational changes proceeding as far as the final hairpin structure (Golding et al., 2002).

2. Coreceptor-Induced Conformational Changes

The formation of Env–CD4–CR complexes is thought to free up the gp41 to undergo refolding into the final hairpin structure. Interestingly, CD4-induced conformational changes in gp120 appear to proceed in the cold, both in the context of monomeric gp120 (Doranz, Baik, & Doms, 1999) and membrane-expressed Env (Kassa et al., 2009). At intermediate temperatures (18–23 °C), Env can slowly engage both CD4 and CR, forming ternary complexes that are resistant to inhibitors of receptor (Henderson & Hope, 2006) and coreceptor binding (Miyauchi, Kim, Latinovic, Morozov, & Melikyan, 2009; Mkrtchyan et al., 2005) (Fig. 2C). Raising the temperature from this temperature-arrested stage induces fast and efficient fusion. The prime-wash experiments using the N- and C-peptides demonstrated that gp41 formed stable prehairpin intermediates following the Env–CD4–CR interactions at reduced temperature (Melikyan et al., 2000).

By definition, the N- and C-peptides would recognize all possible prehairpin intermediates exposing their complementary gp41 domains. Further insight into the temperature-arrested conformation of gp41 was gained using a target membrane-anchored C-peptide, mCP (Hildinger et al., 2001; Melikyan, Egelhofer, & von Laer, 2006). The short linker anchoring the mCP to the target membrane limits its binding to a subset of gp41 prehairpins that directly engage the target membrane, most likely through the insertion of the fusion peptides (Fig. 2C). These bridging prehairpins position their HR1 domains sufficiently close to the target membrane to permit interactions with the mCP. Since the bridging prehairpins are transient structures, the potency of membrane-anchored peptides should be enhanced upon stabilizing these intermediates, which would allow ample time for gp41-mCP interactions. However, preincubation at reduced temperature prior to shifting to 37 °C did not diminish the extent of fusion with mCP-expressing cells compared to control experiments when the fusion reaction was not interrupted (Melikyan et al., 2006). This result implied that gp41 prehairpins formed at reduced temperature did not directly engage the target membrane (Fig. 2C). At physiological temperature, however, gp41 formed bridging prehairpins, as manifested by the enhanced inhibitory activity of the mCP following a brief incubation at 37 °C in the presence of lysophosphatidylcholine (Melikyan et al., 2006), a micelle-forming lipid known to inhibit disparate fusion reactions by disfavoring hemifusion (Fig. 1B).

To conclude, gp41 conformations resulting from Env–CD4–CR interactions at a reduced temperature are distinct from those formed at a physiological temperature upstream of membrane merger. Thus, unlike the early receptor/coreceptor binding steps leading to the formation of early prehairpins, late conformational changes in gp41 are highly temperature dependent (Miyauchi, Kim, et al., 2009; Mkrtchyan et al., 2005), suggesting the existence of a large activation energy barrier at the post-coreceptor binding stage. This energy barrier might be related to the requirement for the noncovalently associated gp120 subunit to disengage gp41, thereby relieving restrictions on its refolding (Abrahamyan et al., 2003).

The notion that gp41 prehairpins are reversible and can sample distinct conformations prior to folding into the final hairpin structure is also supported by the following data. The trimeric HR1 (coiled coil) domain appears to dissociate, at least transiently, during the fusion reaction, as manifested by the ability of the HR1-derived peptide with mutated HR2-interacting residues to inhibit HIV fusion (Bewley, Louis, Ghirlando, & Clore, 2002; Dimitrov et al., 2005; Gustchina, Bewley, & Clore, 2008; Wexler-Cohen & Shai, 2009). These results suggest that the mutant peptide insinuates itself between the HR1 monomers of the gp41 coiled coil and inhibits fusion by forming nonfunctional "mixed" trimers (Bewley et al., 2002) (Fig. 3C). The existence of distinct prehairpin structures is also supported by the markedly different kinetics of HIV inactivation by C-peptides and 5-helix (Kahle et al., 2009).

B. Lipid Mixing Stage

There is evidence that fusion induced by diverse viral proteins progresses through a hemifusion intermediate defined as the merger of contacting leaflets of two bilayers with their distal leaflets forming a shared bilayer (referred to as a hemifusion diaphragm, Fig. 1B) (Melikyan, Brener, Ok, & Cohen, 1997). The so called unrestricted hemifusion (Chernomordik, Frolov, Leikina, Bronk, & Zimmerberg, 1998) is manifested by lipid mixing between the fusing entities. It is believed that lateral diffusion of lipids across the hemifusion junction can be restricted by densely packed fusion proteins forming an oligomeric prefusion complex (Chernomordik et al., 1998). Since restricted hemifusion is not associated with lipid exchange, it cannot be detected by conventional means. However, the existence of such structures can be deduced by disrupting the hemifusion diaphragm by chlorpromazine (CPZ, Fig. 1B). Chlorpromazine and some other cationic amphipathic membrane-permeable compounds appear to selectively partition into and destabilize the inner lipid leaflets comprising the hemifusion diaphragm, thereby promoting full fusion and mixing of both membrane and content markers (Melikyan et al., 1997).

Both restricted and unrestricted hemifusion phenotypes have been reported for HIV Env-induced cell–cell fusion. Unrestricted hemifusion has been detected in the presence of partially inhibitory concentrations of gp41-derived peptides (Munoz-Barroso, Durell, Sakaguchi, Appella, & Blumenthal, 1998). Lipid mixing between Env- and CD4/CR-expressing cells was reportedly less sensitive to peptide inhibitors compared to content transfer. The Env mutant (designated SOS (Binley et al., 2000)), in which gp120 and gp41 were covalently linked by a disulfide bond, was unable to induce fusion, unless this bond was reduced after the coculture of SOS Env-expressing cells with target cells (Abrahamyan et al., 2003; Binley et al., 2003). When this disulfide bond was reduced at lower temperature, the mutant formed a restricted hemifusion-like intermediate, which could be converted to full fusion by chlorpromazine (Abrahamyan et al., 2003).

Recent studies of the role of cellular signaling in HIV fusion and infection suggested the requirement for the Abl kinase activity for Env-mediated fusion (Harmon, Campbell, & Ratner, 2010). In this work, the Abl kinase and Wave2 signaling inhibitors arrested Env-induced cell–cell and virus–cell fusion at a hemifusion stage, implicating active cellular processes in promoting the formation and/or enlargement of a fusion pore. Moreover, single virus imaging experiment demonstrated that HIV particles undergo unrestricted hemifusion on the cell surface without progressing to full fusion (Miyauchi, Kim, et al., 2009). These results support the notion that HIV fusion can proceed through hemifusion and that the progression beyond this stage may rely on yet unidentified cellular factors.

C. Conformational Changes Downstream of Membrane Merger

In the final hairpin structure, the fusion peptides and TMDs are positioned at the same end of a helical hairpin (Fig. 2B). If, by analogy to other viral proteins

(Eckert & Kim, 2001) and as suggested by experiments with mCP (Melikyan et al., 2006), HIV Env first engages the target membrane by inserting its fusion peptides (Fig. 2C), the hairpin structure should bring membranes in contact and promote their fusion. This final structure is often envisioned to form prior to membrane merger (e.g., Chan & Kim, 1998). An obvious problem with this model is that it does not explain how the thermodynamically stable hairpins can drive the subsequent, energy-costly steps of pore formation and enlargement (Borrego-Diaz, Peeples, Markosyan, Melikyan, & Cohen, 2003; Chernomordik et al., 1998; Markosyan, Cohen, & Melikyan, 2000; Zaitseva, Mittal, Griffin, & Chernomordik, 2005). Indeed, real-time fusion experiments showed that gp41 did not complete refolding into the hairpin conformation until after pore opening (Markosyan et al., 2003). This conclusion is supported by the sensitivity of cold-stabilized nascent fusion pores to C- or N-peptides: these structures collapsed immediately upon the addition of inhibitory peptides. Hence, gp41 refolding intermediates, but not the final hairpins on which the peptide binding sites are occupied, are responsible for creating a fusion pore (Fig. 2C).

At physiological temperature, fusogenic gp41 prehairpins converted to hairpins shortly after the pore formation, as manifested by the acquisition of full resistance to N- and C-peptides (Markosyan et al., 2003). This resistance demonstrates that hairpins formed after pore opening are irreversible structures that no longer interact with the inhibitory peptides. By contrast, cold-arrested pores remain sensitive to N- and C-peptides and slowly collapsed, most likely due to gp41 prehairpins regressing back to a nonfusogenic conformation. These and other findings (e.g., Markosyan et al., 2000) imply that (i) fusion pores are energetically unfavorable structures that cannot be sustained without proteins in an active fusogenic conformation and (ii) the formation of gp41 hairpins is contingent on pore formation, suggesting a mechanical coupling between the gp41 refolding and membrane merger.

V. WHAT DRIVES THE ENLARGEMENT OF FUSION PORES?

It has been shown that small fusion pores arrested at intermediate temperatures (12–15 °C) acquire resistance to N- and C-peptides (Markosyan et al., 2003). The existence of these long-lived small pores suggests that the hairpin structure itself can stabilize a pore against closure. If the conformational energy of gp41 is spent to form and stabilize a small pore, then what drives pore dilation, which is perhaps the most energy-demanding step of fusion (Borrego-Diaz et al., 2003; Chernomordik et al., 1998; Markosyan et al., 2000; Zaitseva et al., 2005)?

There is evidence that the gp41 domains other than HR1 and HR2 play a role in fusion by stabilizing the final hairpin structure. These regions include the polar segment (Fig. 2A) downstream of the fusion peptide and the membrane-proximal extracellular region (Bellamy-McIntyre et al., 2007; Munoz-Barroso et al., 1999), which may interact with each other (Bellamy-McIntyre et al., 2007; Lay et al., 2004). It is also feasible that the disulfide-linked loop between heptad repeat domains could stabilize the hairpin conformation. Note that interactions between these additional domains may not be detected by probing the resistance of fusion intermediates to C- and N-peptides, since these peptides are expected to lose their activity once the HR1 and HR2 domains coalesce (Kliger & Shai, 2000; Markosyan et al., 2003). It is therefore possible that a fully zippered hairpin that involves coalescence of regions proximal to the FPs and TMDs are not formed around small pores. However, it is difficult to envision how these relatively minor changes in the overall gp41 hairpin fold can drive the pore dilation.

It has been proposed that the pore dilation could be driven by the difference in the chemical potential between gp41 prehairpins outside a pore and the low-energy hairpins around the pore circumference (Markosyan et al., 2003; Melikyan, 2008). This mechanism, however, may not apply to HIV particles, which are reported to carry only a few Env (Zhu et al., 2006). So, even if several Env glycoproteins form a prefusion complex, it remains uncertain how these proteins create a pore that permits the release of the HIV nucleocapsid (~50 nm). Paradoxically, based on the analyses of infectivity of HIV particles bearing variable ratios of wild-type and dominant-negative Env mutant, it has been concluded that a single Env can induce infection (X. Yang, Kurteva, Ren, Lee, & Sodroski, 2005). Regardless of whether or not the model assumptions that lead to this conclusion are valid (Klasse, 2007; Magnus, Rusert, Bonhoeffer, Trkola, & Regoes, 2008; Salzwedel & Berger, 2009), these results indicate that the energy required for full pore dilation may not come solely from Env.

The above results are consistent with the notion that cellular processes, such as signaling and actin rearrangement (A. Chen et al., 2008; Harmon & Ratner, 2008; Harmon et al., 2010; Pontow, Heyden, Wei, & Ratner, 2004; Richard, Leikina, & Chernomordik, 2009; Stantchev, Markovic, Telford, Clouse, & Broder, 2007; Yoder et al., 2008), as well as virus trafficking to permissive cellular compartments (Le Blanc et al., 2005; Miyauchi, Kim, et al., 2009; Zaitseva, Yang, Melikov, Pourmal, & Chernomordik, 2010), might be coopted by the virus to help enlarge a fusion pore.

VI. HIV ENTRY PATHWAYS

HIV has long been thought to enter permissive cells by fusing directly with the plasma membrane, while endocytosis has been regarded as a nonproductive pathway leading to the virus degradation (Maddon et al., 1988; McClure, Marsh,

& Weiss, 1988). This notion is based on several lines of evidence that, however, are somewhat indirect and/or subject to alternative interpretations. First, ectopic expression of HIV Env on the cell surface promotes fusion with susceptible target cells at neutral pH (Ashorn, Berger, & Moss, 1990). Second, HIV fusion with the plasma membrane of human T cells was captured by electron microscopy (Stein et al., 1987). However, these images were obtained within 2 min of virus–cell coincubation at 4 °C, a temperature that is not permissive for fusion (Melikyan et al., 2000; Miyauchi, Kim, et al., 2009), and could thus be a fixation artifact. On the other hand, HIV-endosome fusion has also been observed by electron microscopy (Marechal et al., 2001; Pauza & Price, 1988). Third, unlike viruses that constitutively enter cells via endocytosis, HIV fusion does not require low pH (McClure et al., 1988; Miyauchi, Kim, et al., 2009; Stein et al., 1987). It is worth stressing, however, that the lack of low pH dependence does not rule out endocytic entry and simply means that this virus is not confined to fusing with acidic intracellular compartments. Fourth, HIV infection of cells expressing wild-type and mutant CD4 does not correlate with this receptor's ability to undergo constitutive or regulated endocytosis (Bedinger et al., 1988; Maddon et al., 1988; Pelchen-Matthews, Clapham, & Marsh, 1995). Here, the lack of biochemical evidence for ligand- or virus-induced CD4 uptake was taken to indicate that infection did not involve receptor-driven HIV internalization and fusion with endosomes. However, biochemical assays may not be sufficiently sensitive to detect a small fraction of virions entering through a productive receptor-mediated pathway (Miyauchi, Kim, et al., 2009). In addition, the CD4-dependent HIV transfer between infected and uninfected cells followed by virus internalization occurs irrespective of the CT of CD4 (Blanco et al., 2004). Fifth, HIV infection of resting $CD4^+$ T cells appears to be associated with CXCR4-mediated signaling that activates an actin-depolymerizing cellular factor, cofilin (Yoder et al., 2008). This finding was interpreted as evidence for direct HIV fusion with the plasma membrane that, in the absence of proper signaling, does not lead to infection due to the viral core entrapment within a dense layer of cortical actin. An alternative possibility supported by recent data (Pace, Agosto, & O'Doherty, 2010) is that cellular signaling is simply required to promote HIV uptake and fusion in resting T cells.

Another approach to infer the sites of HIV entry is based on the comparison of fusion/infection induced by Env and VSV G glycoproteins, which mediated low pH-dependent endosomal entry. Whereas HIV infection is enhanced by its Nef (negative factor) accessory protein (Aiken & Trono, 1995; Miller, Warmerdam, Gaston, Greene, & Feinberg, 1994), infection by the VSV G-pseudotyped HIV cores is Nef-independent (Aiken, 1997; Tobiume, Lineberger, Lundquist, Miller, & Aiken, 2003). Since Nef does not affect the Env fusion activity (Campbell, Nunez, & Hope, 2004; Tobiume et al.,

2003) but can regulate actin polymerization (Fackler, Luo, Geyer, Alberts, & Peterlin, 1999; Fackler et al., 1997), it is thought to facilitate the viral core penetration through the cortical actin layer. Consistent with this notion, actin-depolymerizing drugs have been reported to complement the Nef defect (Campbell et al., 2004). Hence, the ability of VSV G to bypass the Nef requirement for infection may be indicative of distinct entry pathways of these and HIV Env-pseudotyped viruses.

The strikingly different outcomes for the HIV Env- and VSV G-mediated entry have been observed in resting $CD4^+$ T cells. In these cells, HIV established latent infection, while the VSV G pseudotyped HIV cores failed to infect (Agosto et al., 2009; Yu, Wang, Yoder, Spear, & Wu, 2009). According to one group (Yu et al., 2009), the VSV G-mediated entry is blocked at a post-fusion stage, suggesting that endocytic entry into resting T cells is not productive, whereas Env-mediated fusion with the plasma membrane leads to infection. However, a recent study (Pace et al., 2010) demonstrated that VSV G pseudotypes did not infect naïve $CD4^+$ T cells simply because these particles were not internalized. Importantly, the ability of VSV G to mediate low pH-dependent entry into resting T cells could be rescued by incorporating a CCR5-tropic HIV Env into the same virions. Although this Env itself was incapable of mediating fusion due to the low CCR5 expression in these cells, it promoted CD4-dependent virus uptake into acidic compartments where VSV G induced fusion.

Aside from the above arguments for and against endocytic entry, two additional lines of evidence support the endocytic route of HIV entry. In contrast to low pH-dependent viruses, HIV infection is often enhanced in cells pretreated with drugs that raise endosomal pH (Fredericksen, Wei, Yao, Luo, & Garcia, 2002; Schaeffer, Soros, & Greene, 2004; Wei et al., 2005). A likely explanation for this phenomenon is that these conditions extend the time window for endosomal fusion by sparing the virus from degradation. More importantly, HIV fusion and infection in HeLa-derived cells are markedly attenuated upon inhibiting clathrin- and dynamin-dependent endocytosis (Daecke, Fackler, Dittmar, & Krausslich, 2005), implying that endocytosis plays an essential role in productive HIV entry. However, in order to unambiguously determine the sites of HIV entry, one needs to employ noninvasive real-time assays that do not involve blocking essential cellular processes, such as endocytosis.

Miyauchi, Kim, et al. (2009) have implemented two independent functional approaches to deduce the sites of HIV entry. The first approach is based on the kinetic analyses of HIV escape from fusion inhibitors targeting either external viruses (C-peptide) or both external and internalized viruses (low temperature). The comparison of the extents of HIV fusion upon introducing these inhibitors at varied times of incubation at 37 °C revealed that HIV acquired resistance to the membrane-impermeant peptide much earlier than to a low temperature

(Fig. 3A). Thus, the majority of virions appear to escape from the inhibitory peptide by undergoing endocytosis and fusing with intracellular compartments at later times. Another, more direct approach to pinpoint the sites of HIV fusion relies on time-resolved imaging of single virions colabeled with membrane and content markers. Here, different extents of dilution of a lipophilic probe upon its transfer into the cell and endosomal membranes, combined with monitoring of the viral content release, provided spatial information regarding the fusion sites. Briefly, the fluorescence signal from a membrane dye was lost upon HIV fusion with the plasma membrane, whereas this dye remained within a small recipient endosome as a result of fusion with intracellular compartments (Fig. 3B and C). This strategy revealed that, surprisingly, HIV entered permissive cell lines by fusing with endosomes in low pH-independent manner. In contrast, fusion initiated at the cell surface resulted in lipid mixing but failed to form functional fusion pores through which a relatively small (GFP-based) content marker could be released into the cytosol (Miyauchi, Kim, et al., 2009).

To conclude, HIV-plasma membrane fusion appears to be blocked at a hemifusion-like intermediate downstream of lipid mixing. This could occur because the plasma membrane is inherently resistant to viral fusion and/or because HIV requires endosome-resident factors that promote the formation and dilation of fusion pores. Consistent with both possibilities, blocking HIV endocytosis by a small molecule dynamin inhibitor, dynasore, permitted the viral lipid transfer to the plasma membrane without content release (Miyauchi, Kim, et al., 2009). Interestingly, inhibition of cellular signaling has been recently reported to arrest HIV fusion at a hemifusion-like stage that could be converted to full fusion by treatment with chlorpromazine (Harmon et al., 2010), an agent that is believed to selectively destabilize a hemifusion diaphragm (Melikyan et al., 1997). These findings suggest the possibility that, in the absence of active cellular processes, HIV Env glycoprotein(s) may not be able to drive pore dilation.

HIV endocytosis has also been reported to play a role in the productive transfer of viruses across the virological synapses formed between dendritic cells and $CD4^+$ T cells and between infected and uninfected T cells (Clotet-Codina et al., 2009; Hubner et al., 2009). Monitoring HIV movement across the cell–cell junction by electron and confocal microscopy showed that virions internalized by T cells could establish productive infection. These findings imply that endocytosis may be the major entry pathway for both cell-free and cell-associated HIV. Note, however, that other groups did not observe HIV in intracellular compartments of recipient T cells following its transfer across the virological synapse (Felts et al., 2010; Jolly, Kashefi, Hollinshead, & Sattentau, 2004; Martin et al., 2010). The reason for these discrepant results is not clear, but it is likely to be related to differences in cell types (primary cells vs. cell lines) and the degree of T-cell activation (Permanyer, Ballana, & Este, 2010).

HIV entry through endocytosis may offer several advantages for viral replication. First, internalized virions are protected from antibodies and most commonly used entry inhibitors (Miyauchi, Kim, et al., 2009; Miyauchi, Kozlov, & Melikyan, 2009). Second, HIV fusion with endosomal compartments near the perinuclear space can release the nucleocapsids in the vicinity of the nuclear pores, thereby minimizing the cytoplasmic trafficking of viral preintegration complexes and thus the chances for their neutralization by the cellular restriction factors.

VII. CONCLUDING REMARKS

HIV has evolved strategies to evade and destroy the host immune system. Studies of the structural and functional properties of HIV Env uncovered striking features of this glycoprotein that minimize its recognition by neutralizing antibodies while maintaining the ability to mediate membrane fusion. These features include the reduced number and low stability of HIV Env compared to SIV Env in the viral membrane (Zhu et al., 2006), which appears to guide immune responses to antigenically distinct, inactive forms of Env. Whereas most SIV strains can infect cells in CD4-independent manner, the HIV's reliance on CD4 binding protects the conserved coreceptor binding sites that are transiently exposed following the receptor engagement. It is also possible that HIV utilizes an endocytic entry route to minimize the exposure of its cryptic gp120 and gp41 epitopes and perhaps to exploit the cellular machinery to promote enlargement of fusion pores.

In spite of the selective pressure imposed by the host immune responses, perhaps leading to poor expression and stability HIV Env, this glycoprotein seems to be highly efficient at inducing membrane fusion. Indeed, a handful or perhaps even a single Env (X. Yang et al., 2005) incorporated into virions can mediate infection. The slow, stepwise progression of Env through the fusion reaction and the availability of inhibitors targeting distinct steps of HIV fusion facilitated the mechanistic studies of this process. Functional studies identified key intermediates of Env refolding and their coupling to lipid rearrangements associated with membrane fusion. It appears that gp41 progresses through at least three distinct prehairpin intermediates (early, bridging, and fusogenic prehairpins) *en route* to the final hairpin conformation and that the latter post-fusion structure stabilizes a fusion pore.

Although much has been learnt about the HIV Env-mediated fusion in the past two decades, a number of important questions remain unanswered: (i) gp120 signaling that triggers the gp41 refolding; (ii) regulation of the fusion activity through the Gag–Env interactions; (iii) the minimal number of receptors, coreceptors, and Env glycoproteins required for fusion; and (iv) the identity of cellular factors required to fully expand a fusion pore. The application of cutting-edge biophysical techniques to solving the above problems should provide new insights into the HIV entry process and suggest a new means to block infection.

Acknowledgment

This work was supported by the NIH R01 GM054787 grant.

References

Abrahamyan, L. G., Markosyan, R. M., Moore, J. P., Cohen, F. S., & Melikyan, G. B. (2003). Human immunodeficiency virus type 1 Env with an intersubunit disulfide bond engages coreceptors but requires bond reduction after engagement to induce fusion. *Journa ofl Virology, 77*(10), 5829–5836.

Agosto, L. M., Yu, J. J., Liszewski, M. K., Baytop, C., Korokhov, N., & Humeau, L. M., et al. (2009). The CXCR4-tropic human immunodeficiency virus envelope promotes more-efficient gene delivery to resting CD4+ T cells than the vesicular stomatitis virus glycoprotein G envelope. *Journal of Virology, 83*(16), 8153–8162 doi: JVI. 00220-09.

Aiken, C. (1997). Pseudotyping human immunodeficiency virus type 1 (HIV-1) by the glycoprotein of vesicular stomatitis virus targets HIV-1 entry to an endocytic pathway and suppresses both the requirement for Nef and the sensitivity to cyclosporin A. *Journal of Virology, 71*(8), 5871–5877.

Aiken, C., & Trono, D. (1995). Nef stimulates human immunodeficiency virus type 1 proviral DNA synthesis. *Journal of Virology, 69*(8), 5048–5056.

Aloia, R. C., Tian, H., & Jensen, F. C. (1993). Lipid composition and fluidity of the human immunodeficiency virus envelope and host cell plasma membranes. *Proceedings of the National Academy of Sciences of the United States of America, 90*(11), 5181–5185.

Ashorn, P. A., Berger, E. A., & Moss, B. (1990). Human immunodeficiency virus envelope glycoprotein/CD4-mediated fusion of nonprimate cells with human cells. *Journal of Virology, 64*(5), 2149–2156.

Backovic, M., & Jardetzky, T. S. (2009). Class III viral membrane fusion proteins. *Current Opinion in Structural Biology, 19*(2), 189–196.

Bedinger, P., Moriarty, A., von Borstel 2nd, R. C., Donovan, N. J., Steimer, K. S., & Littman, D. R. (1988). Internalization of the human immunodeficiency virus does not require the cytoplasmic domain of CD4. *Nature, 334*(6178), 162–165. doi: 10.1038/334162a0.

Bellamy-McIntyre, A. K., Lay, C. S., Baar, S., Maerz, A. L., Talbo, G. H., & Drummer, H. E., et al. (2007). Functional links between the fusion peptide-proximal polar segment and membrane-proximal region of human immunodeficiency virus gp41 in distinct phases of membrane fusion. *Journal of Biological Chemistry, 282*(32), 23104–23116.

Benaroch, P., Billard, E., Gaudin, R., Schindler, M., & Jouve, M. (2010). HIV-1 assembly in macrophages. *Retrovirology, 7*, 29 1742-4690-7-29.

Benjamin, J., Ganser-Pornillos, B. K., Tivol, W. F., Sundquist, W. I., & Jensen, G. J. (2005). Three-dimensional structure of HIV-1 virus-like particles by electron cryotomography. *Journal of Molecular Biology, 346*(2), 577–588.

Bennett, A., Liu, J., Van Ryk, D., Bliss, D., Arthos, J., & Henderson, R. M., et al. (2007). Cryoelectron tomographic analysis of an HIV-neutralizing protein and its complex with native viral gp120. *Journal of Biological Chemistry, 282*(38), 27754–27759.

Berger, E. A., Murphy, P. M., & Farber, J. M. (1999). Chemokine receptors as HIV-1 coreceptors: Roles in viral entry, tropism, and disease. *Annual Review of Immunology, 17*, 657–700.

Bewley, C. A., Louis, J. M., Ghirlando, R., & Clore, G. M. (2002). Design of a novel peptide inhibitor of HIV fusion that disrupts the internal trimeric coiled-coil of gp41. *Journal of Biological Chemistry, 277*(16), 14238–14245.

Bieniasz, P. D. (2009). The cell biology of HIV-1 virion genesis. *Cell Host Microbe, 5*(6), 550–558 doi: S1931-3128(09)00182-6.

Binley, J. M., Cayanan, C. S., Wiley, C., Schulke, N., Olson, W. C., & Burton, D. R. (2003). Redox-triggered infection by disulfide-shackled human immunodeficiency virus type 1 pseudovirions. *Journal of Virology, 77*(10), 5678–5684.

Binley, J. M., Sanders, R. W., Clas, B., Schuelke, N., Master, A., & Guo, Y., et al. (2000). A recombinant human immunodeficiency virus type 1 envelope glycoprotein complex stabilized by an intermolecular disulfide bond between the gp120 and gp41 subunits is an antigenic mimic of the trimeric virion-associated structure. *Journal of Virology, 74*(2), 627–643.

Blanco, J., Bosch, B., Fernandez-Figueras, M. T., Barretina, J., Clotet, B., & Este, J. A. (2004). High level of coreceptor-independent HIV transfer induced by contacts between primary CD4 T cells. *Journal of Biological Chemistry, 279*(49), 51305–51314. doi: 10.1074/jbc.M408547200.

Borrego-Diaz, E., Peeples, M. E., Markosyan, R. M., Melikyan, G. B., & Cohen, F. S. (2003). Completion of trimeric hairpin formation of influenza virus hemagglutinin promotes fusion pore opening and enlargement. *Virology, 316*(2), 234–244.

Briggs, J. A., Wilk, T., Welker, R., Krausslich, H. G., & Fuller, S. D. (2003). Structural organization of authentic, mature HIV-1 virions and cores. *EMBO Journal, 22*(7), 1707–1715.

Buzon, V., Natrajan, G., Schibli, D., Campelo, F., Kozlov, M. M., & Weissenhorn, W. (2010). Crystal structure of HIV-1 gp41 including both fusion peptide and membrane proximal external regions. *PLoS Pathogens, 6*(5), e1000880.

Caffrey, M., Cai, M., Kaufman, J., Stahl, S. J., Wingfield, P. T., & Covell, D. G., et al. (1998). Three-dimensional solution structure of the 44 kDa ectodomain of SIV gp41. *EMBO Journal, 17*(16), 4572–4584.

Campbell, E. M., Nunez, R., & Hope, T. J. (2004). Disruption of the actin cytoskeleton can complement the ability of Nef to enhance human immunodeficiency virus type 1 infectivity. *Journal of Virology, 78*(11), 5745–5755. doi: 10.1128/JVI.78.11.5745-5755.2004.

Chan, D. C., & Kim, P. S. (1998). HIV entry and its inhibition. *Cell, 93*(5), 681–684.

Chan, R., Uchil, P. D., Jin, J., Shui, G., Ott, D. E., & Mothes, W., et al. (2008). Retroviruses human immunodeficiency virus and murine leukemia virus are enriched in phosphoinositides. *Journal of Virology, 82*(22), 11228–11238 doi: JVI. 00981-08.

Chen, A., Leikina, E., Melikov, K., Podbilewicz, B., Kozlov, M. M., & Chernomordik, L. V. (2008). Fusion-pore expansion during syncytium formation is restricted by an actin network. *J Cell Sci, 121*(Pt 21), 3619–3628 121/21/3619.

Chen, B., Vogan, E. M., Gong, H., Skehel, J. J., Wiley, D. C., & Harrison, S. C. (2005). Structure of an unliganded simian immunodeficiency virus gp120 core. *Nature, 433*(7028), 834–841.

Chernomordik, L. V., Chanturiya, A. N., Suss-Toby, E., Nora, E., & Zimmerberg, J. (1994). An amphipathic peptide from the C-terminal region of the human immunodeficiency virus envelope glycoprotein causes pore formation in membranes. *Journal of Virology, 68*(11), 7115–7123.

Chernomordik, L. V., Frolov, V. A., Leikina, E., Bronk, P., & Zimmerberg, J. (1998). The pathway of membrane fusion catalyzed by influenza hemagglutinin: Restriction of lipids, hemifusion, and lipidic fusion pore formation. *Journal of Cell Biology, 140*(6), 1369–1382.

Chien, MP., Jiang, S., & Chang, D. K. (2008). The function of coreceptor as a basis for the kinetic dissection of HIV type 1 envelope protein-mediated cell fusion. *Faseb Journal, 22*(4), 1179–1192.

Chu, H., Wang, J. J., & Spearman, P. (2009). Human immunodeficiency virus type-1 gag and host vesicular trafficking pathways. *Current Topics in Microbiology and Immunology, 339*, 67–84. doi: 10.1007/978-3-642-02175-6_4.

Clotet-Codina, I., Bosch, B., Senserrich, J., Fernandez-Figueras, M. T., Pena, R., & Ballana, E., et al. (2009). HIV endocytosis after dendritic cell to T cell viral transfer leads to productive virus infection. *Antiviral Research, 83*(1), 94–98. doi: S0166-3542(09)00291-5.

Costin, J. M. (2007). Cytopathic mechanisms of HIV-1. *Virology Journal, 4*, 100 doi: 1743-422X-4-100.

Daecke, J., Fackler, O. T., Dittmar, M. T., & Krausslich, H. G. (2005). Involvement of clathrin-mediated endocytosis in human immunodeficiency virus type 1 entry. *Journal of Virology, 79*(3), 1581–1594.

Decker, J. M., Bibollet-Ruche, F., Wei, X., Wang, S., Levy, D. N., & Wang, W., et al. (2005). Antigenic conservation and immunogenicity of the HIV coreceptor binding site. *Journal of Experimental Medicine, 201*(9), 1407–1419.

Dimitrov, A. S., Jacobs, A., Finnegan, C. M., Stiegler, G., Katinger, H., & Blumenthal, R. (2007). Exposure of the membrane-proximal external region of HIV-1 gp41 in the course of HIV-1 envelope glycoprotein-mediated fusion. *Biochemistry, 46*(5), 1398–1401.

Dimitrov, A. S., Louis, J. M., Bewley, C. A., Clore, G. M., & Blumenthal, R. (2005). Conformational changes in HIV-1 gp41 in the course of HIV-1 envelope glycoprotein-mediated fusion and inactivation. *Biochemistry, 44*(37), 12471–12479.

Doranz, B. J., Baik, S. S., & Doms, R. W. (1999). Use of a gp120 binding assay to dissect the requirements and kinetics of human immunodeficiency virus fusion events. *Journal of Virology, 73*(12), 10346–10358.

Eckert, D. M., & Kim, P. S. (2001). Mechanisms of Viral Membrane Fusion and Its Inhibition. *Annual Review of Biochemistry, 70*, 777–810.

Edwards, T. G., Hoffman, T. L., Baribaud, F., Wyss, S., LaBranche, C. C., & Romano, J., et al. (2001). Relationships between CD4 independence, neutralization sensitivity, and exposure of a CD4-induced epitope in a human immunodeficiency virus type 1 envelope protein. *Journal of Virology, 75*(11), 5230–5239.

Fackler, O T., Kienzle, N., Kremmer, E., Boese, A., Schramm, B., & Klimkait, T., et al. (1997). Association of human immunodeficiency virus Nef protein with actin is myristoylation dependent and influences its subcellular localization. *European Journal of Biochemistry, 247*(3), 843–851.

Fackler, O. T., Luo, W., Geyer, M., Alberts, A. S., & Peterlin, B. M. (1999). Activation of Vav by Nef induces cytoskeletal rearrangements and downstream effector functions. *Molecular Cell, 3*(6), 729–739 doi: S1097-2765(01)80005-8.

Felts, R. L., Narayan, K., Estes, J. D., Shi, D., Trubey, C. M., & Fu, J., et al. (2010). 3D visualization of HIV transfer at the virological synapse between dendritic cells and T cells. *Proceedings of the National Academy of Sciences of the United States of America, 107*(30), 13336–13341 doi: 1003040107.

Finnegan, C. M., Berg, W., Lewis, G. K., & DeVico, A. L. (2002). Antigenic properties of the human immunodeficiency virus transmembrane glycoprotein during cell-cell fusion. *Journal of Virology, 76*(23), 12123–12134.

Fischer, P. B., Karlsson, G. B., Dwek, R. A., & Platt, F. M. (1996). *N*-Butyldeoxynojirimycin-mediated inhibition of human immunodeficiency virus entry correlates with impaired gp120 shedding and gp41 exposure. *Journal of Virology, 70*(10), 7153–7160.

Fredericksen, B. L., Wei, B. L., Yao, J., Luo, T., & Garcia, J. V. (2002). Inhibition of endosomal/lysosomal degradation increases the infectivity of human immunodeficiency virus. *Journal of Virology, 76*(22), 11440–11446.

Freed, E. O., & Martin, M. A. (1995). The role of human immunodeficiency virus type 1 envelope glycoproteins in virus infection. *Journal of Biological Chemistry, 270*(41), 23883–23886.

Freed, E. O., & Mouland, A. J. (2006). The cell biology of HIV-1 and other retroviruses. *Retrovirology, 3*, 77.

Fu, Y. K., Hart, T. K., Jonak, Z. L., & Bugelski, P. J. (1993). Physicochemical dissociation of CD4-mediated syncytium formation and shedding of human immunodeficiency virus type 1 gp120. *Journal of Virology, 67*(7), 3818–3825.

Furuta, R. A., Wild, C. T., Weng, Y., & Weiss, C. D. (1998). Capture of an early fusion-active conformation of HIV-1 gp41. *Nature Structural & Molecular Biology, 5*(4), 276–279.

Gallo, S. A., Finnegan, C. M., Viard, M., Raviv, Y., Dimitrov, A., & Rawat, S. S., et al. (2003). The HIV Env-mediated fusion reaction. *Biochimica et Biophysica Acta, 1614*(1), 36–50.

Gallo, S. A., Puri, A., & Blumenthal, R. (2001). HIV-1 gp41 six-helix bundle formation occurs rapidly after the engagement of gp120 by CXCR4 in the HIV-1 Env-mediated fusion process. *Biochemistry, 40*(41), 12231–12236.

Golding, H., Zaitseva, M., de Rosny, E., King, L. R., Manischewitz, J., & Sidorov, I., et al. (2002). Dissection of human immunodeficiency virus type 1 entry with neutralizing antibodies to gp41 fusion intermediates. *Journal of Virology, 76*(13), 6780–6790.

Gustchina, E., Bewley, C. A., & Clore, G. M. (2008). Sequestering of the prehairpin intermediate of gp41 by peptide N36Mut(e,g) potentiates the human immunodeficiency virus type 1 neutralizing activity of monoclonal antibodies directed against the N-terminal helical repeat of gp41. *Journal of Virology, 82*(20), 10032–10041.

Haim, H., Si, Z., Madani, N., Wang, L., Courter, J. R., & Princiotto, A., et al. (2009). Soluble CD4 and CD4-mimetic compounds inhibit HIV-1 infection by induction of a short-lived activated state. *PLoS Pathogens, 5*(4), e1000360.

Harmon, B., Campbell, N., & Ratner, L. (2010). Role of Abl kinase and the Wave2 signaling complex in HIV-1 entry at a post-hemifusion step. *PLoS Pathogens, 6*(6), e1000956. doi: 10.1371/journal.ppat.1000956.

Harmon, B., & Ratner, L. (2008). Induction of the Galpha(q) signaling cascade by the human immunodeficiency virus envelope is required for virus entry. *Journal of Virology, 82*(18), 9191–9205 doi: JVI. 00424-08.

Harrison, S. C. (2008). Viral membrane fusion. *Nature Structural & Molecular Biology, 15*(7), 690–698.

Henderson, H. I., & Hope, T. J. (2006). The temperature arrested intermediate of virus-cell fusion is a functional step in HIV infection. *Virology Journal, 3*, 36.

Hildinger, M., Dittmar, M. T., Schult-Dietrich, P., Fehse, B., Schnierle, B. S., & Thaler, S., et al. (2001). Membrane-anchored peptide inhibits human immunodeficiency virus entry. *Journal of Virology, 75*(6), 3038–3042.

Hubner, W., McNerney, G. P., Chen, P., Dale, B. M., Gordon, R. E., & Chuang, F. Y., et al. (2009). Quantitative 3D video microscopy of HIV transfer across T cell virological synapses. *Science, 323*(5922), 1743–1747.

Jiang, J., & Aiken, C. (2006). Maturation of the viral core enhances the fusion of HIV-1 particles with primary human T cells and monocyte-derived macrophages. *Virology, 346*(2), 460–468.

Jiang, J., & Aiken, C. (2007). Maturation-dependent human immunodeficiency virus type 1 particle fusion requires a carboxyl-terminal region of the gp41 cytoplasmic tail. *Journal of Virology, 81* (18), 9999–10008.

Jolly, C., Kashefi, K., Hollinshead, M., & Sattentau, Q. J. (2004). HIV-1 cell to cell transfer across an Env-induced, actin-dependent synapse. *Journal of Experimental Medicine, 199*(2), 283–293.

Kahle, K M., Steger, H. K., & Root, M. J. (2009). Asymmetric deactivation of HIV-1 gp41 following fusion inhibitor binding. *PLoS Pathogens, 5*(11), e1000674. doi: 10.1371/journal.ppat.1000674.

Kassa, A., Finzi, A., Pancera, M., Courter, J. R., Smith 3rd, A. B., & Sodroski, J. (2009). Identification of a human immunodeficiency virus type 1 envelope glycoprotein variant resistant to cold inactivation. *Journal of Virology, 83*(9), 4476–4488 doi: JVI. 02110-08.

Kielian, M., & Rey, F. A. (2006). Virus membrane-fusion proteins: More than one way to make a hairpin. *Nature Reviews Microbiology, 4*(1), 67–76.

Kilgore, N. R., Salzwedel, K., Reddick, M., Allaway, G. P., & Wild, C. T. (2003). Direct evidence that C-peptide inhibitors of human immunodeficiency virus type 1 entry bind to the gp41 N-helical domain in receptor-activated viral envelope. *Journal of Virology, 77*(13), 7669–7672.

Klasse, P. J. (2007). Modeling how many envelope glycoprotein trimers per virion participate in human immunodeficiency virus infectivity and its neutralization by antibody. *Virology, 369*(2), 245–262.

Kliger, Y., & Shai, Y. (2000). Inhibition of HIV-1 entry before gp41 folds into its fusion-active conformation. *Journal of Molecular Biology, 295*(2), 163–168.

Kozak, S. L., Platt, E. J., Madani, N., Ferro Jr., F. E., Peden, K., & Kabat, D. (1997). CD4, CXCR-4, and CCR-5 dependencies for infections by primary patient and laboratory-adapted isolates of human immunodeficiency virus type 1. *Journal of Virology, 71*(2), 873–882.

Kwong, P D., Wyatt, R., Robinson, J., Sweet, R. W., Sodroski, J., & Hendrickson, W. A. (1998). Structure of an HIV gp120 envelope glycoprotein in complex with the CD4 receptor and a neutralizing human antibody [see comments]. *Nature, 393*(6686), 648–659.

Labrijn, A. F., Poignard, P., Raja, A., Zwick, M. B., Delgado, K., & Franti, M., et al. (2003). Access of antibody molecules to the conserved coreceptor binding site on glycoprotein gp120 is sterically restricted on primary human immunodeficiency virus type 1. *Journal of Virology, 77* (19), 10557–10565.

Lamb, R. A., & Jardetzky, T. S. (2007). Structural basis of viral invasion: Lessons from paramyxovirus F. *Current Opinion in Structural Biology, 17*(4), 427–436.

Lay, C. S., Wilson, K. A., Kobe, B., Kemp, B. E., Drummer, H. E., & Poumbourios, P. (2004). Expression and biochemical analysis of the entire HIV-2 gp41 ectodomain: Determinants of stability map to N- and C-terminal sequences outside the 6-helix bundle core. *FEBS Letter, 567* (2–3), 183–188. doi: 10.1016/j.febslet.2004.04.054.

Le Blanc, I., Luyet, P. P., Pons, V., Ferguson, C., Emans, N., & Petiot, A., et al. (2005). Endosome-to-cytosol transport of viral nucleocapsids. *Nature Cell Biology, 7*(7), 653–664.

Lee, S. F., Ko, C. Y., Wang, C. T., & Chen, S. S. (2002). Effect of point mutations in the N terminus of the lentivirus lytic peptide-1 sequence of human immunodeficiency virus type 1 transmembrane protein gp41 on Env stability. *Journal of Biological Chemistry, 277*(18), 15363–15375. doi: 10.1074/jbc.M201479200.

Liu, J., Bartesaghi, A., Borgnia, M. J., Sapiro, G., & Subramaniam, S. (2008). Molecular architecture of native HIV-1 gp120 trimers. *Nature, 455*(7209), 109–113 doi: nature07159.

Maddon, P. J., McDougal, J. S., Clapham, P. R., Dalgleish, A. G., Jamal, S., & Weiss, R. A., et al. (1988). HIV infection does not require endocytosis of its receptor. *CD4. Cell, 54*(6), 865–874.

Magnus, C., Rusert, P., Bonhoeffer, S., Trkola, A., & Regoes, R. R. (2008). Estimating the stoichiometry of HIV entry. *Journal of Virology, 83*(3), 1523–1531.

Marechal, V., Prevost, M. C., Petit, C., Perret, E., Heard, J. M., & Schwartz, O. (2001). Human immunodeficiency virus type 1 entry into macrophages mediated by macropinocytosis. *Journal of Virology, 75*(22), 11166–11177.

Markosyan, R. M., Cohen, F. S., & Melikyan, G. B. (2000). The lipid-anchored ectodomain of influenza virus hemagglutinin (GPI-HA) is capable of inducing nonenlarging fusion pores. *Molecular Biology of the Cell, 11*(4), 1143–1152.

Markosyan, R. M., Cohen, F. S., & Melikyan, G. B. (2003). HIV-1 envelope proteins complete their folding into six-helix bundles immediately after fusion pore formation. *Molecular Biology of the Cell, 14*(3), 926–938.

Martin, N., Welsch, S., Jolly, C., Briggs, J. A., Vaux, D., & Sattentau, Q. J. (2010). Virological synapse-mediated spread of human immunodeficiency virus type 1 between T cells is sensitive to entry inhibition. *Journal of Virology, 84*(7), 3516–3527 doi: JVI. 02651-09.

McClure, M. O., Marsh, M., & Weiss, R. A. (1988). Human immunodeficiency virus infection of CD4-bearing cells occurs by a pH-independent mechanism. *EMBO Journal, 7*(2), 513–518.

Melikyan, G. B. (2008). Common principles and intermediates of viral protein-mediated fusion: The HIV-1 paradigm. *Retrovirology, 5*(1), 111.

Melikyan, G. B., Brener, S. A., Ok, D. C., & Cohen, F. S. (1997). Inner but not outer membrane leaflets control the transition from glycosylphosphatidylinositol-anchored influenza hemagglutinin-induced hemifusion to full fusion. *Journal of Cell Biology, 136*, 995–1005.

Melikyan, G. B., Egelhofer, M., & von Laer, D. (2006). Membrane-anchored inhibitory peptides capture human immunodeficiency virus type 1 gp41 conformations that engage the target membrane prior to fusion. *Journal of Virology, 80*(7), 3249–3258.

Melikyan, G. B., Markosyan, R. M., Hemmati, H., Delmedico, M. K., Lambert, D. M., & Cohen, F. S. (2000). Evidence that the transition of HIV-1 gp41 into a six-helix bundle, not the bundle configuration, induces membrane fusion. *Journal of Cell Biology, 151*(2), 413–424.

Miller, M. D., Warmerdam, M. T., Gaston, I., Greene, W. C., & Feinberg, M. B. (1994). The human immunodeficiency virus-1 nef gene product: A positive factor for viral infection and replication in primary lymphocytes and macrophages. *Journal of Experimental Medicine, 179*(1), 101–113.

Mische, C. C., Yuan, W., Strack, B., Craig, S., Farzan, M., & Sodroski, J. (2005). An alternative conformation of the gp41 heptad repeat 1 region coiled coil exists in the human immunodeficiency virus (HIV-1) envelope glycoprotein precursor. *Virology, 338*(1), 133–143.

Miyauchi, K., Kim, Y., Latinovic, O., Morozov, V., & Melikyan, G. B. (2009). HIV enters cells via endocytosis and dynamin-dependent fusion with endosomes. *Cell, 137*(3), 433–444.

Miyauchi, K., Kozlov, M. M., & Melikyan, G. B. (2009b). Early steps of HIV-1 fusion define the sensitivity to inhibitory peptides that block 6-helix bundle formation. *PLoS Pathogens, 5*(9), e1000585. doi: 10.1371/journal.ppat.1000585.

Mkrtchyan, S. R., Markosyan, R. M., Eadon, M. T., Moore, J. P., Melikyan, G. B., & Cohen, F. S. (2005). Ternary complex formation of human immunodeficiency virus type 1 Env, CD4, and chemokine receptor captured as an intermediate of membrane fusion. *Journal of Virology, 79*(17), 11161–11169.

Montero, M., van Houten, N. E., Wang, X., & Scott, J. K. (2008). The membrane-proximal external region of the human immunodeficiency virus type 1 envelope: Dominant site of antibody neutralization and target for vaccine design. *Microbiology and Molecular Biology Reviews, 72*(1), 54–84, table of contents.

Morita, E., & Sundquist, W. I. (2004). Retrovirus budding. *Annual Review of Cell and Developmental Biology, 20*, 395–425.

Munoz-Barroso, I., Durell, S., Sakaguchi, K., Appella, E., & Blumenthal, R. (1998). Dilation of the human immunodeficiency virus-1 envelope glycoprotein fusion pore revealed by the inhibitory action of a synthetic peptide from gp41. *Journal of Cell Biology, 140*(2), 315–323.

Munoz-Barroso, I., Salzwedel, K., Hunter, E., & Blumenthal, R. (1999). Role of the membrane-proximal domain in the initial stages of human immunodeficiency virus type 1 envelope glycoprotein-mediated membrane fusion. *Journal of Virology, 73*(7), 6089–6092.

Murakami, T., Ablan, S., Freed, E. O., & Tanaka, Y. (2004). Regulation of human immunodeficiency virus type 1 Env-mediated membrane fusion by viral protease activity. *Journal of Virology, 78* (2), 1026–1031.

Murakami, T., & Freed, E. O. (2000). The long cytoplasmic tail of gp41 is required in a cell type-dependent manner for HIV-1 envelope glycoprotein incorporation into virions [In Process Citation]. *Proceedings of the National Academy of Sciences of the United States of America, 97*(1), 343–348.

Ohno, H., Aguilar, R. C., Fournier, M. C., Hennecke, S., Cosson, P., & Bonifacino, J. S. (1997). Interaction of endocytic signals from the HIV-1 envelope glycoprotein complex with members of the adaptor medium chain family. *Virology, 238*(2), 305–315. doi: S0042-6822(97) 98839-5.

Orloff, S. L., Kennedy, M. S., Belperron, A. A., Maddon, P. J., & McDougal, J. S. (1993). Two mechanisms of soluble CD4 (sCD4)-mediated inhibition of human immunodeficiency virus type 1 (HIV-1) infectivity and their relation to primary HIV-1 isolates with reduced sensitivity to sCD4. *Journal of Virology, 67*(3), 1461–1471.

Pace, M.J., Agosto, L., & O'Doherty, U. (2010). R5 HIV Env and VSV-G Cooperate to Mediate Fusion to Naive CD4+T Cells. *Journal of Virology*. doi: JVI. 01851-10.

Pauza, C. D., & Price, T. M. (1988). Human immunodeficiency virus infection of T cells and monocytes proceeds via receptor-mediated endocytosis. *Journal of Cell Biology, 107*(3), 959–968.

Pelchen-Matthews, A., Clapham, P., & Marsh, M. (1995). Role of CD4 endocytosis in human immunodeficiency virus infection. *Journal of Virology, 69*(12), 8164–8168.

Pelchen-Matthews, A., Kramer, B., & Marsh, M. (2003). Infectious HIV-1 assembles in late endosomes in primary macrophages. *J Cell Biol, 162*(3), 443–455.

Permanyer, M., Ballana, E., & Este, J. A. (2010). Endocytosis of HIV: Anything goes. *Trends Microbiol, 18*(12), 543–551. doi: S0966-842X(10)00171-X.

Pontow, S. E., Heyden, N. V., Wei, S., & Ratner, L. (2004). Actin cytoskeletal reorganizations and coreceptor-mediated activation of rac during human immunodeficiency virus-induced cell fusion. *Journal of Virology, 78*(13), 7138–7147.

Richard, J. P., Leikina, E., & Chernomordik, L. V. (2009). Cytoskeleton reorganization in influenza hemagglutinin-initiated syncytium formation. *Biochim Biophys Acta, 1788*(2), 450–457. doi: S0005-2736(08)00308-8.

Salzwedel, K., & Berger, E. A. (2009). Complementation of diverse HIV-1 Env defects through cooperative subunit interactions: A general property of the functional trimer. *Retrovirology, 6*, 75. doi: 1742-4690-6-75.

Salzwedel, K., Smith, E. D., Dey, B., & Berger, E. A. (2000). Sequential CD4-coreceptor interactions in human immunodeficiency virus type 1 Env function: Soluble CD4 activates Env for coreceptor-dependent fusion and reveals blocking activities of antibodies against cryptic conserved epitopes on gp120. *Journal of Virology, 74*(1), 326–333.

Sattentau, Q. J., & Moore, J. P. (1993). The role of CD4 in HIV binding and entry. *Philosophical Transactions of the Royal Society B: Biological Sciences, 342*(1299), 59–66.

Sattentau, Q. J., Moore, J. P., Vignaux, F., Traincard, F., & Poignard, P. (1993). Conformational changes induced in the envelope glycoproteins of the human and simian immunodeficiency viruses by soluble receptor binding. *Journal of Virology, 67*(12), 7383–7393.

Schaeffer, E., Soros, V. B., & Greene, W. C. (2004). Compensatory link between fusion and endocytosis of human immunodeficiency virus type 1 in human CD4 T lymphocytes. *Journal of Virology, 78*(3), 1375–1383.

Skehel, J. J., & Wiley, D. C. (1998). Coiled coils in both intracellular vesicle and viral membrane fusion. *Cell, 95*(7), 871–874.

Stantchev, T. S., Markovic, I., Telford, W. G., Clouse, K. A., & Broder, C. C. (2007). The tyrosine kinase inhibitor genistein blocks HIV-1 infection in primary human macrophages. *Virus Research, 123*(2), 178–189.

Stein, B. S., Gowda, S. D., Lifson, J. D., Penhallow, R. C., Bensch, K. G., & Engleman, E. G. (1987). pH-independent HIV entry into CD4-positive T cells via virus envelope fusion to the plasma membrane. *Cell, 49*(5), 659–668.

Sullivan, N., Sun, Y., Sattentau, Q., Thali, M., Wu, D., & Denisova, G., et al. (1998). CD4-Induced conformational changes in the human immunodeficiency virus type 1 gp120 glycoprotein: Consequences for virus entry and neutralization. *Journal of Virology, 72*(6), 4694–4703.

Thali, M., Furman, C., Helseth, E., Repke, H., & Sodroski, J. (1992). Lack of correlation between soluble CD4-induced shedding of the human immunodeficiency virus type 1 exterior envelope glycoprotein and subsequent membrane fusion events. *Journal of Virology, 66*(9), 5516–5524.

Thali, M., Moore, J. P., Furman, C., Charles, M., Ho, D. D., & Robinson, J., et al. (1993). Characterization of conserved human immunodeficiency virus type 1 gp120 neutralization epitopes exposed upon gp120-CD4 binding. *Journal of Virology, 67*(7), 3978–3988.

Tobiume, M., Lineberger, J. E., Lundquist, C. A., Miller, M. D., & Aiken, C. (2003). Nef does not affect the efficiency of human immunodeficiency virus type 1 fusion with target cells. *Journal of Virology, 77*(19), 10645–10650.

Vishwanathan, S. A., & Hunter, E. (2008). Importance of the membrane-perturbing properties of the membrane-proximal external region of human immunodeficiency virus type 1 gp41 to viral fusion. *Journal of Virology, 82*(11), 5118–5126.

Waheed, A. A., & Freed, E. O. (2010). The Role of Lipids in Retrovirus Replication. *Viruses, 2*(5), 1146–1180. doi: 10.3390/v2051146.

Wei, B. L., Denton, P. W., O'Neill, E., Luo, T., Foster, J. L., & Garcia, J. V. (2005). Inhibition of lysosome and proteasome function enhances human immunodeficiency virus type 1 infection. *Journal of Virology, 79*(9), 5705–5712.

Wexler-Cohen, Y., & Shai, Y. (2009). Membrane-anchored HIV-1 N-heptad repeat peptides are highly potent cell fusion inhibitors via an altered mode of action. *PLoS Pathogens, 5*(7), e1000509. doi: 10.1371/journal.ppat.1000509.

White, J. M., Delos, S. E., Brecher, M., & Schornberg, K. (2008). Structures and mechanisms of viral membrane fusion proteins: Multiple variations on a common theme. *Critical Reviews in Biochemistry and Molecular Biology, 43*(3), 189–219.

Willey, R. L., Bonifacino, J. S., Potts, B. J., Martin, M. A., & Klausner, R. D. (1988). Biosynthesis, cleavage, and degradation of the human immunodeficiency virus 1 envelope glycoprotein gp160. *Proceedings of the National Academy of Sciences of the United States of America, 85* (24), 9580–9584.

Willey, R. L., Martin, M. A., & Peden, K. W. (1994). Increase in soluble CD4 binding to and CD4-induced dissociation of gp120 from virions correlates with infectivity of human immunodeficiency virus type 1. *Journal of Virology, 68*(2), 1029–1039.

Wu, L., Zhou, T., Yang, Z. Y., Svehla, K., O'Dell, S., & Louder, M. K., et al. (2009). Enhanced exposure of the CD4-binding site to neutralizing antibodies by structural design of a membrane-anchored human immunodeficiency virus type 1 gp120 domain. *Journal of Virology, 83*(10), 5077–5086. doi: JVI. 02600-08.

Wu, S. R., Loving, R., Lindqvist, B., Hebert, H., Koeck, P. J., & Sjoberg, M., et al. (2010). Single-particle cryoelectron microscopy analysis reveals the HIV-1 spike as a tripod structure. *Proceedings of the National Academy of Sciences of the United States of America, 107*(44), 18844–18849. doi: 1007227107.

Wyatt, R., Kwong, P. D., Desjardins, E., Sweet, R. W., Robinson, J., & Hendrickson, W. A., et al. (1998). The antigenic structure of the HIV gp120 envelope glycoprotein. *Nature, 393*(6686), 705–711.

Wyatt, R., & Sodroski, J. (1998). The HIV-1 envelope glycoproteins: Fusogens, antigens, and immunogens. *Science, 280*(5371), 1884–1888.

Wyma, D. J., Jiang, J., Shi, J., Zhou, J., Lineberger, J. E., & Miller, M. D., et al. (2004). Coupling of human immunodeficiency virus type 1 fusion to virion maturation: A novel role of the gp41 cytoplasmic tail. *Journal of Virology, 78*(7), 3429–3435.

Wyma, D. J., Kotov, A., & Aiken, C. (2000). Evidence for a stable interaction of gp41 with Pr55 (Gag) in immature human immunodeficiency virus type 1 particles. *Journal of Virology, 74*(20), 9381–9387.

Wyss, S., Dimitrov, A. S., Baribaud, F., Edwards, T. G., Blumenthal, R., & Hoxie, J. A. (2005). Regulation of human immunodeficiency virus type 1 envelope glycoprotein fusion by a membrane-interactive domain in the gp41 cytoplasmic tail. *Journal of Virology, 79*(19), 12231–12241 doi: 79/19/12231.

Yang, P., Ai, L. S., Huang, S. C., Li, H. F., Chan, W. E., & Chang, C. W., et al. (2010). The cytoplasmic domain of human immunodeficiency virus type 1 transmembrane protein gp41 harbors lipid raft association determinants. *Journal of Virology, 84*(1), 59–75. doi: JVI. 00899-09.

Yang, X., Kurteva, S., Ren, X., Lee, S., & Sodroski, J. (2005). Stoichiometry of envelope glycoprotein trimers in the entry of human immunodeficiency virus type 1. *Journal of Virology, 79* (19), 12132–12147.

Yoder, A., Yu, D., Dong, L., Iyer, S. R., Xu, X., & Kelly, J., et al. (2008). HIV envelope-CXCR4 signaling activates cofilin to overcome cortical actin restriction in resting CD4 T cells. *Cell, 134* (5), 782–792.

Yu, D., Wang, W., Yoder, A., Spear, M., & Wu, Y. (2009). The HIV envelope but not VSV glycoprotein is capable of mediating HIV latent infection of resting CD4 T cells. *PLoS Pathogens, 5*(10), e1000633. doi: 10.1371/journal.ppat.1000633.

Yuan, W., Bazick, J., & Sodroski, J. (2006). Characterization of the multiple conformational States of free monomeric and trimeric human immunodeficiency virus envelope glycoproteins after fixation by cross-linker. *Journal of Virology, 80*(14), 6725–6737.

Zaitseva, E., Mittal, A., Griffin, D. E., & Chernomordik, L. V. (2005). Class II fusion protein of alphaviruses drives membrane fusion through the same pathway as class I proteins. *J Cell Biol, 169*(1), 167–177.

Zaitseva, E., Yang, S. -T., Melikov, K., Pourmal, S., & Chernomordik, L. V. (2010). Dengue virus ensures its fusion in late endosomes using compartment-specific lipids. *PLoS Pathogens, 6*(10), e1001131. doi: 10.1371/journal.ppat.1001131.

Zanetti, G., Briggs, J. A., Grunewald, K., Sattentau, Q. J., & Fuller, S. D. (2006). Cryo-electron tomographic structure of an immunodeficiency virus envelope complex in situ. *PLoS Pathogens, 2*(8), e83.

Zhu, P., Liu, J., Bess Jr., J., Chertova, E., Lifson, J. D., & Grise, H., et al. (2006). Distribution and three-dimensional structure of AIDS virus envelope spikes. *Nature, 441*(7095), 847–852.

Zhu, P., Winkler, H., Chertova, E., Taylor, K. A., & Roux, K. H. (2008). Cryoelectron tomography of HIV-1 envelope spikes: Further evidence for tripod-like legs. *PLoS Pathogens, 4*(11), e1000203. doi: 10.1371/journal.ppat.1000203.

CHAPTER 5

The Reovirus Fusion-Associated Small Transmembrane (FAST) Proteins: Virus-Encoded Cellular Fusogens

Julie Boutilier and Roy Duncan
Department of Microbiology and Immunology, Dalhousie University, Halifax, Nova Scotia, Canada

I. OVERVIEW

The reovirus fusion-associated small transmembrane (FAST) proteins are a singular family of viral membrane fusion proteins that function as dedicated cell–cell fusogens. Each of the FAST proteins contains a unique repertoire and arrangement of functional motifs, assembled into three small fusion modules (the ectodomain, transmembrane domain (TMD), and endodomain). These fusion modules are interchangeable, but specific combinations of membrane

Current Topics in Membranes, Volume 68
0065-230X/10 $35.00

DOI: 10.1016/B978-0-12-385891-7.00005-2

interaction motifs are required to generate a functional fusogen, implying the fusion modules function in a coordinated manner to induce syncytium formation. The cell–cell fusion reaction is best viewed as three distinct, multistep stages: the prefusion, membrane merger, and postfusion stages. The FAST proteins have delegated the membrane attachment step of the prefusion reaction to cellular adhesins but likely use their ecto- and TMDs to promote the final close apposition of tethered membranes. All three fusion modules are required for the membrane merger stage, functioning from both sides of the plasma membrane to alter the lamellar membrane structure and promote lipid mixing and pore formation. During the postfusion stage, proteolytic processing of the FAST proteins generates a soluble endodomain fusion module that increases the rate of pore expansion and syncytium formation mediated by diverse viral and cellular fusogens, implying recruitment of cellular partners involved in the general process of syncytium formation. The use of simple fusogens to mediate the merger of closely apposed membranes and accessory factors for the pre- and postfusion stages is well suited to the evolution of the FAST proteins as virus-encoded cellular fusogens and may have relevance to diverse, naturally occurring cell–cell fusion events.

II. INTRODUCTION

The lipid membrane that surrounds enveloped viruses is studded with multiple copies of a fusion protein complex. These fusion complexes mediate intracellular delivery of the viral genome by fusing the viral membrane with a cellular membrane (J. M. White, Delos, Brecher, & Schornberg, 2008). Enveloped viruses that assemble by budding through the plasma membrane localize their fusion proteins on the cell surface, which can result in cell–cell fusion and syncytium formation and contribute to the pathogenic potential of these enveloped viruses (Cathomen et al., 1998; Steain, Wang, & Saksena, 2008). However, syncytiogenesis is not an evolutionary imperative of these fusogens that have evolved to mediate virus–cell fusion, not cell–cell fusion. Cell–cell membrane fusion resulting in multinucleated syncytium formation is, however, an essential feature of diverse biological events ranging from fertilization to the formation and maintenance of muscle, bone, and the placenta (Oren-Suissa & Podbilewicz, 2010; Sapir, Avinoam, Podbilewicz, & Chernomordik, 2008). Similar to virus–cell fusion, it is widely accepted that these physiological cell–cell fusion events require specific protein fusogens.

Nonenveloped viruses lack a lipid membrane surrounding the virus particle and thus virus entry is not dependent on a membrane fusion event. Consequently, nonenveloped viruses do not encode membrane fusion proteins and syncytium formation is not a typical cytopathic effect induced by these viruses. The only exceptions to this general statement are several viruses from the family Reoviridae, a diverse group of nonenveloped viruses whose genomes consist of 10–12 segments of double-stranded RNA (Mertens, Attoui, Duncan, & Dermody, 2005). Within the genera *Orthoreovirus* and *Aquareovirus*, numerous virus isolates induce syncytium formation, both in cell culture and in virus-infected animals (Duncan, 1999; Duncan, Corcoran, Shou, & Stoltz, 2004; Kibenge & Dhillon, 1987; Lamirande, Nichols, Owens, Gaskin, & Jacobson, 1999; Vieler, Baumgartner, Herbst, & Kohler, 1994). Over the past few years, the viral proteins responsible for the syncytium-inducing phenotype of these fusogenic reoviruses have been identified (Corcoran & Duncan, 2004; Dawe & Duncan, 2002; Racine et al., 2009; Shmulevitz & Duncan, 2000; Thalmann et al., 2010). These FAST proteins define a new family of virus-encoded membrane fusion proteins, distinct both structurally and functionally from the paradigmatic enveloped virus fusion proteins.

This review summarizes our current understanding of FAST protein structure–function relationships and how these remarkable viral fusogens mediate cell–cell fusion and syncytium formation. To provide a context for this discussion, a brief overview is first provided of the general membrane fusion reaction and how structural features of the archetypal enveloped virus fusion proteins are envisioned to mediate this process. Readers are referred to several excellent recent reviews (Harrison, 2008; Melikyan, 2008; J. M. White et al., 2008), and to other chapters in this text, for more detailed descriptions of the biophysics of membrane fusion and the enveloped virus fusogens.

A. Protein-Mediated Membrane Fusion

Studies in numerous different systems suggest there may be a universal five-step pathway for protein-mediated membrane fusion that involves membrane-binding, close membrane apposition, hemifusion (i.e., lipid mixing of only the outer leaflets of the two membranes), pore formation, and pore expansion (Chernomordik & Kozlov, 2003). The initial step in protein-mediated membrane fusion requires adhesion molecules to tether the donor membrane to the target membrane. Prior to the onset of fusion, additional effort is required to reduce the interbilayer distance to <3 nm, the critical repulsive range of hydrated phospholipid headgroups (Chernomordik & Kozlov, 2005; L. V. Chernomordik, Zimmerberg, & Kozlov, 2006; Cohen & Melikyan, 2004). Establishment of close membrane contact at the critical repulsive range is insufficient to fuse biological membranes, and additional energy is needed to generate the lipid rearrangements required during the multistep membrane merger process (Chernomordik & Kozlov, 2003, 2008). The prevailing view is that shallow insertion of hydrophobic amino acid side chains into the outer leaflet of a bilayer induces the formation of a highly curved membrane dimple (Kozlov, McMahon, & Chernomordik, 2010; McMahon, Kozlov, & Martens, 2010). The small diameter at the tip of the dimple lowers the hydration energy at the point of contact to allow the bilayers to bridge the final intermembrane gap. The high curvature of the dimple also increases the spacing between lipid headgroups, creating membrane stress and energetically unfavorable hydrophobic defects (Chernomordik & Kozlov, 2005; Cohen & Melikyan, 2004). These stresses are partially relieved by lipid mixing between only the outer leaflets of the two bilayers and creation of a hemifusion stalk structure. Stresses induced by the corresponding curvature changes in the inner leaflet during stalk formation are subsequently relieved by lipid mixing between the inner leaflets and formation of the fusion pore (L. V. Chernomordik et al., 2006). An inherent feature of this model is that protein fusogens function by deforming membranes, generating elastic stresses that are relieved by fusion. Moreover, the membrane merger reaction must go energetically "downhill" and the energies of the transient intermediate membrane structures must be low enough to be biologically feasible (Kozlov et al., 2010).

Formation of a stable fusion pore represents completion of the membrane merger reaction. If the ultimate endpoint is syncytium formation, then these small stable pores must laterally expand to a micron-sized lumenal opening. Little is known about this postfusion stage of syncytiogenesis and the role, if any, of protein fusogens in the process. Various scenarios for how this might be achieved include relaxing the bending energy of the pore rim, membrane removal by vesiculation or lateral membrane displacement, lateral membrane tension, direct or indirect effects of the fusion protein, or actin-driven forces that push the pore rim apart (Chen et al., 2008; Chernomordik & Kozlov, 2003; L. V. Chernomordik et al., 2006; Eitzen, 2003; Mohler, Simske, Williams-Masson, Hardin, & White et al., 1998; Nolan, Cowan, Koppel, Jin, & Grote, 2006).

B. Model of Membrane Fusion Mediated by Enveloped Virus Fusogens

A major goal in the fusion field is to determine how protein fusogens mediate the lipid rearrangements involved in the multistep membrane merger process. Enveloped viruses mediate virus–cell fusion using a fixed quantity of one or more envelope glycoproteins embedded in the viral envelope. These fusion protein complexes must therefore be able to function autonomously to mediate all steps in the fusion reaction. Despite considerable diversity in the architecture of the enveloped virus fusion proteins, there is a remarkable conservation in the structural remodelling events these proteins undergo during the fusion reaction (Harrison, 2008; Kielian & Rey, 2006; Lamb & Jardetzky, 2007; Melikyan, 2008; Rey, 2006; Roche, Albertini, Lepault, Bressanelli, & Gaudin, 2008; J. M.

White et al., 2008). Initial virus attachment to cell receptors anchors the virus to the cell at intermembrane distances >10–15 nm. Receptor binding and/or the low pH in endosomes triggers conformational changes that convert the fusion protein into an extended prefusion structure. This prefusion intermediate exposes and projects a sequestered fusion peptide (FP) away from the viral envelope and toward the target membrane. The FPs are short (~20–30 residues), predominantly apolar sequences that exist as either N-terminal amphipathic helices or as internal loop structures containing hydrophobic residues at the apex of the loop (Tamm, Han, Li, & Lai, 2002). The precise role of the viral FPs remains unclear. One possibility is that insertion of the hydrophobic residues into the outer leaflet of the target cell membrane displaces lipid headgroups, promoting dimple formation and creation of the hemifusion stalk (Campelo, McMahon, & Kozlov, 2008). Insertion of the FP into the target membrane may also serve to anchor the fusion protein in both membranes, via the TMD at one end and the FP at the other end of the extended prefusion structure. Subsequent folding back of the prefusion structure into a postfusion hairpin would pull the two membranes into close proximity to allow membrane merger to proceed.

The mechanism of transition from hemifusion to stable pore formation remains largely speculative. Pore formation is likely the rate-limiting event in the fusion reaction, requiring increased energy expenditure provided by an increased number of fusion proteins (L. V. Chernomordik, Frolov, Leikina, Bronk, & Zimmerberg, 1998; Markosyan, Melikyan, & Cohen, 2001; Mittal, Leikina, Chernomordik, & Bentz, 2003). The elastic coupling model suggests that the dramatic structural changes in the fusion protein ectodomain result in tilting of the TMD and membrane stress (Melikyan, White, & Cohen, 1995). A similar force transmission model has been proposed for the SNARE proteins involved in intracellular vesicle fusion, where the force generated by folding of the SNARE complex into a four-helix bundle is converted into membrane bending stress by the TMD (Kozlov et al., 2010; Stein, Weber, Wahl, & Jahn, 2009; Weber et al., 1998). The lipid rearrangements required for pore formation may also arise via specific FP–TMD interactions, since hairpin formation leads to the juxtaposition of these two motifs in the final postfusion protein structure (Tamm, 2003).

Although the structural transitions that occur in numerous enveloped virus fusion proteins have been determined, the precise relationships between protein remodelling and the different steps in the fusion reaction remain unclear. Despite this caveat, the enveloped virus fusion model has dominated our view of protein-mediated membrane fusion, resulting in the expectation that fusion proteins, in general, may function via triggered FP exposure, membrane insertion, and hairpin formation. Within the field, this model also instilled the concepts that a single fusion protein complex mediates all steps of the fusion reaction and that the energy released from dramatic structural rearrangement of complex ectodomain structures serves to pull membranes together and drive membrane merger. While these concepts are well suited to virus–cell membrane fusion, they are unlikely to apply to either the FAST proteins or to many of the physiological processes that require cell–cell fusion.

III. THE DIVERSITY OF FAST PROTEINS

The first description of a FAST protein occurred a decade ago (Shmulevitz & Duncan, 2000), with the two most recent members being added to the family in the past year (Racine et al., 2009; Thalmann et al., 2010). All of the identified FAST proteins are encoded by fusogenic members of the *Orthoreovirus* and *Aquareovirus* genera of the Reoviridae family. At 95–198 amino acids in size, the FAST proteins are the smallest known membrane fusion proteins. Unlike the enveloped virus fusion proteins, the FAST proteins are nonstructural viral proteins (i.e., they are not components of the virus particle) and play no role in virus entry. Upon reovirus infection of a cell, viral transcripts are generated from the parental dsRNA genome segments and translated to produce both structural and nonstructural proteins (Benavente & Martinez-Costas, 2007). For the fusogenic reoviruses, one of these genome segments encodes a polycistronic mRNA whose 5'-proximal open reading frame encodes a FAST protein. The use of suboptimal translation start codons delays the onset of FAST protein-induced cell–cell fusion, thereby coordinating the rate of syncytiogenesis with the remainder of the virus replication cycle (Shmulevitz, Yameen, Dawe, Shou, & O'Hara, 2002). Translation of the FAST proteins occurs on the endoplasmic reticulum, with the TMD serving as a reverse signal-anchor to direct a bitopic, $N_{exoplasmic}/C_{cytoplasmic}$ topology in the membrane (Corcoran & Duncan, 2004; Dawe, Corcoran, Clancy, Salsman, & Duncan, 2005; Shmulevitz & Duncan, 2000; Spiess, 1995). The FAST proteins are then trafficked through the secretory pathway to the plasma membrane where they induce cell–cell fusion with neighboring uninfected cells. These nonenveloped virus fusogens have therefore specifically evolved to mediate cell–cell rather than virus–cell fusion and are potent inducers of syncytiogenesis in a wide variety of adherent cell types. Syncytiogenesis promotes localized cell–cell transmission of the infection, with subsequent apoptotic disruption of large syncytia resulting in progeny virus release and systemic dissemination of the infection (Duncan, Chen, Walsh, & Wu, 1996; Salsman, Top, Boutilier, & Duncan, 2005). Syncytium formation correlates with the pathogenic potential of these fusogenic nonenveloped viruses (C. W. Brown et al., 2009; Duncan & Sullivan, 1998), which induce an array of pathologies in virus-infected animals including meningoencephalitis, arthritis, cardiomyopathies, and respiratory symptoms (Chua et al., 2007, 2008; Lamirande et al., 1999; Leland, Hubbard, Sentmore, Soike, & Hilliard, 2000; Palacios et al., 2010).

Since their discovery, significant progress has been made toward defining the structural and biological features of the FAST proteins. To date, there are five confirmed members of the FAST protein family, each named for their approximate size in kilodaltons (Fig. 1). Multiple isolates of the p10 FAST protein segregate into two distinct clades; those encoded by the avian reoviruses (ARV) and those encoded by several reoviruses isolated from bats, including the prototype Nelson Bay reovirus (NBV) (Chua et al., 2007, 2008; Gard & Compans, 1970; Pritchard et al., 2006; Shmulevitz & Duncan, 2000). At present, there are only single isolates of each of the other four FAST proteins. The p14 FAST protein is encoded by a reptilian reovirus (RRV) isolated from a ball python (Ahne, Thomsen, Winton, 1987; Corcoran & Duncan, 2004), baboon reovirus (BRV) encodes the p15 FAST protein (Dawe & Duncan, 2002; Duncan, Murphy, & Mirkovic, 1995), and the largest of the FAST proteins, p22, is encoded by Atlantic salmon reovirus (AtSRV), a member of the *Aquareovirus* genus (Racine et al., 2009). The most recent addition to the FAST protein family is p13 encoded by Broome reovirus (BroV). Although this orthoreovirus was isolated from a bat, the p13 FAST protein is unrelated to the p10 protein of other bat-borne reoviruses and instead shares some features with the RRV p14 FAST

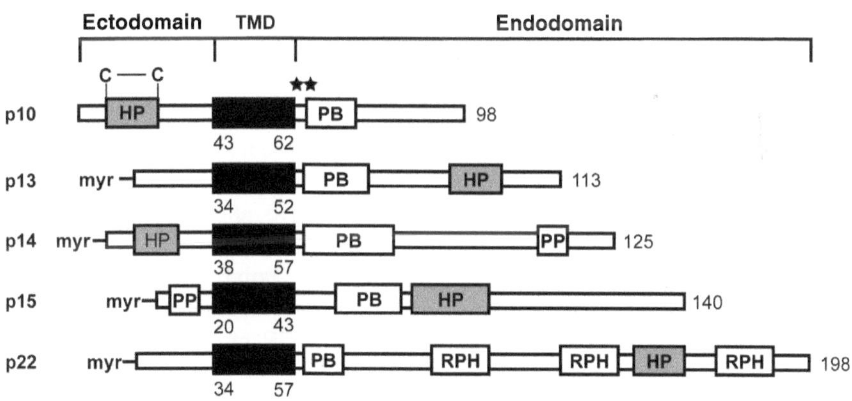

FIGURE 1. The fusion-associated small transmembrane (FAST) proteins. Schematic of the FAST proteins drawn to approximate scale. Each of the FAST proteins has an N-terminal ectodomain and C-terminal endodomain separated by a single TMD. The amino acid numbers defining the approximate boundaries of the transmembrane domains (TMDs) are indicated. All of the FAST proteins are modified by acylation, either with an N-terminal myristic acid (myr) or a palmitoylated dicysteine motif (★) in p10 (myristoylation of p13 and p22 is inferred). Two additional conserved cysteines in the p10 ectodomain form an intramolecular disulphide bond. Each of the FAST proteins contains a hydrophobic patch (HP), and a membrane-proximal polybasic (PB) motif in their endodomain. The p14 and p15 proteins contain polyproline motifs (PP), while p22 contains a polycysteine (PC) motif in the TMD and three endodomain motifs enriched in arginine, proline, and histidine (RPH).

protein (Thalmann et al., 2010). There is every reason to believe that additional unique members of the FAST protein family exist. For instance, a predicted gene product in a recently isolated aquatic orthoreovirus may represent an additional p13 FAST protein unrelated to the p13 FAST protein of BroV (Palacios et al., 2010). Similarly, predicted p16 gene products in the syncytiogenic genogroup A and C aquareoviruses share certain features with the FAST proteins, as does the VP12 protein of Colorado tick fever virus, a member of the *Coltivirus* genus in the family Reoviridae (Mohd Jaafar et al., 2008).

The different FAST proteins display no direct sequence conservation, but they do share a number of common features. The FAST proteins are all basic proteins with isoelectric points ranging from 8.80 to 9.84, and all are single-pass transmembrane proteins. The asymmetric membrane topology of the FAST proteins results in unexpectedly small N-terminal ectodomains of only ~19–42 residues (the approximation reflects the lack of precision in predicting the precise boundaries of a TMD) and considerably larger C-terminal endodomains of ~36–141 residues. In contrast, almost all enveloped virus fusogens position the majority of their mass external to the membrane. Each of the FAST proteins is modified by acylation; either palmitoylation of a membrane-proximal dicysteine motif in the endodomain of p10 or N-terminal myristoylation of the ectodomain of p14 and p15. Acylation-minus mutants of the p10, p14, and p15 FAST proteins display no cell–cell fusion activity (Corcoran & Duncan, 2004; Dawe et al., 2005; Shmulevitz, Salsman, & Duncan, 2003). The p13 and p22 FAST proteins have N-terminal myristoylation consensus sequences, but their acylation status has not been confirmed. In addition to their TMDs, each of the FAST proteins contains one additional region that is more hydrophobic than hydrophilic. These hydrophobic patches (HPs) are found in the ectodomains of the p10 and p14 FAST proteins, but in the endodomains of the p13, p15, and p22 FAST proteins (Fig. 1). All of the FAST proteins also contain a small cluster of membrane-proximal basic residues in their endodomains. These polybasic (PB) motifs are 6–10 residues in size, they have no conserved primary sequence, and they vary considerably in the number of basic residues; the PB motifs in p10 and p22 contain only 3–4 basic residues while p13, p14, and p15 contain 6–7 basic residues. Mutational analysis of the p10 (Shmulevitz et al., 2003) and p14 (Barry, Parmar, Kai, & Duncan, unpublished) PB motifs indicate the positively charged residues in both regions are essential for fusion activity.

The FAST proteins contain several additional motifs, with each FAST protein possessing its own unique repertoire and arrangement of these structural motifs (Fig. 1). The p14 endodomain and p15 ectodomain contain polyproline (PP) motifs. Truncations that remove the C-proximal p14 PP motif reduce the rate, but not the final extent, of syncytium formation (Corcoran & Duncan, 2004), while mutations within the p15 PP motif eliminate fusion activity (Dawe & Duncan, unpublished). The ectodomains of the p10 proteins contain two conserved cysteine residues that form an intramolecular disulfide bond, creating an 11-residue cystine loop (Barry, Key, Haddad, & Duncan, 2010). Both the formation of the cystine loop and specific residues within the loop are essential for p10 fusion activity (Barry et al., 2010; Cheng, Plemper, & Compans, 2005; Shmulevitz, Epand, Epand, & Duncan, 2004). In addition to the unique cystine loop, the ectodomains of over 50 different isolates of the p10 protein contain a conserved region (CR) of nine residues (AGGDLQATS). The absolute conservation of this sequence, despite an overall 29% sequence identity of the p10 proteins, is highly suggestive of an important, though undefined, function. As predicted, even very conservative amino acid substitutions at any of the nine positions in the CR eliminate p10 fusion activity (Key & Duncan, unpublished). Lastly, the AtSRV p22 FAST protein contains a polycysteine (PC) motif within its TMD (CCCCCICC) and three regions in its 141-residue endodomain that are rich in arginine, proline, and histidine residues (RPH motifs) (Racine et al., 2009). The importance of these motifs to p22 fusion activity has not been investigated.

IV. THE FAST PROTEINS ARE NECESSARY AND SUFFICIENT TO INDUCE MEMBRANE FUSION

The exceptionally small size of the FAST protein ectodomains is incompatible with the concept that dramatic structural rearrangement of complex ectodomains serves to pull membranes together and drive membrane merger. While transient transfection studies established that the FAST proteins are the only reovirus proteins required to induce syncytium formation, it was unclear whether they might be functioning through undefined cellular fusion proteins. This issue was addressed by demonstrating that the purified p14 FAST protein, when reconstituted into pure lipid bilayers, induces liposome–cell fusion (Top et al., 2005). The fusion reaction results in transfer of an aqueous fluorescent probe entrapped within p14-liposomes into the cytosol of target cells and of a lipidic probe from the liposome membrane into the cell membrane. Transfer of these probes was inhibited by lysophosphatidylcholine (LPC), an inhibitor of the hemifusion stage of the fusion reaction (Chernomordik & Kozlov, 2003), but not by various endocytosis inhibitors, consistent with p14 mediating liposome–cell fusion at the plasma membrane. The density of p14 in liposome membranes was $\sim 1 \times 10^4$ p14 molecules/μm^2, in the same range as influenza HA molecules on the surface of transfected cells and of reconstituted liposomes containing the SNARE proteins involved in intracellular vesicle fusion (Danieli, Pelletier, Henis, & White et al., 1996; Weber et al., 1998), suggesting p14 fusion activity was not due to excessive protein:lipid ratios. The liposome studies, coupled with heterotypic cell–cell fusion assays, also established that the FAST proteins need only be present in one of the two membranes being fused

(Corcoran et al., 2006; Shmulevitz et al., 2004; Top et al., 2005). The FAST proteins are therefore both necessary and sufficient to mediate membrane fusion, and unlike the SNARE proteins and the recently discovered *Caenorhabditis elegans* fusogens, Aff-1 and Eff-1 (Jahn & Scheller, 2006; Sapir et al., 2008), the FAST proteins do not need to interact in *trans* to form a functional fusion complex.

Studies over the past few years have focused on defining the roles of the various FAST protein structural motifs in the cell–cell fusion reaction. These results indicate the general five-step membrane fusion reaction is better viewed as three distinct stages; prefusion events (membrane binding and apposition), membrane merger (lipid mixing and pore formation), and postfusion events (syncytiogenesis). Each FAST protein can also be viewed as an assembly of three separate fusion modules (the ecto- endo-, and transmembrane domains) that exert complementary or overlapping roles on various stages of the fusion process. The following discussion examines the roles of the FAST protein fusion modules and their structural motifs in this three-stage process of cell–cell fusion and syncytiogenesis.

V. THE FAST PROTEINS AND PREFUSION EVENTS

Atomic force microscopy measurements of the p14 FAST protein in liposomes revealed the ~37-residue ectodomain extends <1.5 nm from the membrane (Corcoran et al., 2006), far less than the ~20–40 nm distance that normally separates the membranes of two adjacent cells. Binding studies also indicated low-level, nonsaturable binding of p14-liposomes to cells, implying p14 lacks high-affinity receptor-binding activity (Salsman, Top, Barry, & Duncan, 2008). It therefore seemed unlikely that the FAST proteins would directly contribute to tethering the two membranes together, or to decreasing the intermembrane separation down to the critical ~2–3 nm distance needed to allow membrane fusion to proceed. Recent studies confirmed the first prediction, revealing the FAST proteins rely on cellular adhesion factors to mediate cell–cell attachment. However, additional studies at least suggest the latter prediction may not be true, and that the FAST proteins may directly contribute to close membrane apposition during the prefusion stage.

A. Surrogate Adhesins and Membrane Attachment

Cadherins are a ubiquitous family of cell adhesion proteins involved in the calcium-dependent formation of adherens junctions, a widespread form of cell–cell adhesion (Perez-Moreno, Jamora, & Fuchs, 2003). Using calcium depletion, siRNA knockdown and ectopic cadherin expression, the formation of adherens junctions was shown to be particularly important for optimal FAST protein-induced cell–cell fusion (Salsman et al., 2008). Fluorescence microscopy results indicate the FAST proteins colocalize with cadherins, but they do so randomly rather than through specific cadherin interactions. Other adhesins can also substitute for cadherins in promoting FAST protein-induced syncytium formation, implying the FAST proteins rely on generic adhesion factors for the membrane attachment step of the prefusion reaction. However, cell–cell fusion is maximized by the actin remodelling triggered specifically by *trans*-cadherin interactions, which results in extended regions of stronger cell–cell contact and the formation of adherens junctions. Interestingly, some of the cell–cell fusion events during development in *C. elegans* occur only at or near adherens junctions (Mohler et al., 1998). In the absence of *trans*-cadherin interactions for membrane tethering, nonspecific adsorption of p14-liposomes to cells is presumably sufficient to allow low levels of liposome–cell fusion, although this is not a particularly efficient event since the majority of liposomes do not interact strongly with target cells (Salsman et al., 2008). The FAST proteins therefore rely on surrogate, generic adhesion factors to mediate the earliest stages of the fusion reaction, retaining within their rudimentary structures a minimal assembly of motifs capable of converting preexisting cellular junctions into sites of cell–cell membrane fusion.

B. The FAST Protein Ectodomains and Membrane Apposition

Estimates of the intermembrane distance at adherens junctions range from ~28 to 38 nm (Koch, Manzur, & Shan, 2004; Zhu et al., 2003), well beyond the ~1.5 nm distance the p14 ectodomain extends from the membrane surface. Factors other than *trans*-cadherin interactions alone must therefore contribute to the second step (i.e., close membrane apposition) of the prefusion stage. Cadherin-induced actin remodelling that leads to adherens junctions provides an extended region of stable cell–cell contact, increasing the probability that out-of-plane thermal fluctuations or actin-driven membrane undulations would allow the two membranes to "probe" each other (Salsman et al., 2008). Since close, even transient, approach of the two membranes may be enough to allow the FAST proteins to exert their fusion activity, adherens junctions would increase the likelihood that localized regions of the two membranes approach the intermembrane distances needed to trigger FAST protein-induced membrane fusion.

The FAST proteins themselves may also directly contribute to promoting close membrane apposition following the initial cadherin-dependent tethering event. For enveloped virus fusogens, close membrane apposition is presumably mediated by folding back the extended prefusion intermediate into the postfusion hairpin structure. Considering the structural limitations of their ectodomains, the FAST proteins cannot use mechanical energy provided by hairpin formation to juxtapose the two membranes. Instead, structural motifs in the ectodomain fusion modules may promote membrane approach to the critical repulsive distance, either by altering membrane curvature or by reducing the effects of hydration repulsion. In the case of p14, the ectodomain contains the HP and an essential N-terminal myristate moiety. As determined by NMR spectroscopy, a nonmyristoylated p14 ectodomain peptide is disordered at the C-terminus but contains an N-proximal proline-hinged loop in the HP, both in aqueous and membrane mimetic environments (Corcoran et al., 2004). Phenylalanine and valine residues are located at the apex of the loop (Fig. 2), positioned for shallow insertion into the outer leaflet of the membrane in which p14 resides (i.e., the donor membrane). Membrane insertion of these residues could force apart lipid headgroups, resulting in positive curvature of the outer leaflet, dimple formation, and decreased intermembrane separation.

NMR and CD spectroscopy also indicate the amphiphilic p14 ectodomain exhibits considerable structural plasticity. Addition of the N-terminal myristate to the p14 ectodomain peptide promotes membrane interactions and triggers a structural transition, such that the proline-hinged loop becomes disordered while the disordered C-terminus assumes an amphipathic helix-kink-helix structure (Fig. 2). A structurally dynamic, amphiphilic p14 ectodomain could therefore reversibly interact with the donor membrane, transiently exposing myristate and hydrophobic amino acid side chains to the solvent phase. The chaotropic effects of such interactions on the lamellar structure of the outer leaflets of the bilayers and/or on their hydration state would lower hydration repulsion. Therefore, the combination of random membrane undulations in adherens junctions and dimple formation due to hydrophobic side chain insertions might partially decrease intermembrane separation to <10 nm, with the chaotropic effects of the structurally dynamic, amphiphilic p14 ectodomain promoting the final close approach of the two membranes to the critical repulsive distance of 2–3 nm.

A similar situation, but with significant variations, may apply to the other FAST protein ectodomains. The $\sim$42-residue p10 ectodomain contains a HP but lacks the N-terminal myristate present on p14. Instead, the two conserved cysteines in the p10 ectodomain form an intramolecular disulfide bond that generates a cystine loop (Fig. 2). The formation of the cystine loop is essential for p10 cell–cell fusion activity (Barry et al., 2010), and substitutions of hydrophobic residues in the loop adversely affect fusion activity (Cheng et al., 2005; Shmulevitz et al., 2004). The disulfide-stabilized loop may function as a cystine noose, an unusual structure where the steric effects of the extreme ϕ and ψ angles force solvent exposure of hydrophobic side chains (Lapthorn, Janes, Isaacs, & Wallace, 1995). Inhibitors of cell surface-localized thiol-disulfide oxidoreductase activity also dramatically reduce p10 fusion activity (Barry et al., 2010), suggesting the fusion activity of p10 is dependent on

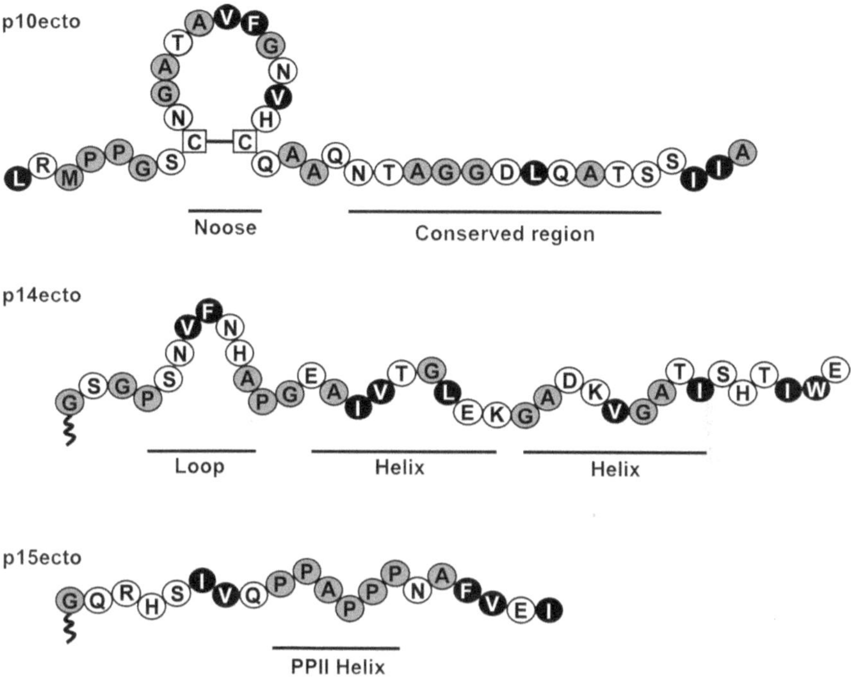

FIGURE 2. Structural motifs in the FAST protein ectodomains. The amino acid sequences of the p10, p14, and p15 FAST proteins ectodomains are depicted. Residues are color coded black (hydrophobic), gray (apolar), or white (polar or charged). The N-terminal myristic acid moieties in p14 and p15 are shown (squiggles). The indicated structural motifs (underlined and labelled) are present in the NMR structures of synthetic ectodomain peptides. In the case of p14, the loop structure is present in a nonmyristoylated peptide present in an aqueous environment and the helix-kink-helix structure is disordered. Conversely, the helix-kink-helix structure is present when a myristoylated peptide is analyzed in a membrane mimetic environment while the loop structure becomes disordered (see text for details).

dynamic changes in the small, amphiphilic cystine loop. As with p14, the p10 ectodomain could therefore use a combination of membrane curvature and chaotropic effects to promote close membrane apposition.

In the case of p15, the myristoylated ~19-residue p15 ectodomain lacks a HP or loop structure (Dawe et al., 2005), containing instead a central polyproline type II (PPII) helix flanked by disordered, amphiphilic N- and C-termini (Fig. 2). PPII helices are extended, left-handed helices that result in solvent exposure of both the hydrophobic prolyl side chains and the polar carbonyl oxygens of the peptide bonds (Stapley & Creamer, 1999). No single proline in the p15 PP motif is required for fusion activity, but substitutions that disrupt the PPII helix eliminate p15-induced syncytium formation (Dawe & Duncan, unpublished). Phenylalanine and valine residues on the C-terminal side of the PPII helix are also essential for fusion activity (Read & Duncan, unpublished). Although PPII helices are rigid structures when formed, the helix is not particularly stable due to the absence of backbone hydrogen bonds (Reiersen & Rees, 2001; Stapley & Creamer, 1999). All of the FAST proteins therefore contain small, structurally dynamic, amphiphilic ectodomains that could alter membrane curvature and/or the hydration state of the donor membrane, contributing to the final close approach of the two membranes.

C. The FAST Protein Transmembrane Domains and Membrane Apposition

In addition to the ectodomains, the FAST protein TMDs may also promote close membrane apposition. Despite an absence of direct sequence similarity and predicted TMD lengths that vary from ~19 to 23 residues, the FAST protein TMDs are functionally interchangeable in a p14 backbone (Clancy & Duncan, 2009). Conversely, substitution of the p14 TMD with the TMDs of heterologous enveloped virus fusion proteins or cellular membrane proteins eliminates cell–cell fusion activity. Molecular modeling suggests the FAST protein TMD helices have a funnel-shaped architecture, with a broader C-terminal cross-sectional area due to the clustering of aromatic residues at this end (Clancy & Duncan, 2009). Since the C-proximal end of the TMD resides within the inner leaflet of the bilayer, the bulky conical shape would induce positive curvature in this leaflet (i.e., the lipid headgroups would curve away from each other) and negative curvature in the corresponding outer leaflet (i.e., the lipid headgroups curve toward each other). Clustering of the FAST proteins due to multimer formation or association with membrane microdomains (Clancy & Duncan, 2011; Corcoran, Clancy, & Duncan, 2011; Corcoran et al., 2006) would amplify these membrane deformations. The FAST proteins may therefore employ two fusion modules to decrease the intermembrane separation, using the complementary curvature effects of the TMD on the inner leaflet and insertion of hydrophobic ectodomain residues in the outer leaflet to cause the donor membrane to bulge toward the target membrane.

Driven by distinct evolutionary imperatives to function as virus–cell fusogens versus cell–cell fusogens, the enveloped virus fusion proteins and the FAST proteins have developed unique approaches to deal with membrane attachment and close apposition as a prelude to membrane merger. As opportunistic, virus-encoded cellular fusogens, the FAST proteins have consigned the membrane attachment step to surrogate cellular adhesion factors. Although a similar situation can occur with enveloped virus fusogens, either in contrived systems where surrogate adhesins have been employed or with mutant fusogens (Paterson, Russell, & Lamb, 2000; Zaitseva, Mittal, Griffin, &

Chernomordik, 2005), the FAST proteins have specifically evolved to use surrogate adhesins. The prefusion events required for cell–cell fusion therefore do not need to be directly coupled to the membrane fusion reaction, nor do they need to be mediated by the fusogen itself or by a cognate receptor binding protein. The FAST proteins suggest that physiological cell–cell fusion events may also rely on separate adhesins and less complex fusion proteins. In fact, many of the proteins implicated in numerous cell–cell fusion processes contain immunoglobulin-like domains and function as tethering factors but not as fusion proteins *per se* (Helming & Gordon, 2009; Martens & McMahon, 2008; Oren-Suissa & Podbilewicz, 2010; Richardson, Nowak, & Baylies, 2008). The FAST proteins have also circumvented the need for mechanical energy provided by extensive structural remodelling of the fusion complex to mediate membrane juxtaposition. Motifs present in the FAST protein ectodomains and TMDs may instead alter the lamellar structure and hydration status of the membranes, lowering the energy barrier to close membrane apposition. The culmination of the prefusion stage is two membranes anchored together at distances approaching the critical repulsive range and poised to fuse.

VI. THE FAST PROTEINS AND MEMBRANE MERGER

The observation that the FAST protein TMDs are interchangeable in a p14 backbone led to the suggestion the FAST proteins evolved as "modular" fusogens (Clancy & Duncan, 2009). This proposal is supported by a comprehensive series of chimeric FAST proteins containing various combinations of the fusion modules from the p10, p14, and p15 FAST proteins (Fig. 3), many of which retain syncytiogenic activity. However, several of these chimeric constructs are nonfusogenic, suggesting that specific combinations of motifs are needed to generate a functional FAST protein fusogen. The implication is that the individual fusion modules function in a coordinated manner to drive membrane merger. How these fusion modules induce the structural rearrangements of the lipid bilayer needed to initiate and complete the fusion reaction is still unclear, although mechanical energy provided by extensive structural changes in complex ectodomains is clearly not an option. Recent studies have described several interesting functional features of the FAST protein fusion modules that provide the basis for a working model of the membrane fusion reaction.

A. *Role of the FAST Protein Ectodomains in Lipid Mixing and Hemifusion*

The mechanisms employed by the FAST protein ectodomains to promote close membrane apposition during the prefusion stage (Section V.B) may also be directly involved in lipid mixing during the membrane merger stage. In this

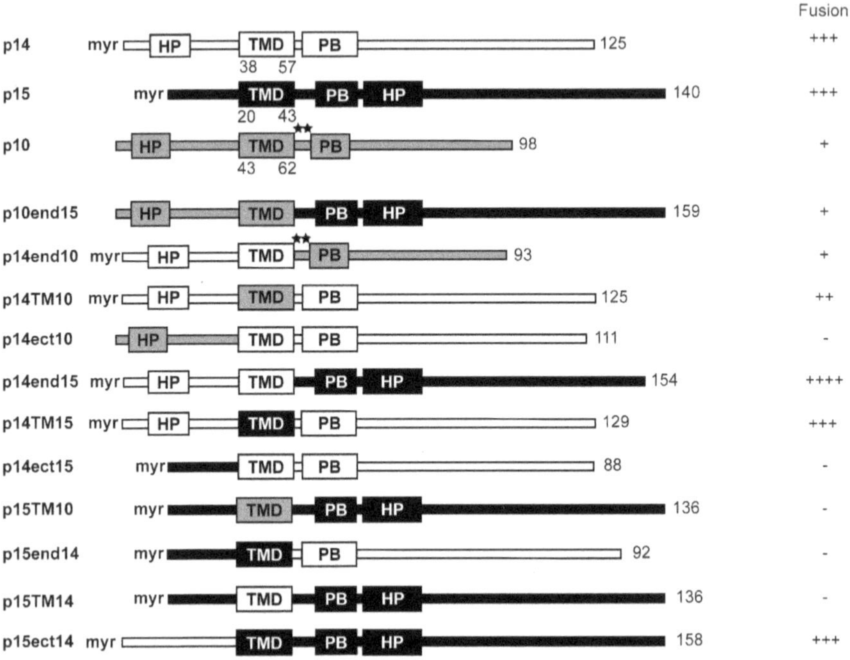

FIGURE 3. Chimeric constructs of the ARV p10, BRV p15, and RRV p14 FAST proteins. Schematic of the authentic FAST proteins and various ecto-, endo-, and transmembrane domain chimeric constructs. myr, Myristic acid; HP, hydrophobic patch; TM, transmembrane domain; ★★, palmitoylated dicysteine motif; PB, polybasic motif. Numbers indicate the number of amino acids in the constructs or the boundaries of the TMD in the parental constructs. Fusion activities were qualitatively scored by observing the extent of syncytia formation in Giemsa-stained monolayers and the relative fusion activities of the constructs are indicated.

regard, FAST protein ectodomain motifs share biophysical and biochemical properties with the FPs present in the enveloped virus fusion proteins (Tamm & Han, 2000). The cystine loop in the p10 ectodomain and the proline-hinged loop in the p14 ectodomain are both amphiphilic structures that position phenylalanine and valine residues at the apex of their loops (Fig. 2). These hydrophobic residues resemble the hydrophobic anchors located at the tips of the fusion loops in the Class II and Class III enveloped virus proteins (Gibbons et al., 2004; Modis, Ogata, Clements, & Harrison, 2004; Roche, Bressanelli, Rey, & Gaudin, 2006). The p10 and p14 HPs are also highly sensitive to amino acid substitutions, and synthetic peptides of these motifs induce lipid mixing between liposomes (Cheng et al., 2005; Corcoran et al., 2004; Shmulevitz et al., 2004), characteristic features of FPs. Shallow membrane insertion of the apical hydrophobic loop residues of the p10 or p14 FPs could induce membrane curvature, with membrane merger relieving the resulting elastic stresses and hydrophobic defects generated by insertion of multiple FAST protein FPs at the fusion site (Campelo et al., 2008). Dimple formation arising from hydrophobic insertion of the p10 and p14 HPs may therefore promote both membrane apposition and lipid mixing.

There are several unique features of the p15 FAST protein ectodomain that may be incompatible with this fusion module functioning as a FP by hydrophobic insertion and curvature-induction. As with p14 (Corcoran et al., 2004), a myristoylated p15 ectodomain peptide induces lipid mixing between liposomes (Top & Duncan, unpublished). The p14 and p15 FPs are the only characterized viral FP motifs that require a myristoylated N-terminus. Since the p14 HP and the p15 polyproline helix are more amphiphilic than hydrophobic, the myristate may be needed to initiate peptide–membrane interactions. Lipid mixing induced by the p15 ectodomain also exhibits an uncharacteristically long lag phase (~200 s) that correlates with extensive peptide-induced liposome aggregation prior to the onset of lipid mixing (Read & Duncan, unpublished). No such lag phase or liposome aggregation occurs during lipid mixing with the myristoylated p14 ectodomain peptide (Corcoran et al., 2004), suggesting that these two FPs may use distinct mechanisms to mediate lipid mixing. It is also difficult to reconcile the structure of the p15 PPII helix with a role as just a curvature-inducing FP, since the helical axis has perfect threefold rotational symmetry and the hydrophobic prolyl rings are equally distributed around the helix (Stapley & Creamer, 1999). This motif therefore does not cluster hydrophobic residues at the apex of a fusion loop or on one side of an amphipathic helix, suggesting it may alter bilayer structure to induce lipid mixing by doing something other than just shallow insertion of hydrophobic side chains.

In addition to the anomalous features of the p15 ectodomain, studies with membrane curvature agents suggest the FAST proteins may not generate an ordered stalk-like structure as a fusion intermediate. LPC induces positive curvature in the outer leaflet of bilayers, which inhibits the negative curvature needed for stalk formation and impedes hemifusion induced by numerous enveloped virus fusion proteins and by the SNARE proteins (L. Chernomordik, Leikina, Cho, & Zimmerberg, 1995; Reese & Mayer, 2005; Russell, Jardetzky, & Lamb, 2004; Zaitseva et al., 2005). LPC also inhibits p14-induced liposome–cell fusion (Top et al., 2005). However, LPC has no inhibitory effect on p14-induced cell–cell pore formation, even when p14 expression is reduced to slow the fusion reaction (Clancy, Barry, Ciechonska, & Duncan, 2010). LPC does, however, inhibit the pore expansion that leads to FAST protein-induced syncytium formation (Ciechonska & Duncan, unpublished). LPC therefore inhibits propagation of the negative curvature in the outer leaflet of the pore rim needed for pore expansion, but its curvature properties have no effect on lipid rearrangements that are required prior to pore formation.

It is unclear whether the differential effects of LPC on liposome–cell versus cell–cell fusion reflects differences in the FAST protein-mediated fusion pathway in these two systems or technical differences between the two assays (Clancy et al., 2010). However, the cell–cell fusion results with LPC at least suggest the possibility that the FAST protein FPs may generate hemifusion lipid intermediates that are not strictly based on formation of an ordered stalk structure with strong negative curvature of the outer leaflets of the bilayers.

It is worth noting that liposome tubulation and fusion studies, such as those conducted on the C2-domains of synaptotagmin-1 (Martens, Kozlov, & McMahon, 2007), have not been conducted with the FPs of the enveloped virus and FAST protein fusogens. The role of viral FPs in inducing membrane curvature is therefore inferred, and other roles have also been posited. For example, as opposed to inducing formation of an ordered stalk structure, dynamic FP–membrane interactions could partially dehydrate the interstitial space at the point of membrane contact. Assuming formation of a point-like protrusion, only minimal dehydration may be required to sufficiently lower the energy barrier of hydration repulsion to allow lipid mixing between the two membranes. Localized reduction of the hydrophobic effect would also decrease the need for an ordered stalk structure to minimize solvent exposure of phospholipid acyl chains, resulting in a more disordered fusion intermediate (Jahn & Grubmuller, 2002; Tamm, 2003; Tamm, Crane, & Kiessling, 2003). A disordered stalk structure is compatible with the inability of LPC to inhibit FAST protein-induced pore formation. As opposed to inducing curvature stresses in the bilayers that are partially relieved by hemifusion, the hydrated membranes and bulk aqueous phase surrounding the region of lipid disorder would provide the thermodynamic potential to resolve this energetically unfavorable lipid intermediate.

Fluid membranes also display a considerable degree of out-of-plane thermal fluctuations of the lipid molecules, such that the combined thickness of the interfacial regions on either side of a bilayer (i.e., the hydration layer associated with the polar headgroup) actually equals the thickness of the hydrocarbon core of the membrane (S. H. White, Ladokhin, Jayasinghe, & Hristova et al., 2001). Membrane interactions of flexible, amphiphilic FPs may further blur the hydrophobic–hydrophilic boundary, increasing the amplitude of the lipid oscillations and promoting a more disordered lipid mixing intermediate. The hydrophobic insertion/curvature model and dehydration/disorder model need not be mutually exclusive. A combination of these two models might be well suited to the unusual p15 PPII FP motif. The hydrophobic prolyl rings alternate with the solvent-exposed polar backbone carbonyl oxygens and this alternating hydrophobic–hydrophilic structure is flanked on one end by a myristic acid and on the other by phenylalanine and valine residues. Insertion of these flanking hydrophobic anchors into the outer leaflet of the bilayer could promote curvature defects while simultaneously drawing the amphiphilic polyproline helix deeper into the interfacial region, altering the properties of this important barrier to lipid mixing.

B. Role of the FAST Protein Transmembrane Domains in Lipid Mixing and Hemifusion

Evidence suggests that the FAST protein TMDs also promote lipid mixing, and that properties other than just the funnel shape may be important for this function. Although the FAST protein TMDs are functionally interchangeable in a p14 backbone (Clancy & Duncan, 2009), this is not the case in a p15 backbone (Clancy & Duncan, 2011). Mutational analysis indicates the p15 TMD has stringent sequence requirements when functioning in its natural context; a tri-serine motif, two N-terminal β-branched residues, and two N-proximal glycines, all unique to the p15 TMD, are required for fusion activity (Clancy & Duncan, 2011). The fusion activities of TMD peptides correlate with their degree of helix instability (Langosch, Hofmann, & Ungermann, 2007; Poschner, Quint, Hofmann, & Langosch, 2009), suggesting the p15 glycine and β-branched residues may destabilize the portion of the TMD helix that resides in the outer leaflet of the bilayer. Altered lipid packing due to helix instability can promote the initial stages of lipid mixing, as suggested by peptide studies showing that a dynamic TMD helix selectively promotes lipid mixing between the outer leaflets of bilayers (Poschner, Fischer, Herrmann, Hofmann, & Langosch, 2010). Interestingly, N-terminal acyl chains stabilize intrinsically dynamic TMD helices (Poschner & Langosch, 2009). This raises the intriguing possibility that the "trigger" for p15 fusion activity is the close approach of a target membrane. In the absence of a closely apposed membrane, insertion of the p15 myristate into the donor membrane may stabilize the TMD helix while transfer of the myristate from the donor to a juxtaposed target membrane would trigger TMD helix instability and the onset of lipid mixing. Lipid mixing might be further promoted by hydrogen bonding between water molecules and the polar tri-serine motif in the p15 TMD, drawing water molecules deeper into the outer leaflet and expanding the interfacial region.

The stringent sequence requirements of the p15 TMD, when functioning in its natural context, may reflect the need for additional fusion attributes of the TMD to compensate for "deficiencies" in the p15 ecto- or endodomains. For example, the limited size of the p15 ectodomain and the atypical features of the p15 PPII FP may necessitate additional roles for the TMD beyond a funnel-shaped architecture. This conjecture is supported by the observation that replacement of the p15 ectodomain with that of p14 generates a functional FAST protein chimera (Clancy & Duncan, 2011), while a p14 chimera containing the p15 ectodomain is nonfunctional (Fig. 3).

C. The FAST Protein Transmembrane Domains and Pore Formation

The transition from hemifusion to pore formation is considered the rate-limiting step in most membrane fusion reactions, although this may not be the case for the FAST proteins. Chlorpromazine (CPZ) is a membrane curvature agent that promotes positive curvature of the inner leaflet, increasing the transition from hemifusion to pore formation (Melikyan, Brener, Ok, & Cohen, 1997). Consistent with this presumed mode of action, CPZ increases the rate of syncytium formation induced by wild-type influenza virus hemagglutinin and restores syncytiogenic activity to a hemifusion-positive, pore formation-negative mutant of hemagglutinin (Clancy et al., 2010). However, parallel experiments with p14-transfected cells indicate CPZ has no effect on the rate or extent of pore formation or syncytiogenesis, even under limiting plasma membrane concentrations of the FAST protein. The absence of a CPZ-sensitive stage led to the proposal of an "all-or-nothing" model for the FAST protein fusion process. Under this scenario, accumulation of the FAST proteins in suitable quantities and at an adhesion site with appropriate intermembrane distances may be sufficient to both initiate lipid mixing and directly transition to stable fusion pore formation (Clancy et al., 2010). By promoting positive curvature of the inner leaflet, the conical FAST protein TMDs mimic the effect of CPZ. The FAST proteins also cluster together in membrane microdomains, so-called lipid rafts (Munro, 2003), as shown by their association with the detergent-resistant membrane fraction of cells and by atomic force microscopy visualization of aqueous lipid membranes containing the p14 FAST protein (Corcoran et al., 2006). Coimmunoprecipitation of untagged and epitope-tagged p14, under detergent conditions that solubilize lipid rafts, also indicates that the FAST proteins form homotypic interactions (Corcoran et al., 2011). Clustering of the FAST proteins would therefore accentuate the membrane curvature effects of their conical TMDs, resulting in the direct transition from lipid mixing to pore formation and the absence of a CPZ-sensitive step in the fusion process.

D. The FAST Protein Endodomains and Pore Formation

A striking feature of the FAST proteins is the disproportionate size of their cytoplasmic, C-terminal endodomains (Fig. 1), implying an important role in the fusion process. Truncation and substitution studies indicate the membrane-distal regions of the FAST protein endodomains are required for stable pore formation (Barry & Duncan, 2009). Furthermore, the C-terminal amino acid sequences of the p10 and p14 endodomains can be scrambled and retain their fusion promoting function, but replacement of these sequences with heterologous amino acids eliminates membrane fusion activity (Barry & Duncan, 2009). The C-termini of the FAST protein endodomains therefore function in an amino acid content-dependent, linear sequence-independent manner. The cell–cell fusion kinetics of the scrambled sequence constructs are, however, delayed relative to the authentic sequence constructs, implying that linear sequence motifs exert a second, nonessential role in promoting pore formation. While the basis for these fusion phenotypes has not been determined, there is a direct correlation between predicted structural disorder of their C-termini and the functionality of the various FAST protein constructs (Barry & Duncan, 2009). The inherent flexibility of intrinsically disordered domains facilitates interaction with multiple partners, both proteins and membranes (Cortese, Uversky, & Dunker, 2008; Popovic, De Biasio, Pintar, & Pongor, 2007). The disordered C-termini of the FAST protein endodomains might therefore interact with the inner leaflet of the bilayer to induce or stabilize membrane curvature in the rim of the fusion pore. Similar effects on the fusion pore rim could also be generated by the disordered C-termini recruiting cellular protein partners that alter membrane curvature or actin–membrane interactions at adherens junctions. If the C-termini recruit protein partners needed for pore formation, this would appear to be a specific requirement during cell–cell fusion since the p14 FAST protein is sufficient to induce liposome–cell fusion (Top et al., 2005).

In addition to the extreme C-termini of the FAST proteins, additional membrane-proximal endodomain motifs also play an active role in the membrane fusion process. Mutational analysis indicates the p10 and p14 PB motifs are directly involved in the fusion process (Barry & Duncan, unpublished). The clustering of PB motifs due to homotypic FAST protein interactions and/or association with lipid rafts could recruit anionic phospholipids in the inner leaflet to the fusion site, altering the local lipid composition of the inner monolayer and creating membrane deformations that promote stable pore formation. A somewhat similar mechanism has been suggested for membrane fission during yeast endocytosis, where localized phosphatidylinositol 4,5-biphosphate concentrations create lipid phase boundaries that promote membrane scission (Liu, Sun, Oster, & Drubin, 2010).

One of the more surprising features of the p13, p15, and p22 FAST proteins is the presence of a HP in their endodomains, rather than in their ectodomains where one would expect to find a FP. Mutational studies indicate the p15 HP is essential for stable pore formation (Clancy & Duncan, unpublished). Furthermore, a chimeric p14 FAST protein containing the endodomain of p15 is the only FAST protein mutant with higher fusion activity than the authentic FAST proteins (p14end15 construct in Fig. 3). It is noteworthy that this construct contains the HPs from both the p14 ectodomain and the p15 endodomain. Most interestingly, the 20-residue p15 HP can be functionally replaced by the 25-residue ALPS (amphipathic lipid packing sensor) motif from the ArfGap1 protein (Read & Duncan, unpublished). As opposed to FPs that insert hydrophobic side chains into planar bilayers to induce membrane curvature, ALPS motifs sense altered lipid headgroup packing in highly curved membranes. Insertion of the hydrophobic residues of ALPS motifs into these membrane defects stabilizes the highly curved membrane structure (Bigay, Casella, Drin, Mesmin, & Antonny, 2005; Drin & Antonny, 2010). The same ALPS motif can also functionally replace a previously unrecognized potential amphipathic helix located in the p14 endodomain, just downstream of the PB motif (Read & Duncan, unpublished). Assuming an interbilayer distance of ~10–20 nm between apposed membranes undergoing fusion, the rim of a nascent fusion pore is a highly curved and energetically unfavorable membrane structure. The FAST protein endodomain HPs or other membrane-proximal amphipathic helices may therefore function as "internal" FPs, lowering the energy barrier to stable pore formation by masking the hydrophobic defects generated in the rim of a nascent fusion pore.

VII. THE FAST PROTEINS AND POSTFUSION EVENTS

Cell–cell membrane fusion designed to generate multinucleated syncytia requires the nanometer-sized micropores formed at the completion of the membrane merger process expand to micron-sized openings. We refer to this latter event as macropore formation, a process that is enhanced by the extended cytoplasmic tails of the FAST proteins. In transfected or virus-infected cells expressing the p14 FAST protein, a soluble p14 endodomain fragment is broadly distributed throughout the cytosol and nucleus (Top, Barry, Racine, Ellis, & Duncan, 2009). Size estimates suggest this fragment arises by regulated intramembrane proteolysis of the full-length p14 protein, a process used to generate bioactive signaling peptides from certain membrane receptors following ligand binding (M. S. Brown, Ye, Rawson, & Goldstein, 2000). Ectopic expression of the FAST protein endodomains substantially enhances the rate of syncytiogenesis, and fluorescent pore formation assays indicate this effect occurs downstream of micropore formation (Top et al., 2009). The ability of the soluble endodomain to enhance the rate of macropore formation is sequence-dependent (i.e., scrambling the endodomain or alanine substitutions abrogate activity) and does not involve direct interactions with either the fusogen or the membranes being fused. Most importantly, the bioactive endodomains function to increase the syncytiogenic activity of not only the homologous FAST protein but also of heterologous FAST proteins, influenza virus hemagglutinin, and the unidentified fusogen responsible for the differentiation-dependent fusion of muscle cells into multinucleated myotubes (Top et al.). The FAST protein endodomains therefore function in *trans* to recruit an intracellular factor or pathway broadly involved in the conversion of micropores into macropores.

In keeping with the concept that the endodomains may be intrinsically disordered domains designed to interact with multiple partners, recent yeast two-hybrid screens with the FAST protein endodomains identified numerous potential protein partners. Several of these protein partners are involved in regulating the actin cytoskeleton (Boutilier & Duncan, unpublished). In view of the inhibitory role of the actin cytoskeleton on formation of macropores (Chen et al., 2008; Wurth et al., 2010), the FAST protein endodomains could promote macropore formation by altering actin dynamics, either indirectly by engaging transcription factors that regulate the cytoskeletal network or directly via interactions with proteins that mediate actin polymerization. Additional endodomain protein partners are involved in adherens junction formation or stability. This is an intriguing observation considering that adherens junctions are required for the prefusion stage of FAST protein-induced cell–cell fusion (Salsman et al., 2008), but these junctions clearly need to be disrupted to allow macropore formation to occur. Other potential endodomain partners are proteins that recognize or alter membrane curvature, which could function to reduce the line tension of the highly curved pore rim to promote lateral expansion of the fusion pore (Chen et al., 2008). Future studies are needed to define the roles, if any, of these cellular proteins in the general process of macropore formation.

Pore expansion leading to syncytium formation is also augmented by the *cis*-acting activity of the FAST protein endodomains (i.e., when the endodomain functions as part of the intact FAST protein). C-terminal truncations that modestly reduce syncytium formation have no discernible effect on micropore formation (Barry & Duncan, 2009), suggesting these truncations affect the creation of macropores. Furthermore, reextending a fusion-dead truncation mutant with a heterologous sequence has no effect on restoring syncytiogenesis but partially restores pore formation. Since the *trans*-acting activity of the endodomain is sequence-dependent, these results cannot be explained by effects on the soluble endodomain fragment. The *cis*-acting FAST protein endodomains therefore exert overlapping or independent effects on generating both micropores and macropores. These effects involve the membrane-distal region of the endodomain, and therefore function in addition to the membrane-proximal PB regions or HPs to efficiently complete the process of syncytium formation.

Given the small size of the FAST protein ectodomains, the disproportionate size of their endodomains, and the biological imperative of the endodomains to function from within the cytosol and not from within a virus particle, it is not surprising that these fusion modules evolved to serve more than one purpose in mediating cell–cell fusion. The multifunctional role of the FAST protein endodomains in the membrane merger and postfusion stages of syncytiogenesis is therefore perfectly suited to the evolution of the FAST proteins as virus-encoded cellular fusogens. Along with the role of surrogate adhesins in mediating the prefusion stage of cell–cell fusion, the multifaceted nature of the endodomains helps to explain how these diminutive fusogens efficiently induce syncytium formation in the absence of energy provided by complex ectodomain rearrangements.

VIII. MODEL OF THE FAST PROTEIN-MEDIATED CELL–CELL FUSION REACTION

The above-mentioned studies of the various FAST proteins have culminated in a working model of FAST protein-induced cell–cell membrane fusion and syncytium formation (Fig. 4). The FAST proteins are comprised of three fusion modules, positioned within and on either side of the membrane in which the FAST proteins reside, and each module contains its own unique series of membrane interaction motifs. In the prefusion stage, the conical TMD and flexible, amphiphilic ectodomain may function together to promote dimple formation in the donor membrane (Fig. 4, panel I). Under this scenario, the

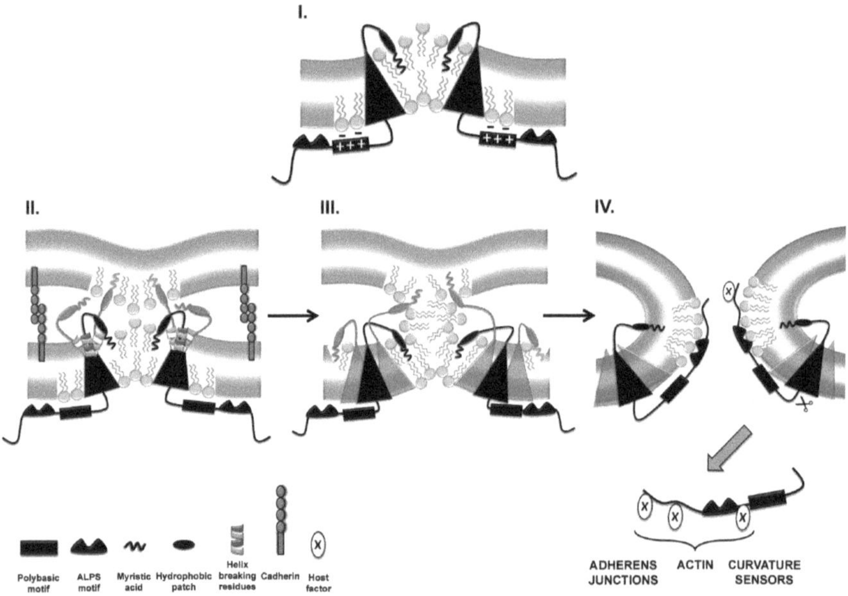

FIGURE 4. Model of the FAST protein membrane fusion reaction. The influence of the FAST proteins on the prefusion state of the donor membrane is depicted (panel I). The close membrane apposition (panel II), lipid mixing (panel III), and pore formation (panel IV) steps of the membrane reaction are shown. The FAST protein fusion motifs (shaded black) and accessory factors are indicated in the legend. The gray shading of the fusion motifs in panels II and III reflect either clustering of the FAST proteins or dynamic membrane interactions of the fusion motifs. The soluble endodomain fragment generated by proteolytic cleavage of the FAST proteins is depicted in panel IV (see text for details).

funnel-shaped TMD induces positive curvature in the inner leaflet and negative curvature in the corresponding outer leaflet. These curvature effects may be amplified by insertion of the myristic acid and hydrophobic amino acid side chains of the ectodomain into the outer leaflet of the donor membrane, and by clustering of the FAST proteins due to reversible homotypic FAST protein interactions or dynamic association with membrane microdomains. The lamellar structure at the inflection point between the planar membrane bilayer and the emerging dimple may be further altered by the endodomain PB motif recruiting phosphatidylinositol lipids present in the inner leaflet of the membrane. The unusual features of the FAST protein FPs also suggests that reversible membrane association of the amphiphilic, flexible ectodomains may alter the interfacial region, blurring the membrane-solvent boundary by increasing vertical fluctuations of membrane lipids. We view this nascent, prefusion dimple as a highly dynamic structure, similar to wave undulations. The model suggests that clustering of the FAST proteins leads to increased dimple curvature resulting in energetically unfavorable hydrophobic defects at the tip of the dimple. The energy penalty for creation of these hydrophobic defects presumably exceeds the energy that can be accumulated in the membrane by curvature stress and/or lipid disorder created by the FAST proteins. Thus, the wave continually forms and recedes. The cumulative effects of the FAST protein ectodomain and TMD fusion modules on the structure of the donor membrane in effect primes the fusion site for interaction with a closely apposed target membrane.

Colocalization of the FAST proteins at adherens junctions creates a favorable environment for close membrane apposition, which triggers initiation of membrane merger (Fig. 4, panel II). A region of extended, stable cell–cell contact may allow random membrane fluctuations and actin-driven protrusions to bring the target membrane into the sphere of influence of the FAST protein ectodomains. At this intermembrane distance (~5–10 nm), transient FAST protein-induced dimpling of the donor membrane would help to bridge the final gap between the two membranes. The previously unfavorable thermodynamics that forced ectodomain hydrophobic moieties to interact with the donor membrane are now reduced, increasing the likelihood for these molecules to reversibly interact with both the donor and target membranes. Flickering membrane interactions of the ectodomain could alter the hydration state, lamellar structure, and interfacial boundaries of the membranes at the point of contact, breaching the critical repulsive barrier and initiating fusion (Fig. 4, panel II). For the majority of FAST proteins, exit of their N-terminal myristate from the donor membrane may also trigger activation of other fusion motifs, such as TMD helix instability. The net result of these FAST protein-induced changes to the lamellar membrane structure is a reduction in the activation energy for lipid mixing.

The membrane merger reaction, once initiated, proceeds directly from lipid mixing to pore formation via the coordinated actions of all three fusion modules working from both sides of the donor membrane, and possibly from the outer leaflet of the target membrane. The model suggests that lipid mixing may involve a more disordered stalk-like intermediate that is less sensitive to inhibition by LPC (Fig. 4, panel III). Increased disorder of the interfacial regions and partial dehydration of lipid headgroups in the zone of membrane contact may make a disordered intermediate more energetically feasible. However, hydration pressure surrounding the fusion site would still make this a high-energy intermediate. Unique features of the FAST protein TMDs and endodomains may lower the activation energy of pore formation, making pore formation a kinetically feasible means to allow resolution of the energetically unfavorable lipid mixing intermediate (Fig. 4, panel IV). These features include the curvature effects of the TMD, lipid packing sensor motifs in the endodomain, and membrane interactions of the intrinsically disordered C-termini of the FAST proteins, which together promote and stabilize positive curvature in the inner leaflet. The C-termini may also recruit host factors to alter the actin–membrane interactions that impede pore formation. The interchangeable nature of the FAST protein fusion modules suggests that various combinations of the fusion motifs present in these different modules can work together to mediate membrane merger. However, certain motifs in some modules, such as the PPII helix in the small p15 ectodomain module, may necessitate specific motifs in other modules, such as increased helix instability and a tri-serine motif in the TMD, to successfully promote the membrane remodelling needed for fusion.

Following completion of the membrane merger stage, the FAST proteins promote syncytiogenesis by enhancing the conversion of micropores into macropores. FAST protein-induced syncytiogenesis presumably occurs in the absence of the *trans*-acting endodomain, as occurs during syncytium formation induced by enveloped virus fusion proteins. As dedicated cell–cell fusogens, however, the FAST proteins employ a *trans*-acting signaling peptide to specifically enhance this postfusion stage of syncytium formation. The ability of the *trans*-acting FAST protein endodomains to promote macropore formation induced by diverse viral and cellular fusogens indicates the soluble endodomain recruits cellular partners generally involved in pore expansion (Fig. 4, panel IV). Likely candidates for such partners are cellular proteins that alter actin remodelling, cell–cell junctions, or highly curved membrane structures.

IX. SUMMARY AND FUTURE DIRECTIONS

The substantial discrepancies between the structural features of the enveloped virus fusogens and the FAST proteins reflect the fact that these fusogens are charged with very different tasks. Enveloped viruses use synchronous activation of a predetermined quantity of fusion protein complexes to mediate all steps of the virus–cell fusion reaction. The FAST proteins use expression-dependent accumulation in specific regions of the plasma membrane to mediate cell–cell fusion, exploiting preexisting cellular cofactors during the pre- and postfusion stages of syncytium formation. It is therefore unclear whether the features that are shared between the FAST proteins and the enveloped virus fusogens function analogously. For example, do their FPs and TMDs serve a similar purpose in the fusion reaction or did the HA FP and TMD evolve specifically to function in conjunction with hairpin formation? If triggering HA fusion activity in transfected cells leads to membrane fusion and syncytium formation, even when the very short cytoplasmic tail is deleted, then why do the FAST proteins require an extended endodomain to complete merger? Further study of these minimal fusogens should continue to provide alternative perspectives on the general process of protein-mediated membrane fusion.

Numerous issues regarding the mechanism of action of the FAST proteins still need to be resolved. Additional mutational analysis is needed to define how specific structural motifs function. Do the FAST protein FPs function according to the curvature-inducing model? Do all of the internal HPs function as ALPS motifs, and does their proximity to the membrane affect this activity? Do portions of the endodomains function as intrinsically disordered domains? What is the specific role of myristoylation in the membrane fusion reaction? What are the roles of the palmitoylated dicysteine motif and CR in p10, and does the cystine loop function as a cystine noose? How do the PB motifs promote membrane fusion? This information should provide important insights into the factors constraining the modularity of the FAST proteins, with the intriguing possibility of creating "designer fusogens" based on assembling the correct repertoire of fusion motifs.

There are also broader issues that need to be addressed. Considerable analysis of the multifunctional endodomains is required to tease out the complementary *cis/trans* functions of this fusion module. These studies should reveal the role of undefined cellular pathways in the general process of syncytium formation, and whether macropore formation reflects a cellular response to micropore formation. Solving the structure of a full-length FAST protein in a membrane would considerably advance our understanding of the prefusion state. If coupled with real-time atomic force microscopy or neutron diffraction studies with fluid membranes, these studies can reveal the effects of the FAST proteins on thermal membrane fluctuations. NMR analysis of the TMD coupled with TMD peptide-liposome studies should address whether the predicted conical shape of the TMD and/or TMD helix instability are important for fusion activity. It would be particularly informative to determine whether the FAST proteins induce hemifusion through a more disordered lipid intermediate, or do they induce formation of an ordered stalk structure and if so, why is this not inhibited by LPC? Similarly, if close membrane apposition is the rate limiting event in the FAST protein membrane fusion reaction, how do p14-liposomes fuse to cells and could increased liposome–cell attachment result in increased fusion? This is highly relevant to ongoing development of FAST–liposomes as an intracellular drug delivery platform. In addition, although the FAST proteins are viral proteins, they evolved to function as cellular fusogens, which raises the question of whether other cellular fusogens resemble the FAST proteins or mimic their mode of action. For example, does physiological syncytium formation rely on specific cell–cell adhesion sites created during cell differentiation that function analogously to the generic adherens junctions involved in FAST protein-induced syncytiogenesis? Additional insights into the *modus operandi* of the FAST proteins may therefore be relevant to understanding how cell–cell fusion occurs during such biological processes as muscle, bone, and placenta formation and during fertilization.

Abbreviations

ARV	avian reovirus
AtSRV	Atlantic salmon aquareovirus
BroV	Broome reovirus
BRV	baboon reovirus
CPZ	chlorpromazine
CR	conserved region
FAST	fusion-associated small transmembrane
FP	fusion peptide
HP	hydrophobic patch
LPC	lysophosphatidylcholine
NBV	Nelson Bay reovirus
PB	polybasic
PC	polycysteine
PP	polyproline
PPII	polyproline type II helix
RPH	arginine–proline–histidine
RRV	reptilian reovirus
TMD	transmembrane domain

References

Ahne, W., Thomsen, I., & Winton, J. (1987). Isolation of a reovirus from the snake, *Python regius*. Brief report. *Archives of Virology, 94*, 135–139.

Barry, C., & Duncan, R. (2009). Multifaceted sequence-dependent and -independent roles for reovirus FAST protein cytoplasmic tails in fusion pore formation and syncytiogenesis. *Journal of Virology, 83*, 12185–12195.

Barry, C., Key, T., Haddad, R., & Duncan, R. (2010). Features of a spatially constrained cystine loop in the p10 FAST protein ectodomain define a new class of viral fusion peptides. *Journal of Biological Chemistry, 285*, 16424–16433.

Benavente, J., & Martinez-Costas, J. (2007). Avian reovirus: Structure and biology. *Virus Research, 123*, 105–119.
Bigay, J., Casella, J. F., Drin, G., Mesmin, B., & Antonny, B. (2005). ArfGAP1 responds to membrane curvature through the folding of a lipid packing sensor motif. *EMBO Journal, 24*, 2244–2253.
Brown, C. W., Stephenson, K. B., Hanson, S., Kucharczyk, M., Duncan, R., & Bell, J. C., et al. (2009). The p14 FAST protein of reptilian reovirus increases vesicular stomatitis virus neuropathogenesis. *Journal of Virology, 83*, 552–561.
Brown, M. S., Ye, J., Rawson, R. B., & Goldstein, J. L. (2000). Regulated intramembrane proteolysis: A control mechanism conserved from bacteria to humans. *Cell, 100*, 391–398.
Campelo, F., McMahon, H. T., & Kozlov, M. M. (2008). The hydrophobic insertion mechanism of membrane curvature generation by proteins. *Biophysical J., 95*, 2325–2339.
Cathomen, T., Mrkic, B., Spehner, D., Drillien, R., Naef, R., & Pavlovic, J., et al. (1998). A matrix-less measles virus is infectious and elicits extensive cell fusion: Consequences for propagation in the brain. *EMBO Journal, 17*, 3899–3908.
Chen, A., Leikina, E., Melikov, K., Podbilewicz, B., Kozlov, M. M., & Chernomordik, L. V. (2008). Fusion-pore expansion during syncytium formation is restricted by an actin network. *Journal of Cell Science, 121*, 3619–3628.
Cheng, L. T., Plemper, R. K., & Compans, R. W. (2005). Atypical fusion peptide of Nelson Bay virus fusion-associated small transmembrane protein. *Journal of Virology, 79*, 1853–1860.
Chernomordik, L., Leikina, E., Cho, M. S., & Zimmerberg, J. (1995). Control of baculovirus gp64-induced syncytium formation by membrane lipid composition. *Journal of Virology, 69*, 3049–3058.
Chernomordik, L. V., Frolov, V. A., Leikina, E., Bronk, P., & Zimmerberg, J. (1998). The pathway of membrane fusion catalyzed by influenza hemagglutinin: Restriction of lipids, hemifusion, and lipidic fusion pore formation. *The Journal of Cell Biology, 140*, 1369–1382.
Chernomordik, L. V., & Kozlov, M. M. (2003). Protein-lipid interplay in fusion and fission of biological membranes. *Annual Review of Biochemistry, 72*, 175–207.
Chernomordik, L. V., & Kozlov, M. M. (2005). Membrane hemifusion: Crossing a chasm in two leaps. *Cell, 123*, 375–382.
Chernomordik, L. V., & Kozlov, M. M. (2008). Mechanics of membrane fusion. *Nature Structural & Molecular Biology, 15*, 675–683.
Chernomordik, L. V., Zimmerberg, J., & Kozlov, M. M. (2006). Membranes of the world unite!. *The Journal of Cell Biology, 175*, 201–207.
Chua, K. B., Crameri, G., Hyatt, A., Yu, M., Tompang, M. R., & Rosli, J., et al. (2007). A previously unknown reovirus of bat origin is associated with an acute respiratory disease in humans. *Proceedings of the National Academy of Sciences United States of America, 104*, 11424–11429.
Chua, K. B., Voon, K., Crameri, G., Tan, H. S., Rosli, J., & McEachern, J. A., et al. (2008). Identification and characterization of a new orthoreovirus from patients with acute respiratory infections. *PLoS One, 3*, e3803.
Clancy, E. K., Barry, C., Ciechonska, M., & Duncan, R. (2010). Different activities of the reovirus FAST proteins and influenza hemagglutinin in cell-cell fusion assays and in response to membrane curvature agents. *Virology, 397*, 119–129.
Clancy, E. K., & Duncan, R. (2009). Reovirus FAST protein transmembrane domains function in a modular, primary sequence-independent manner to mediate cell-cell membrane fusion. *Journal of Virology, 83*, 2941–2950.
Clancy, E. K., & Duncan, R. (2011). Helix-destabilizing, β-branched and polar residues in the baboon reovirus p15 transmembrane domain influence the modularity of the FAST proteins. *Journal of Virology, 85*, 4707–4719.

Cohen, F. S., & Melikyan, G. B. (2004). The energetics of membrane fusion from binding, through hemifusion, pore formation, and pore enlargement. *Journal of Membrane Biology, 199*, 1–14.
Corcoran, J A., Clancy, E. K., & Duncan, R. (2011). Homomultimerization of the reovirus p14 fusion-associated small transmembrane protein during transit through the ER-Golgi complex secretory pathway. *Journal of General Virology, 92*, 162–166.
Corcoran, J. A., & Duncan, R. (2004). Reptilian reovirus utilizes a small type III protein with an external myristylated amino terminus to mediate cell-cell fusion. *Journal of Virology, 78*, 4342–4351.
Corcoran, J. A., Salsman, J., de Antueno, R., Touhami, A., Jericho, M. H., & Clancy, E. K., et al. (2006). The p14 fusion-associated small transmembrane (FAST) protein effects membrane fusion from a subset of membrane microdomains. *The Journal of Biological Chemistry, 281*, 31778–31789.
Corcoran, J. A., Syvitski, R., Top, D., Epand, R. M., Epand, R. F., & Jakeman, D., et al. (2004). Myristoylation, a protruding loop, and structural plasticity are essential features of a nonenveloped virus fusion peptide motif. *The Journal of Biological Chemistry, 279*, 51386–51394.
Cortese, M. S., Uversky, V. N., & Dunker, A. K. (2008). Intrinsic disorder in scaffold proteins: Getting more from less. *Progress in Biophysics and Molecular Biology, 98*, 85–106.
Danieli, T., Pelletier, S. L., Henis, Y. I., & White, J. M. (1996). Membrane fusion mediated by the influenza virus hemagglutinin requires the concerted action of at least three hemagglutinin trimers. *The Journal of Cell Biology, 133*, 559–569.
Dawe, S., Corcoran, J. A., Clancy, E. K., Salsman, J., & Duncan, R. (2005). Unusual topological arrangement of structural motifs in the baboon reovirus fusion-associated small transmembrane protein. *Journal of Virology, 79*, 6216–6226.
Dawe, S., & Duncan, R. (2002). The S4 genome segment of baboon reovirus is bicistronic and encodes a novel fusion-associated small transmembrane protein. *Journal of Virology, 76*, 2131–2140.
Drin, G., & Antonny, B. (2010). Amphipathic helices and membrane curvature. *FEBS Letters, 584*, 1840–1847.
Duncan, R. (1999). Extensive sequence divergence and phylogenetic relationships between the fusogenic and nonfusogenic orthoreoviruses: A species proposal. *Virology, 260*, 316–328.
Duncan, R., Chen, Z., Walsh, S., & Wu., S. (1996). Avian reovirus-induced syncytium formation is independent of infectious progeny virus production and enhances the rate, but is not essential, for virus-induced cytopathology and virus egress. *Virology, 224*, 453–464.
Duncan, R., Corcoran, J., Shou, J., & Stoltz, D. (2004). Reptilian reovirus: A new fusogenic orthoreovirus species. *Virology, 319*, 131–140.
Duncan, R., Murphy, F. A., & Mirkovic, R. R. (1995). Characterization of a novel syncytium-inducing baboon reovirus. *Virology, 212*, 752–756.
Duncan, R., & Sullivan, K. (1998). Characterization of two avian reoviruses that exhibit strain-specific quantitative differences in their syncytium-inducing and pathogenic capabilities. *Virology, 250*, 263–272.
Eitzen, G. (2003). Actin remodeling to facilitate membrane fusion. *Biochimica et Biophysica Acta, 1641*, 175–181.
Gard, G., & Compans, R. W. (1970). Structure and cytopathic effects of Nelson Bay virus. *Journal of Virology, 6*, 100–106.
Gibbons, D. L., Vaney, M. C., Roussel, A., Vigouroux, A., Reilly, B., & Lepault, J., et al. (2004). Conformational change and protein-protein interactions of the fusion protein of Semliki Forest virus. *Nature, 427*, 320–325.
Harrison, S. C. (2008). Viral membrane fusion. *Nature Structural & Molecular Biology, 15*, 690–698.

Helming, L., & Gordon, S. (2009). Molecular mediators of macrophage fusion. *Trends in Cell Biology, 19*, 514–522.

Jahn, R., & Grubmuller, H. (2002). Membrane fusion. *Current Opinion in Cell Biology, 14*, 488–495.

Jahn, R., & Scheller, R. H. (2006). SNAREs—Engines for membrane fusion. *Nature Reviews Molecular Cell Biology, 7*, 631–643.

Kibenge, F. S., & Dhillon, A. S. (1987). A comparison of the pathogenicity of four avian reoviruses in chickens. *Avian Diseases, 31*, 39–42.

Kielian, M., & Rey, F. A. (2006). Virus membrane-fusion proteins: More than one way to make a hairpin. *Nature Reviews Microbiology, 4*, 67–76.

Koch, A. W., Manzur, K. L., & Shan, W. (2004). Structure-based models of cadherin-mediated cell adhesion: The evolution continues. *Cellular and Molecular Life Sciences, 61*, 1884–1895.

Kozlov, M. M., McMahon, H. T., & Chernomordik, L. V. (2010). Protein-driven membrane stresses in fusion and fission. *Trends in Biochemical Sciences, 35*, 699–706.

Lamb, R. A., & Jardetzky, T. S. (2007). Structural basis of viral invasion: Lessons from paramyxovirus F. *Current Opinion in Structural Biology, 17*, 427–436.

Lamirande, E. W., Nichols, D. K., Owens, J. W., Gaskin, J. M., & Jacobson, E. R. (1999). Isolation and experimental transmission of a reovirus pathogenic in ratsnakes (*Elaphe* species). *Virus Research, 63*, 135–141.

Langosch, D., Hofmann, M., & Ungermann, C. (2007). The role of transmembrane domains in membrane fusion. *Cellular and Molecular Life Sciences, 64*, 850–864.

Lapthorn, A. J., Janes, R. W., Isaacs, N. W., & Wallace, B. A. (1995). Cystine nooses and protein specificity. *Nature Structural Biology, 2*, 266–268.

Leland, M. M., Hubbard, G. B., Sentmore III, H. T., Soike, K. F., & Hilliard, J. K. (2000). Outbreak of Orthoreovirus-induced meningoencephalomyelitis in baboons. *Comparative Medicine, 50*, 199–205.

Liu, J., Sun, Y., Oster, G. F., & Drubin, D. G. (2010). Mechanochemical crosstalk during endocytic vesicle formation. *Current Opinion in Cell Biology, 22*, 36–43.

Markosyan, R. M., Melikyan, G. B., & Cohen, F. S. (2001). Evolution of intermediates of influenza virus hemagglutinin-mediated fusion revealed by kinetic measurements of pore formation. *Biophysical J., 80*, 812–821.

Martens, S., Kozlov, M. M., & McMahon, H. T. (2007). How synaptotagmin promotes membrane fusion. *Science, 316*, 1205–1208.

Martens, S., & McMahon, H. T. (2008). Mechanisms of membrane fusion: Disparate players and common principles. *Nature Reviews Molecular Cell Biology, 9*, 543–556.

McMahon, H. T., Kozlov, M. M., & Martens, S. (2010). Membrane curvature in synaptic vesicle fusion and beyond. *Cell, 140*, 601–605.

Melikyan, G. B. (2008). Common principles and intermediates of viral protein-mediated fusion: The HIV-1 paradigm. *Retrovirology, 5*, 111.

Melikyan, G. B., Brener, S. A., Ok, D. C., & Cohen, F. S. (1997). Inner but not outer membrane leaflets control the transition from glycosylphosphatidylinositol-anchored influenza hemagglutinin-induced hemifusion to full fusion. *The Journal of Cell Biology, 136*, 995–1005.

Melikyan, G. B., White, J. M., & Cohen, F. S. (1995). GPI-anchored influenza hemagglutinin induces hemifusion to both red blood cell and planar bilayer membranes. *The Journal of Cell Biology, 131*, 679–691.

Mertens, P. P. C., Attoui, H., Duncan, R., & Dermody, T. S. (2005). Reoviridae. Elsevier/Academic Press, London 447–454.

Mittal, A., Leikina, E., Chernomordik, L. V., & Bentz, J. (2003). Kinetically differentiating influenza hemagglutinin fusion and hemifusion machines. *Biophysical J., 85*, 1713–1724.

Modis, Y., Ogata, S., Clements, D., & Harrison, S. C. (2004). Structure of the dengue virus envelope protein after membrane fusion. *Nature, 427*, 313–319.

Mohd Jaafar, F., Goodwin, A. E., Belhouchet, M., Merry, G., Fang, Q., & Cantaloube, J. F., et al. (2008). Complete characterisation of the American grass carp reovirus genome (genus Aquareovirus: family Reoviridae) reveals an evolutionary link between aquareoviruses and coltiviruses. *Virology, 373*, 310–321.

Mohler, W. A., Simske, J. S., Williams-Masson, E. M., Hardin, J. D., & White, J. G. (1998). Dynamics and ultrastructure of developmental cell fusions in the *Caenorhabditis elegans* hypodermis. *Current Biology, 8*, 1087–1090.

Munro, S. (2003). Lipid rafts: Elusive or illusive? *Cell, 115*, 377–388.

Nolan, S., Cowan, A. E., Koppel, D. E., Jin, H., & Grote, E. (2006). FUS1 regulates the opening and expansion of fusion pores between mating yeast. *Molecular Biology of the Cell, 17*, 2439–2450.

Oren-Suissa, M., & Podbilewicz, B. (2010). Evolution of programmed cell fusion: Common mechanisms and distinct functions. *Developmental Dynamics, 239*, 1515–1528.

Palacios, G., Lovoll, M., Tengs, T., Hornig, M., Hutchison, S., & Hui, J., et al. (2010). Heart and skeletal muscle inflammation of farmed salmon is associated with infection with a novel reovirus. *PLoS One, 5*, e11487.

Paterson, R. G., Russell, C. J., & Lamb, R. A. (2000). Fusion protein of the paramyxovirus SV5: Destabilizing and stabilizing mutants of fusion activation. *Virology, 270*, 17–30.

Perez-Moreno, M., Jamora, C., & Fuchs, E. (2003). Sticky business: Orchestrating cellular signals at adherens junctions. *Cell, 112*, 535–548.

Popovic, M., De Biasio, A., Pintar, A., & Pongor, S. (2007). The intracellular region of the Notch ligand Jagged-1 gains partial structure upon binding to synthetic membranes. *FEBS Journal, 274*, 5325–5336.

Poschner, B. C., Fischer, K., Herrmann, J. R., Hofmann, M. W., & Langosch, D. (2010). Structural features of fusogenic model transmembrane domains that differentially regulate inner and outer leaflet mixing in membrane fusion. *Molecular Membrane Biology, 27*, 1–10.

Poschner, B. C., & Langosch, D. (2009). Stabilization of conformationally dynamic helices by covalently attached acyl chains. *Protein Science, 18*, 1801–1805.

Poschner, B. C., Quint, S., Hofmann, M. W., & Langosch, D. (2009). Sequence-specific conformational dynamics of model transmembrane domains determines their membrane fusogenic function. *Journal of Molecular Biology, 386*, 733–741.

Pritchard, L. I., Chua, K. B., Cummins, D., Hyatt, A., Crameri, G., & Eaton, B. T., et al. (2006). Pulau virus: A new member of the Nelson Bay orthoreovirus species isolated from fruit bats in Malaysia. *Archives of Virology, 151*, 229–239.

Racine, T., Hurst, T., Barry, C., Shou, J., Kibenge, F., & Duncan, R. (2009). Aquareovirus effects syncytiogenesis by using a novel member of the FAST protein family translated from a noncanonical translation start site. *Journal of Virology, 83*, 5951–5955.

Reese, C., & Mayer, A. (2005). Transition from hemifusion to pore opening is rate limiting for vacuole membrane fusion. *The Journal of Cell Biology, 171*, 981–990.

Reiersen, H., & Rees, A. R. (2001). The hunchback and its neighbours: Proline as an environmental modulator. *Trends in Biochemical Sciences, 26*, 679–684.

Rey, F. A. (2006). Molecular gymnastics at the herpesvirus surface. *EMBO Reports, 7*, 1000–1005.

Richardson, B. E., Nowak, S. J., & Baylies, M. K. (2008). Myoblast fusion in fly and vertebrates: New genes, new processes and new perspectives. *Traffic, 9*, 1050–1059.

Roche, S., Albertini, A. A., Lepault, J., Bressanelli, S., & Gaudin, Y. (2008). Structures of vesicular stomatitis virus glycoprotein: Membrane fusion revisited. *Cellular and Molecular Life Sciences, 65*, 1716–1728.

Roche, S., Bressanelli, S., Rey, F. A., & Gaudin, Y. (2006). Crystal structure of the low-pH form of the vesicular stomatitis virus glycoprotein G. *Science, 313*, 187–191.

Russell, C. J., Jardetzky, T. S., & Lamb, R. A. (2004). Conserved glycine residues in the fusion peptide of the paramyxovirus fusion protein regulate activation of the native state. *Journal of Virology, 78*, 13727–13742.

Salsman, J., Top, D., Barry, C., & Duncan, R. (2008). A virus-encoded cell-cell fusion machine dependent on surrogate adhesins. *PLoS Pathogens, 4*, e1000016.

Salsman, J., Top, D., Boutilier, J., & Duncan, R. (2005). Extensive syncytium formation mediated by the reovirus FAST proteins triggers apoptosis-induced membrane instability. *Journal of Virology, 79*, 8090–8100.

Sapir, A., Avinoam, O., Podbilewicz, B., & Chernomordik, L. V. (2008). Viral and developmental cell fusion mechanisms: Conservation and divergence. *Developmental Cell, 14*, 11–21.

Shmulevitz, M., & Duncan, R. (2000). A new class of fusion-associated small transmembrane (FAST) proteins encoded by the non-enveloped fusogenic reoviruses. *EMBO Journal, 19*, 902–912.

Shmulevitz, M., Epand, R. F., Epand, R. M., & Duncan, R. (2004). Structural and functional properties of an unusual internal fusion peptide in a nonenveloped virus membrane fusion protein. *Journal of Virology, 78*, 2808–2818.

Shmulevitz, M., Salsman, J., & Duncan, R. (2003). Palmitoylation, membrane-proximal basic residues, and transmembrane glycine residues in the reovirus p10 protein are essential for syncytium formation. *Journal of Virology, 77*, 9769–9779.

Shmulevitz, M., Yameen, Z., Dawe, S., Shou, J., & O'Hara, D., et al. (2002). Sequential partially overlapping gene arrangement in the tricistronic S1 genome segments of avian reovirus and Nelson Bay reovirus: Implications for translation initiation. *Journal of Virology, 76*, 609–618.

Spiess, M. (1995). Heads or tails—What determines the orientation of proteins in the membrane. *FEBS Letters, 369*, 76–79.

Stapley, B. J., & Creamer, T. P. (1999). A survey of left-handed polyproline II helices. *Protein Sciences, 8*, 587–595.

Steain, M. C., Wang, B., & Saksena, N. K. (2008). The possible contribution of HIV-1-induced syncytia to the generation of intersubtype recombinants in vitro. *Aids, 22*, 1009–1017.

Stein, A., Weber, G., Wahl, M. C., & Jahn, R. (2009). Helical extension of the neuronal SNARE complex into the membrane. *Nature, 460*, 525–528.

Tamm, L. K. (2003). Hypothesis: Spring-loaded boomerang mechanism of influenza hemagglutinin-mediated membrane fusion. *Biochimica et Biophysica Acta, 1614*, 14–23.

Tamm, L. K., Crane, J., & Kiessling, V. (2003). Membrane fusion: A structural perspective on the interplay of lipids and proteins. *Current Opinion in Structural Biology, 13*, 453–466.

Tamm, L. K., & Han, X. (2000). Viral fusion peptides: A tool set to disrupt and connect biological membranes. *Bioscience Reports, 20*, 501–518.

Tamm, L. K., Han, X., Li, Y., & Lai, A. L. (2002). Structure and function of membrane fusion peptides. *Biopolymers, 66*, 249–260.

Thalmann, C. M., Cummins, D. M., Yu, M., Lunt, R., Pritchard, L. I., & Hansson, E., et al. (2010). Broome virus, a new fusogenic Orthoreovirus species isolated from an Australian fruit bat. *Virology, 402*, 26–40.

Top, D., Barry, C., Racine, T., Ellis, C. L., & Duncan, R. (2009). Enhanced fusion pore expansion mediated by the trans-acting endodomain of the reovirus FAST proteins. *PLoS Pathogens, 5*, e1000331.

Top, D., de Antueno, R., Salsman, J., Corcoran, J., Mader, J., & Hoskin, D., et al. (2005). Liposome reconstitution of a minimal protein-mediated membrane fusion machine. *EMBO Journal, 24*, 2980–2988.

Vieler, E., Baumgartner, W., Herbst, W., & Kohler, G. (1994). Characterization of a reovirus isolate from a rattle snake, *Crotalus viridis*, with neurological dysfunction. *Archives of Virology, 138*, 341–344.

Weber, T., Zemelman, B. V., McNew, J. A., Westermann, B., Gmachl, M., & Parlati, F., et al. (1998). SNAREpins: Minimal machinery for membrane fusion. *Cel, 92*, 759–772.

White, J. M., Delos, S. E., Brecher, M., & Schornberg, K. (2008). Structures and mechanisms of viral membrane fusion proteins: Multiple variations on a common theme. *Critical Reviews in Biochemistry and Molecular Biology, 43*, 189–219.

White, S. H., Ladokhin, A. S., Jayasinghe, S., & Hristova, K. (2001). How membranes shape protein structure. *The Journal of Biological Chemistry, 276*, 32395–32398.

Wurth, M. A., Schowalter, R. M., Smith, E. C., Moncman, C. L., Dutch, R. E., & McCann, F R.O. (2010). The actin cytoskeleton inhibits pore expansion during PIV5 fusion protein-promoted cell-cell fusion. *Virology, 404*, 117–126.

Zaitseva, E., Mittal, A., Griffin, D. E., & Chernomordik, L. V. (2005). Class II fusion protein of alphaviruses drives membrane fusion through the same pathway as class I proteins. *The Journal of Cell Biology, 169*, 167–177.

Zhu, B., Chappuis-Flament, S., Wong, E., Jensen, I. E., Gumbiner, B. M., & Leckband, D. (2003). Functional analysis of the structural basis of homophilic cadherin adhesion. *Biophysical J., 84*, 4033–4042.

PART 3

INTRACELLULAR FUSION

CHAPTER 6

C2 Domains and Membrane Fusion

Sascha Martens[1] and Harvey T. McMahon[2]

[1]Max F. Perutz Laboratories, University of Vienna, Vienna, Austria
[2]MRC Laboratory of Molecular Biology, Cambridge, United Kingdom

I. OVERVIEW

In biology, membrane fusion describes the controlled merger of two initially separated membranes. In this chapter we discuss how cellular proteins mediate the lipid rearrangements that are necessary for fusion to occur. We focus on the Ca^{2+}-dependent fusion of granules and synaptic vesicles in professional secretory cells and neurons, respectively. The key molecules discussed are the SNARE-proteins and C2 domain-containing proteins of the synaptotagmin family and its relatives. Special emphasis is given to recent evidence showing that C2 domains are able to induce membrane curvature in a Ca^{2+}-dependent manner. We discuss how this activity in combination with the energy provided by SNARE complex assembly brings about the extremely fast and controlled fusion of granules and synaptic vesicles.

II. MEMBRANE FUSION

Membrane fusion or the controlled merger of two initially separated membranes is of fundamental importance in biology. This is particularly true for eukaryotic cells that contain a plethora of membrane-bound compartments. The transport of proteins, lipids, and other material between the endoplasmic

Current Topics in Membranes, Volume 68 0065-230X/10 $35.00

DOI: 10.1016/B978-0-12-385891-7.00001-5

reticulum, the Golgi apparatus, the endocytic system, and other intracellular compartments is in large part mediated by small vesicular and tubular carriers (Jahn & Scheller, 2006; Martens & McMahon, 2008; Zerial & McBride, 2001). These carriers bud off from a donor compartment and fuse with their target compartment thereby delivering membrane bound proteins, lipids, and luminal cargo. If these vesicular carriers fuse with the plasma membrane they can deliver molecules such as neurotransmitters and other signaling factors into the extracellular space. Membrane fusion can also occur between whole organelles. Thus, mitochondria undergo constant fusion and fission events (Hoppins, Lackner, & Nunnari, 2007). During cell division the Golgi apparatus fragments in order to allow it to be partitioned between the two daughter cells. After partitioning it reassembles by the controlled fusion of these membranous fragments (Shorter & Warren, 2002). Another fascinating example is the reassembly of the nuclear envelope after mitosis. During open mitosis the nuclear envelope fragments into small vesicular structures. These structures fuse after mitosis in order to reform the nuclear envelope (Guttinger, Laurell, & Kutay, 2009). However, membrane fusion processes are not limited to intracellular events as fusion can also occur between cells. For instance, myoblasts fuse to generate myotubes and trophoblasts fuse to form the syncytiotrophoblast layer (Oren-Suissa & Podbilewicz, 2007). Membrane fusion is also exploited by pathogens such as enveloped viruses that fuse with their target cells in order to gain access to the cell's intracellular space (Skehel & Wiley, 2000).

In all examples of membrane fusion events mentioned above the requirements for the actual fusion event is similar. The fusion event must be tightly controlled and it should be non-leaky, meaning that the membranes must keep their integrity and only the content of the two membrane structures should mix (Chernomordik & Kozlov, 2003; Jahn, Lang, & Sudhof, 2003; Martens & McMahon, 2008).

Most if not all membrane fusion processes are believed to proceed through the same intermediates (Chernomordik & Kozlov, 2008; Martens & McMahon, 2008). Initially the two membranes destined to fuse must be brought into close contact. In most biological fusion events this will require the two membranes to be denuded of proteins in order to allow close contact to occur. Next, energy (e.g., curvature stress) must be injected into both membranes destined to fuse in order to overcome the energy barriers for the subsequent fusion intermediates. If the fusion event is initiated by highly curved membrane intermediates (e.g., buckle-like protrusions) this will facilitate both, close membrane contact and the intermediates to follow (Efrat, Chernomordik, & Kozlov, 2007; Martens, Kozlov, & McMahon, 2007). The next step entails the merger of the two contacting monolayers of each membrane into a structure referred to as hemifusion stalk. During hemifusion the two distal monolayers remain separated. The hemifusion stalk resolves by the subsequent merger of the distal monolayers allowing the formation of the fusion pore (Chernomordik & Kozlov, 2003). This is the first time the contents of the two compartments come into contact and mix. Finally, the fusion pore dilates resulting in the complete intermixing of the two membranes.

During fusion the membranes undergo major conformational changes. However, these changes are controlled and mediated by fusion proteins. The task of the fusion proteins is to reduce the energy barriers separating the highly curved intermediates during the membrane fusion process (Chernomordik & Kozlov, 2003, 2008). Flat membranes do not like to undergo spontaneous fusion, but highly curved membranes have a high spontaneous fusion rate, likely because the energy barriers for the high curvature intermediates is lower. The more unstable the highly curved intermediate the more likely it will want to fuse to relax to a lower energy state (McMahon, Kozlov, & Martens, 2010).

As outlined above the fundamental process of membrane fusion plays crucial roles in a wide range of biological processes. Consequently, it is tightly regulated by a variety of signals that allow the fusion reaction to proceed only when needed (Martens et al., 2008). One of the most fascinating and most investigated fusion reactions is the calcium-dependent fusion of vesicles and granules in professional secretory cells such as neurons and chromaffin cells (Sorensen, 2004; Sudhof, 2004) (Fig. 1).

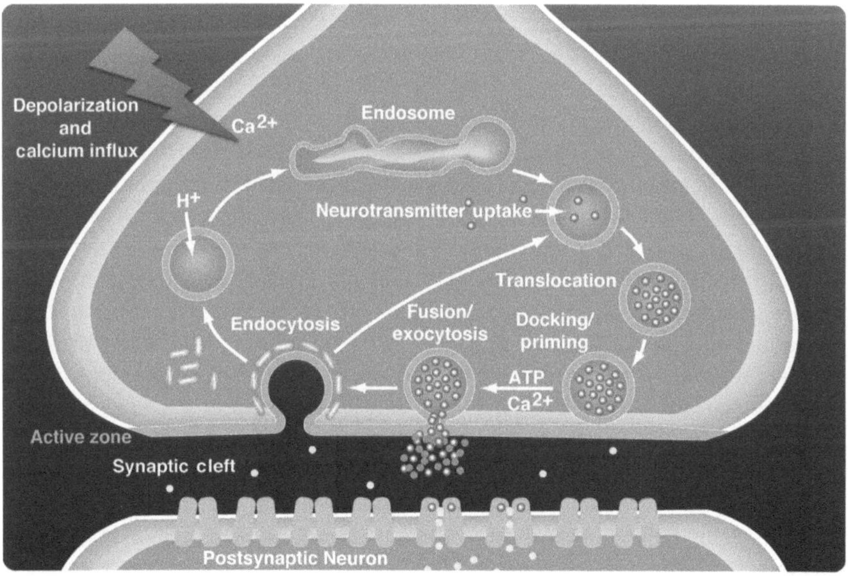

FIGURE 1 The synaptic vesicle exocytic/endocytic cycle. Neurons on depolarization respond to the influx of Ca^{2+} by fusion of vesicles in a region called the active zone. Neurotransmitters are released into the synaptic cleft and bind to postsynaptic receptors to activate/modulate its activity. Vesicle components and the membrane are recycled by endocytosis either directly to form new synaptic vesicles or via an endosome intermediate. These vesicles are then docked/primed and wait on another stimulus for SNARE-dependent fusion. (See Color Insert.)

A. Calcium-Dependent Membrane Fusion

Calcium or more precisely Ca^{2+} ions serve as second messengers for a variety of signaling events. There is a steep gradient of the Ca^{2+} concentration across the limiting membrane of the cell. Thus, the concentration of Ca^{2+} in the extracellular space and some intracellular compartments such as the endoplasmic reticulum is in the μM range whereas the Ca^{2+} concentration in the cytoplasm is in the sub-μM range, at least under resting conditions. The activation of certain signaling receptors and ion channels results in the opening of channels that allow Ca^{2+} to cross the plasma membrane or endoplasmic reticulum membrane. The elevated cytoplasmic Ca^{2+} concentration is sensed by Ca^{2+} binding proteins, which in turn trigger various downstream events. One of the most striking actions is the triggering of exocytic fusion events in neurons, chromaffin cells, and many other professional secretory cells (Fig. 1). In fact many of the hormones controlling our body functions are secreted into the extracellular space by calcium-dependent exocytosis. Furthermore, most of our neurons communicate with each other at chemical synapses. Here the signaling molecules, termed neurotransmitters, are stored in small vesicles and are released by the fusion of these vesicles with the plasma membrane. Upon their release these neurotransmitters bind to receptors on the surface of the neighboring cell and thereby alter its behavior. Its fundamental importance and fascinating biophysics has made synaptic vesicle fusion one of the most heavily studied processes in biology.

1. Synaptic Vesicle Fusion

Synaptic vesicle fusion lies at the core of the communication between the vast majority of our neurons. At chemical synapses one neuron communicates with its neighbor across a specialized site called *synapse* (Fig. 1). Synapses are composed of pre- and postsynaptic structures where the presynaptic part is supplied by the signal-sending neuron and the postsynaptic part is provided by the signal-receiving neuron (DeFelipe, 2010). Within the presynapse the neurotransmitters are packed into small rather uniform 40–50 nm diameter vesicles termed *synaptic vesicles* (Takamori et al., 2006). Dependent on the synapse, the number of synaptic vesicles per presynaptic terminal can vary between a hundred to several thousand. If an action potential arrives at the presynapse it triggers the opening of voltage-gated Ca^{2+} channels resulting in the increase of the intracellular Ca^{2+} concentration. The increased Ca^{2+} concentration triggers the fusion of some synaptic vesicles with the presynaptic plasma membrane resulting in the release of neurotransmitters into the space between the pre- and postsynapse termed *the synaptic cleft* where they can bind to receptors on the postsynaptic plasma membrane (Neher & Sakaba, 2008; Sudhof, 2004). The probability with which an action potential can trigger the fusion of one or more synaptic vesicle is called *release probability* and can vary from 0.1 at some central synapses to 1 at the neuromuscular junction. Changes in this release probability are one of the fundamental mechanisms by which long- and short-term memory are mediated (Citri & Malenka, 2007).

Synaptic vesicles undergo a complex life cycle (Sudhof, 2004) (Fig. 1). Synaptic vesicles initially bud off from a donor compartment and become filled with neurotransmitters. Subsequently some of these vesicles are believed to undergo a complex series of events that are collectively called *docking* and *priming* (Verhage & Sørensen, 2008). The term docking is based on a morphological criterion and describes the phenomenon where some synaptic vesicles are deemed to be close to the presynaptic plasma membrane. The term priming is based on a functional classification and describes the ability of some synaptic vesicles to rapidly fuse after the increase in the intercellular Ca^{2+} concentration. It is commonly believed that during docking and priming the molecular machinery that ultimately brings about the actual fusion event is assembled. Furthermore, the primed vesicles are believed to be a subset of the docked vesicles where docked but not primed vesicles are on some intermediate stage toward the fully fusion competent state. It should be noted that most studies investigating the molecular events during docking and priming have been conducted in chromaffin cells (discussed below) and the strict classification into docked and primed vesicles in neurons should be applied with some caution (de Wit et al., 2009; Sorensen, 2004; Verhage & Sørensen, 2008). After Ca^{2+}-dependent fusion the membrane and proteins of the synaptic vesicles now found in the plasma membrane are recycled by *endocytosis*, after which the synaptic vesicles become re-filled with neurotransmitters and are ready for the next cycle of depolarization and fusion (Sudhof, 2004) (Fig. 1).

It has been observed that upon arrival of an action potential, and subsequent Ca^{2+} channel opening, synaptic vesicles fuse in different phases (Barrett & Stevens, 1972; Sun et al., 2007). *Synchronous release* refers to the neurotransmitters that are released immediately after Ca^{2+} channel opening. For synchronous release, the delay between the increase of the intracellular Ca^{2+} concentration and synaptic vesicle fusion is surprisingly short (less than 1 ms). This short delay has prompted the belief that the molecular release machinery responsible for fusion must be largely preassembled. Synchronous release is followed by *asynchronous release*, which refers to the synaptic vesicles that fuse in a slightly delayed manner. Both synchronous and asynchronous synaptic vesicle fusion are absolutely dependent on Ca^{2+} (Hagler & Goda, 2001). In addition a third form of synaptic vesicle fusion is known, the so-called *spontaneous release*. Spontaneous release is not coupled to an action potential and has long been considered as Ca^{2+} independent. However, there is now compelling evidence that also this type of release is at least to a large extent triggered by Ca^{2+} (Emptage, Reid, & Fine, 2001; Groffen et al., 2010; Xu, Pang, Shin, & Sudhof, 2009).

A large body of evidence suggests that a common molecular mechanism underlies synchronous, asynchronous, as well as spontaneous release (Groffen et al., 2010; Schoch et al., 2001; Verhage et al., 2000). However, before we discuss the molecules that drive these fusion reactions we will briefly discuss Ca^{2+}-dependent fusion in non-neuronal cells.

2. Secretory Granule Fusion

Granules are relatively large secretory organelles (up to several hundred nm in diameter) found in adrenal chromaffin cells, pancreatic beta cells, mast cells, cytotoxic T cells, and many other cell types. Like the fusion of synaptic vesicles in neurons, secretory granule fusion is strictly dependent on Ca^{2+}. In addition, the molecular core machinery mediating the fusion of these granules is the same as the machinery mediating the fusion of synaptic vesicles, although important differences do exist (Sorensen, 2004). However, fusion of secretory granules is about 10 times slower than synaptic vesicle fusion. The most intensely studied system is the exocytosis of secretory granules in chromaffin cells. Analogous to synaptic vesicle fusion, the fusion of secretory granules in chromaffin cells occurs in different phases. Immediately after the rise of the intracellular Ca^{2+} concentration fast fusion is observed. This fast phase is followed by a slow phase, which in turn is followed by a sustained phase (Sorensen, 2004). A key advantage of the study of granule fusion over the study of synaptic vesicle fusion is that the cells are more accessible. Thus, while most measurements of synaptic vesicle fusion are made indirectly by recording the postsynaptic response to the neurotransmitters that are released by these fusion events, individual exocytic events can be detected directly in chromaffin cells. One such method involves measuring plasma membrane capacitance, which allows the detection of an increase in cell surface area upon fusion of secretory granules. Another technique is amperometry that allows the detection of the signaling molecules that are released upon fusion (Haller, Heinemann, Chow, Heidelberger, & Neher, 1998). Both of these techniques are extremely sensitive and accurate and have in combination with the precise manipulation of the molecular release machinery given us deep insights into the core machinery driving Ca^{2+}-dependent exocytosis and membrane fusion in general.

III. THE MOLECULAR MACHINERY MEDIATING CALCIUM-DEPENDENT MEMBRANE FUSION

A wealth of studies over the last decades has unearthed a plethora of molecules that are required for the precisely controlled and fascinating events outlined above. In fact we now have an accurate picture of the molecular composition of a complete synaptic vesicle and although it is not excluded that new molecules playing key roles during Ca^{2+} dependent will be discovered it is likely that the major players are known (Rizo & Rosenmund, 2008; Sudhof, 2004; Takamori et al., 2006). The task now is therefore to assign precise functions and mechanisms to these molecules. In our subsequent discussion we will limit ourselves to the so-called SNARE complex, the synaptotagmins and related C2 domain containing proteins.

A. The SNARE Complex

The SNARE complex is at the core of all Ca^{2+}-dependent exocytic fusion events including synaptic vesicle fusion and secretory granule fusion (Borisovska et al., 2005; Jahn & Scheller, 2006; Schiavo et al., 1992; Schoch et al., 2001; Südhof & Rothman, 2009; Washbourne et al., 2002) (Fig. 2). The

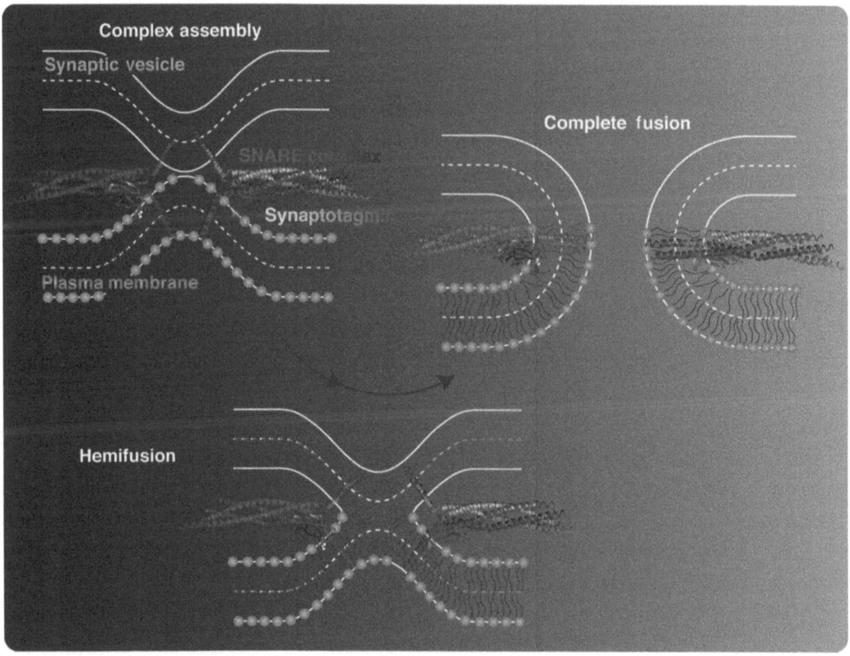

FIGURE 2 SNARE and synaptotagmin-dependent membrane fusion. Zippering of SNARE proteins pulls the synaptic vesicle membrane into close apposition with the plasma membrane. The C2 domains of synaptotagmin/Doc2 proteins bind to the SNARE complex, and so are positioned at the fusion site, where they can dip into the membrane and promote membrane curvature. This will help to reduce the energy barrier that is needed to get to the hemifusion intermediate. The C2 domains of synaptotagmin/Doc2 will still prefer to promote positive membrane curvature and so will help to drive the fusion reaction to complete fusion, where the membranes can relax on collapse of the vesicle. (See Color Insert.)

SNARE complex is an extremely stable four-helical bundle and the fully formed complex likely represents the postfusion state of these proteins. The individual helices are provided by separate SNARE proteins (Stein, Weber, Wahl, & Jahn, 2009; Sutton, Fasshauer, Jahn, & Brunger, 1998). In neurons and chromaffin cells these SNARE proteins are syntaxin1, SNAP25, and synaptobrevin. Each of these SNAREs contains an approximately 60 amino acid long SNARE domain that is largely disordered in isolation (Fasshauer, Otto, Eliason, Jahn, & Brünger, 1997; Stein et al., 2009; Sutton et al., 1998). Syntaxin1 and synaptobrevin are transmembrane domain proteins localized to the plasma membrane and vesicular membrane, respectively (Fig. 2). Syntaxin1 and synaptobrevin each provide one helix to the four-helical SNARE complex. In contrast to syntaxin1 and synaptobrevin, SNAP25 is localized to the plasma membrane via a series of palmitoylated cysteines and provides two helices to the complex (Fasshauer et al., 1997; Stein et al., 2009; Sutton et al., 1998). The discovery that two transmembrane proteins localized to the two membranes destined to fuse have the ability to participate in an extremely stable complex, immediately suggested a mechanism by which these molecules could mediate membrane fusion (Sollner et al., 1993; Weber et al., 1998) (Fig. 2). The SNARE domains of syntaxin1 and synaptobrevin are located to the N-terminus of their transmembrane domains and only a few amino acids separate the SNARE domains from the membrane spanning regions (Stein et al., 2009; Sutton et al., 1998). Structural evidence suggests that upon SNARE complex formation the SNARE helices extend into the transmembrane domains transducing at least some force generated by the SNARE complex formation into the membranes (Stein et al., 2009). In fact it is now widely accepted that SNARE complex formation is the driving force bringing the two membranes destined to fuse into close contact. Furthermore, tetanus and botulinum toxins have been found to specifically cleave SNARE proteins and thereby inhibit synaptic vesicle fusion (Davletov, Bajohrs, & Binz, 2005; Schiavo et al., 1992; Sutton et al., 1998). SNARE proteins are not limited to synapses. Most, but not all, intracellular membrane fusion events are SNARE-dependent and it is widely assumed that the Ca^{2+}-dependent fusion of synaptic vesicles and granules is a special adaptation of a general mechanism (Jahn & Scheller, 2006; Martens & McMahon, 2008; McMahon et al., 1993).

Syntaxin1 and SNAP25 are also called target-SNAREs (tSNAREs) and have been reported to form a tSNARE *acceptor complex* in the plasma membrane for the SNARE motif of synaptobrevin. There is good evidence suggesting a model where synaptobrevin initially makes contact with the preformed tSNARE complex via its N-terminus (Sorensen et al., 2006; Walter, Wiederhold, Bruns, Fasshauer, & Sørensen, 2010). Thus, the regions furthest from the transmembrane domains make contact and as the proteins zipper together the transmembrane domains are pulled closer together. This model is also referred to as N- to C-terminal zippering. It is an appealing model because it explains how the two membranes are brought into proximity and further suggests a mechanism by which the full zippering could be stalled at certain stages. It now appears that accessory molecules regulate the formation of the SNARE complex at various stages (Jahn & Scheller, 2006; Rizo & Rosenmund, 2008; Südhof & Rothman, 2009).

Two lines of evidence point to the SNARE complex as the major driving force during membrane fusion. First, disruption of SNARE function *in vivo* severely inhibits Ca^{2+}-dependent exocytosis. The targeted deletion of individual SNAREs almost completely abolished synaptic vesicle and secretory granule fusion (Borisovska et al., 2005; Schiavo et al., 1992; Schoch et al., 2001; Washbourne et al., 2002). Furthermore, mutations of critical amino acids within the SNARE complex reduce fusion (Kesavan, Borisovska, & Bruns, 2007; Sorensen et al., 2006; Walter et al., 2010). Second, membrane fusion mediated by SNAREs has been successfully reconstituted *in vitro*. Starting with a seminal study by Rothman and colleagues (Weber et al., 1998) membrane fusion mediated by the neuronal SNAREs has been reconstituted and the role of SNAREs has been analyzed (Chen et al., 2006; Domanska, Kiessling, Stein, Fasshauer, & Tamm, 2009; Fix et al., 2004; Kiessling, Domanska, & Tamm, 2010; Liu, Tucker, Bhalla, Chapman, & Weisshaar, 2005; Montecucco, Schiavo, & Pantano, 2005; Pobbati, Stein, & Fasshauer, 2006; van den Bogaart et al., 2010). The results from these *in vitro* systems have led to various and often conflicting interpretations. In some cases the fusion rates observed were extremely low or dependent on non-physiological conditions. It was therefore suggested that the SNAREs are not the minimal machinery for membrane fusion. In other studies it was observed that SNAREs alone suffice to mediate fast fusion and it was thus concluded that they are the minimal machinery mediating membrane fusion *in vivo* (Chen et al., 2006). In addition, the number of SNARE complexes required for an individual fusion event was found to range from 15 to just a few (Domanska et al., 2009; Montecucco et al., 2005). It has recently been suggested that as few as one SNARE complex is sufficient for membrane fusion *in vitro* prompting the question why each synaptic vesicle carries about 70 synaptobrevin molecules (Takamori et al., 2006; van den Bogaart et al., 2010). Titration experiments of wild type and mutant SNAP25 expression in SNAP25 knockout cells have now proposed that at least three SNARE complexes are required for fast fusion but less than three SNARE complexes may be sufficient to drive slow fusion (Mohrmann, de Wit, Verhage, Neher, & Sørensen, 2010).

In all *in vitro* reconstitutions described above one crucial feature of synaptic vesicle and secretory granule fusion was missing, that is the strict Ca^{2+}-dependence of the fusion reaction. Thus, it became clear that although the SNAREs are at the core of the fusion event they are neither Ca^{2+}-sensitive nor sufficient for Ca^{2+}-dependent fusion (Chen, Tang, Sudhof, & Rizo, 2005).

B. Synaptotagmins and Doc2 Proteins

1. Synaptotagmins

The strict Ca^{2+} dependence of synaptic vesicle and secretory granule fusion prompted the search for Ca^{2+} sensors that transmit the Ca^{2+} signal into the actual fusion event. The first such sensor identified was synaptotagmin-1 (Geppert et al., 1994; Matthew, Tsavaler, & Reichardt, 1981; Perin, Fried, Mignery, Jahn, & Sudhof, 1990; Perin et al., 1991). Synaptotagmin-1 is localized to synaptic vesicles and secretory granules by an N-terminal transmembrane domain. Its main functional modules are two C-terminal C2 domains (Chapman, 2002; Fernandez et al., 2001; Sutton, Davletov, Berghuis, Sudhof, & Sprang, 1995). C2 domains are composed of two beta sheets with each sheet composed of four beta strands. The beta strands are connected by variable loops (Rizo & Sudhof, 1998). The C2 domain fold is conserved from yeast to humans. In synaptotagmins the N-terminal C2 domain is called C2A and the C-terminal C2 domain is called C2B domain. In synaptotagmin-1 and several other synaptotagmins, the C2A domain is able to bind three Ca^{2+} ions whereas the C2B domain binds two Ca^{2+} ions (Chapman, 2002). Ca^{2+} binding is mediated by one pocket in each domain located between the loops connecting the beta strands. Acidic residues make these pockets highly negatively charged, allowing the pocket to coordinate Ca^{2+} ions (Cheng et al., 2004; Fernandez et al., 2001; Sutton et al., 1995). Ca^{2+} binding reverses the net negative charge to a net positive charge enabling the C2 domains to bind membranes. In solution the Ca^{2+} affinity of synaptotagmins is very low, in fact too low to be able to serve as a Ca^{2+} sensor *in vivo.* However, in the presence of negatively charged membranes its Ca^{2+} affinity increases dramatically to the low μM range. The C2 domains of synaptotagmin-1 are thus Ca^{2+}-dependent membrane binding modules (Zhang, Rizo, & Sudhof, 1998). Furthermore, as the C2 domains insert into one monolayer of the membrane in a shallow manner they induce membrane curvature, triggering the membrane to bend towards them (Herrick, Sterbling, Rasch, Hinderliter, & Cafiso, 2006; Hui, Bai, & Chapman, 2006; Hui, Johnson, Yao, Dunning, & Chapman, 2009; Martens et al., 2007; McMahon, Kozlov, Martens, 2010). They thereby act analogously to amphipathic helices found in other membrane trafficking proteins such as epsins, arfs, some BAR domain proteins, or HIV Nef (Beck et al., 2008; Ford et al., 2002; Gerlach et al., 2010; Lee et al., 2005; Peter et al., 2004).

In addition to Ca^{2+}-dependent membrane binding, the C2 domains of synaptotagmin-1 can bind to the SNAREs (Chapman, 2002; Martens & McMahon, 2008). Some of the SNARE interactions with the synaptotagmins appear to be Ca^{2+}-dependent (Lai, Huang, Herrick, Epp, & Cafiso, 2011). It is generally observed that the interaction with SNAREs is relatively weak (Choi et al., 2010; Lai et al., 2011; Tang et al., 2006; Vrljic et al., 2010). The C2B domain binds to the tSNARE complex in a Ca^{2+}-independent manner (Rickman et al., 2006), an activity that has been implicated in secretory granule docking (de Wit et al., 2009). It can also bind to the fully assembled SNARE complex but here the interaction appears to be enhanced by Ca^{2+} (Choi et al., 2010; Lai et al., 2011; Tang et al., 2006; Vrljic et al., 2010). The C2A domain has been reported to bind to SNAP-25 within the SNARE complex. This interaction is of low affinity and is dependent on Ca^{2+} (Lynch et al., 2007).

In synaptotagmin-1 knockout neurons the synchronous phase of neurotransmitter release is completely lost (Geppert et al., 1994). In some neurons synaptotagmin-2 and synaptotagmin-9 take over the function of synaptotagmin-1 (Xu, Mashimo, & Sudhof, 2007). When they are deleted in the respective cells the synchronous phase is lost. In these knockout cells asynchronous release is still present and spontaneous release is increased (Fernandez-Chacon et al., 2001; Geppert et al., 1994; Xu et al., 2009). In chromaffin cells deletion of synaptotagmin-1 results in the loss of the fast phase of release whereas the slow and sustained phases are still present (Schonn, Maximov, Lao, Südhof, & Sørensen, 2008; Voets et al., 2001).

In addition, it was observed that deletion or expression of mutant synaptotagmin-1 has a direct effect on the kinetics of release. Effects on the kinetics of release have so far only been observed for certain SNARE mutations and the synaptotagmins (Kesavan et al., 2007; Sorensen, 2004; Sorensen et al., 2006). These results suggest that synaptotagmins-1, -2, and -9 are integral parts of the release machinery and several models have been proposed for the final events leading from the Ca^{2+} signal to membrane fusion. Thus, there is now good evidence that the SNARE complex is at least partially assembled before the Ca^{2+} trigger for release arrives (Sorensen et al., 2006; Walter et al., 2010). Due to its Ca^{2+}-independent interaction via the C2B domain, synaptotagmin-1 may already be bound to this partially assembled SNARE complex (Rickman et al., 2006). In fact it has been proposed that the Ca^{2+}-independent interaction of synaptotagmin-1 with the tSNARE complex composed of syntaxin1 and SNAP-25 may be required for the docking of secretory granules in chromaffin cells (de Wit et al., 2009). Whether synaptotagmin-1 has a similar function in neurons is less clear (Burgalossi et al., 2010). It is thus conceivable that during docking synaptotagmin-1 binds to the tSNARE complex and that during priming synaptobrevin joins this complex to from a partially zippered SNARE complex. This partially assembled SNARE complex may be prevented from complete zippering by accessory molecules such as the complexins (Giraudo et al., 2009; Tang et al., 2006). The preassembly of the fusion apparatus offers a good explanation for the short time delay between the Ca^{2+} trigger and the fusion event. It should be stressed however that most of the studies examining the molecular events during docking and priming have been conducted in chromaffin cells. It is therefore possible that the molecular events in neurons are different. In fact it has recently been suggested that synaptotagmin-1 is not required for the priming of synaptic vesicles (Burgalossi et al., 2010).

Upon Ca^{2+} channel opening triggered by an action potential, the local Ca^{2+} concentration rises to the low μM range (Neher & Sakaba, 2008). At this Ca^{2+} concentration the C2 domains of synaptotagmin-1 are able to bind membranes, which is likely the final trigger for fusion. As the C2 domains insert into the membrane they will locally change the curvature of the bilayer inducing the bending of the membrane towards them (Herrick et al., 2006; Hui et al., 2006, 2009; Martens et al., 2007; McMahon et al., 2010) (Fig. 2). Since the plasma membrane contains more negative charge than the vesicular membrane the C2 domains are likely to bind to the plasma membrane. It is however not excluded that one of the C2 domain, possibly the C2A domains, binds the vesicular membrane (Herrick et al., 2009). As synaptotagmin-1 is associated with the SNARE complex and assuming that several SNARE complexes are located in a ring-like manner close to the future fusion site, the C2 domains are sufficiently concentrated to induce a buckle-like protrusion of the plasma membrane towards the vesicular membrane (Martens et al., 2007). Due to the interaction with the SNAREs the C2 domains are confined to the circumference of the buckle and thus the end caps are free from C2 domains (Choi et al., 2010; Lai et al., 2011; Tang et al., 2006; Vrljic et al., 2010) (Fig. 2). As the membrane in this end cap is highly curved in the absence of insertions the lipids are under curvature stress in this area. This curvature stress reduces the energy barriers for the later stages of the fusion process (Hui et al., 2009; Martens et al., 2007; McMahon et al., 2010). Concomitant with membrane binding the C2 domains may change their mode of binding to the SNARE complex and thereby induce the complete zippering of the complex (McMahon et al., 2010). The C2 domains of synaptotagmin-1 may also displace a hypothetical fusion clamp, such as complexin (Giraudo, Eng, Melia, & Rothman, 2006; Tang et al., 2006). The complete zippering of the SNARE complexes will further bring the membranes into close contact (Stein et al., 2009). As the helical SNARE complex is likely to extend into the membrane via syntaxin1 and synaptobrevin the complete zippering will act synergistically with the curvature induced by synaptatogmin-1 (McMahon et al., 2010; Stein et al., 2009) (Fig. 2). This model offers an explanation for the Ca^{2+}-dependent induction of membrane fusion by synaptotagmin-1 and is supported by *in vitro* and *in vivo* evidence. However, the lack of structural data on key intermediates as well as structural information about the fusion apparatus prior to fusion precludes a complete understanding of the events during fusion.

As mentioned above synaptotagmin-2 and synaptotagmin-9 act analogously to synaptotagmin-1 during synchronous release. Given the central role these synaptotagmins play during synchronous release it was assumed that other synaptotagmins or related molecules trigger asynchronous release. An interesting candidate is synaptotagmin-7, which is a major Ca^{2+} sensor for granule fusion in chromaffin cells (Schonn et al., 2008). However, in neurons it does not seem to play a role during synaptic vesicle fusion (Maximov et al., 2008). Until now the Ca^{2+} sensor for asynchronous synaptic vesicle fusion has remained elusive. In contrast, significant progress has been made in understanding spontaneous release, the third form of release at the synapse.

2. Double C2 Domain Proteins (Doc2)

Spontaneous release has long been regarded as some oddity, which may be the result of an accidental trigger of the release machinery. In synaptotagmin-1 knockout neurons it was observed that the frequency of spontaneous release was increased (Littleton, Stern, Perin, & Bellen, 1994; Maximov & Sudhof, 2005; Pang, Sun, Rizo, Maximov, & Sudhof, 2006; Xu et al., 2009). In recent years it has emerged that spontaneous release is at least to a large extent Ca^{2+} dependent (Emptage et al., 2001; Llano et al., 2000; Groffen et al., 2010). Recently, the Doc2A and Doc2B proteins have been identified as Ca^{2+} sensors that are responsible for a large fraction of the Ca^{2+}-dependent release events (Groffen et al., 2010). Doc2 comprises a small protein family of three proteins in mammals (Martens, 2010; Martens & McMahon, 2008). Like the synaptotagmins, the Doc2 proteins contain two C-terminal C2 domains called C2A and C2B domains. Unlike the synaptotagmins the Doc2 proteins contain no transmembrane domain at their N-terminus. The C2 domains of Doc2A and Doc2B are able to bind Ca^{2+} (Groffen, Friedrich, Brian, Ashery, & Verhage, 2006; Groffen et al., 2004, 2010). However, their Ca^{2+} affinity is in the high nM range and thus significantly higher than for the C2 domains of synaptotagmin-1 (Groffen et al., 2004, 2006, 2010). The expression of Doc2A and Doc2B is restricted to neuronal cells and other professional secreting cells. Within the brain Doc2A and Doc2B are expressed in an overlapping manner (Verhage et al., 1997). In cells in which the two proteins are co-expressed, they appear to function in a redundant manner (Groffen et al., 2010). Biochemically the Doc2 proteins show striking analogies to synaptotagmin-1 and the other Ca^{2+}-dependent synaptotagmins. Thus, both the C2A and C2B domain of Doc2A and Doc2B are able to bind to negatively charged membranes (Groffen et al., 2010). Interestingly, the C2B domain of Doc2B shows a strong phosphatidylinositol 4,5 bisphosphate dependence for Ca^{2+}-dependent membrane binding (Groffen et al., 2010). By analogy to synaptotagmin-1, a fragment containing both, the C2A and C2B domains of Doc2B (C2AB) is a potent inducer of membrane curvature (Groffen et al., 2010). Furthermore, the C2AB domain of Doc2B strongly promotes SNARE-dependent membrane fusion in a reconstituted system. Both, membrane curvature induction and fusion promotion are Ca^{2+} dependent (Groffen et al., 2010). In addition, the C2 domains of Doc2B have been shown to bind to the neuronal SNARE complex. When expressed in Doc2B knockout hippocampal neurons, Doc2B that harbors mutations in the C2 domains, that interfere with Ca^{2+}-dependent membrane binding and SNARE complex binding, are unable to restore wild-type frequency of spontaneous release. Furthermore, a mutant Doc2B that is able to bind membranes at lower Ca^{2+} concentrations compared to the wild-type protein supports a higher frequency of spontaneous release compared to the wild type protein under conditions where Ca^{2+} is limiting (Groffen et al., 2010). These results strongly suggest that Doc2A and Doc2B function as Ca^{2+} sensors for spontaneous neurotransmitter release and furthermore that they act in the same mechanistic manner as synaptotagmin-1 during synchronous release. The finding that Doc2B affects the kinetics of individual fusion events in chromaffin cells further supports the hypothesis that it is an integral part of the fusion machinery (Friedrich et al., 2008).

Synaptotagmin-1 and Doc2B compete with each other for binding to the neuronal SNARE complex *in vitro* (Groffen et al., 2010). Therefore, a complex picture is emerging where different Ca^{2+} sensors that act with the same general mechanism but differ in their kinetics, Ca^{2+} affinity and precise regulation, coexist at the synapse. In an extreme view, many different Ca^{2+} sensors coexist that function in an overlapping manner. The reason no sensor for asynchronous release has been identified so far may be due to several such sensors functioning in a redundant manner. Among the Ca^{2+} sensors, synaptotagmin-1 may be uniquely positioned to rapidly respond to the rise in Ca^{2+} on depolarization (de Wit et al., 2009), whereas Doc2A/B with its high Ca^{2+} affinity may be ideally positioned to respond to very low fluctuations in Ca^{2+} concentrations (Groffen et al., 2010).

IV. CONCLUSION

While the SNARE proteins and synaptotagmins are both implicated in the kinetics of membrane fusion and thus in the fusion event itself there are many other accessory proteins that likely regulate the placement of the fusion site, the assembly of fusion complexes and other proteins may also participate directly in the fusion event. Thus, proteins showing phenotypes in docking or priming of vesicle fusion may not be excluded from playing a role in later stages. Indeed, synaptotagmin-1 has now been proposed to play a role in docking, fusion, and in the post fusion opening of the fusion pore (collapse of the vesicle). Finally, while synaptic vesicle fusion is Ca^{2+}-dependent many other fusion events are constitutive and do not involve a Ca^{2+} trigger, but yet are still likely to require membrane identity selection and membrane priming by accessory proteins of the SNARE complex in an analogous manner to C2 domains.

References

Barrett, E. F., & Stevens, C. F. (1972). The kinetics of transmitter release at the frog neuromuscular junction. *The Journal of Physiology, 227*(3), 691–708.

Beck, R., Sun, Z., Adolf, F., Rutz, C., Bassler, J., Wild, K., et al. (2008). Membrane curvature induced by Arf1-GTP is essential for vesicle formation. *Proceedings of the National Academy of Sciences, 105*(33), 11731–11736.

Borisovska, M., Zhao, Y., Tsytsyura, Y., Glyvuk, N., Takamori, S., Matti, U., et al. (2005). v-SNAREs control exocytosis of vesicles from priming to fusion. *EMBO Journal, 24*(12), 2114–2126.

Burgalossi, A., Jung, S., Meyer, G., Jockusch, W. J., Jahn, O., Taschenberger, H., et al. (2010). SNARE protein recycling by ±SNAP and ^{2}SNAP supports synaptic vesicle priming. *Neuron, 68* (3), 473–487.

Chapman, E. R. (2002). Synaptotagmin: A Ca(2+) sensor that triggers exocytosis? *Nature Reviews Molecular Cell Biology, 3*(7), 498–508.

Chen, X., Arac, D., Wang, T. M., Gilpin, C. J., Zimmerberg, J., & Rizo, J. (2006). SNARE-mediated lipid mixing depends on the physical state of the vesicles. *Biophysical Journal, 90*(6), 2062–2074.

Chen, X., Tang, J., Sudhof, T. C., & Rizo, J. (2005). Are neuronal SNARE proteins Ca^{2+} sensors? *Journal of Molecular Biology, 347*(1), 145–158.

Cheng, Y., Sequeira, S. M., Malinina, L., Tereshko, V., Söllner, T. H., & Patel, D. J. (2004). Crystallographic identification of Ca^{2+} and Sr^{2+} coordination sites in synaptotagmin I C2B domain. *Protein Science, 13*(10), 2665–2672.

Chernomordik, L. V., & Kozlov, M. M. (2003). Protein-lipid interplay in fusion and fission of biological membranes. *Annual Review of Biochemistry, 72*, 175–207.

Chernomordik, L. V., & Kozlov, M. (2008). Mechanics of membrane fusion. *Nature Structural & Molecular Biology, 15*(7), 675–683.

Choi, U. B., Strop, P., Vrljic, M., Chu, S., Brunger, A. T., & Weninger, K. R. (2010). Single-molecule FRET-derived model of the synaptotagmin 1-SNARE fusion complex. *Nature Structural & Molecular Biology, 17*(3), 318–324.

Citri, A., & Malenka, R. C. (2007). Synaptic plasticity: Multiple forms, functions, and mechanisms. *Neuropsychopharmacology, 33*(1), 18–41.

Davletov, B., Bajohrs, M., & Binz, T. (2005). Beyond BOTOX: Advantages and limitations of individual botulinum neurotoxins. *Trends in Neurosciences, 28*(8), 446–452.

DeFelipe, J. (2010). From the connectome to the synaptome: An epic love story. *Science, 330*(6008), 1198–1201.

de Wit, H., Walter, A. M., Milosevic, I., Gulyás-Kovács, A., Riedel, D., Sørensen, J. B., et al. (2009). Synaptotagmin-1 docks secretory vesicles to syntaxin-1/SNAP-25 acceptor complexes. *Cell, 138*(5), 935–946.

Domanska, M. K., Kiessling, V., Stein, A., Fasshauer, D., & Tamm, L. K. (2009). Single vesicle millisecond fusion kinetics reveals number of SNARE complexes optimal for fast SNARE-mediated membrane fusion. *Journal of Biological Chemistry, 284*(46), 32158–32166.

Efrat, A., Chernomordik, L. V., & Kozlov, M. M. (2007). Point-like protrusion as a prestalk intermediate in membrane fusion pathway. *Biophysical Journal, 92*(8), L61–L63.

Emptage, N. J., Reid, C. A., & Fine, A. (2001). Calcium stores in hippocampal synaptic boutons mediate short-term plasticity, store-operated Ca^{2+} entry, and spontaneous transmitter release. *Neuron, 29*(1), 197–208.

Fasshauer, D., Otto, H., Eliason, W. K., Jahn, R., & Brünger, A. T. (1997). Structural changes are associated with soluble *N*-ethylmaleimide-sensitive fusion protein attachment protein receptor complex formation. *Journal of Biological Chemistry, 272*(44), 28036–28041.

Fernandez, I., Arac, D., Ubach, J., Gerber, S. H., Shin, O., Gao, Y., et al. (2001). Three-dimensional structure of the synaptotagmin 1 C2B-domain: Synaptotagmin 1 as a phospholipid binding machine. *Neuron, 32*(6), 1057–1069.

Fernandez-Chacon, R., Konigstorfer, A., Gerber, S. H., Garcia, J., Matos, M. F., Stevens, C. F., et al. (2001). Synaptotagmin I functions as a calcium regulator of release probability. *Nature, 410* (6824), 41–49.

Fix, M., Melia, T. J., Jaiswal, J. K., Rappoport, J. Z., You, D., Söllner, T. H., et al. (2004). Imaging single membrane fusion events mediated by SNARE proteins. *Proceedings of the National Academy of Sciences of the United States of America, 101*(19), 7311–7316.

Ford, M., Mills, I., Peter, B., Vallis, Y., Praefcke, G., Evans, P., et al. (2002). Curvature of clathrin-coated pits driven by epsin. *Nature, 419*(6905), 361–366.

Friedrich, R., Groffen, A. J., Connell, E., van Weering, J., Gutman, O., Henis, Y., et al. (2008). DOC2B acts as a calcium switch and enhances vesicle fusion. *Journal of Neuroscience, 28*(27), 6794–6806.

Geppert, M., Goda, Y., Hammer, R. E., Li, C., Rosahl, T. W., Stevens, C. F., et al. (1994). Synaptotagmin I: A major Ca^{2+} sensor for transmitter release at a central synapse. *Cell, 79*(4), 717–727.

Gerlach, H., Laumann, V., Martens, S., Becker, C. F. W., Goody, R. S., & Geyer, M. (2010). HIV-1 Nef membrane association depends on charge, curvature, composition and sequence. *Nature Chemical Biology, 6*(1), 46–53.

Giraudo, C. G., Eng, W. S., Melia, T. J., & Rothman, J. E. (2006). A clamping mechanism involved in SNARE-dependent exocytosis. *Science, 313*(5787), 676–680.

Giraudo, C. G., Garcia-Diaz, A., Eng, W. S., Chen, Y., Hendrickson, W. A., Melia, T. J., et al. (2009). Alternative zippering as an on-off switch for SNARE-mediated fusion. *Science, 323*(5913), 512–516.

Groffen, A. J., Brian, E., Dudok, J., Kampmeijer, J., Toonen, R., & Verhage, M. (2004). Ca(2 +)-induced recruitment of the secretory vesicle protein DOC2B to the target membrane. *Journal of Biological Chemistry, 279*(22), 23740–23747.

Groffen, A. J., Friedrich, R., Brian, E., Ashery, U., & Verhage, M. (2006). DOC2A and DOC2B are sensors for neuronal activity with unique calcium-dependent and kinetic properties. *Journal of Neurochemistry, 97*(3), 818–833.

Groffen, A. J., Martens, S., Arazola, R. D., Cornelisse, L. N., Lozovaya, N., de Jong, A. P. H., et al. (2010). Doc2b is a high-affinity Ca^{2+} sensor for spontaneous neurotransmitter release. *Science, 327*(5973), 1614–1618.

Guttinger, S., Laurell, E., & Kutay, U. (2009). Orchestrating nuclear envelope disassembly and reassembly during mitosis. *Nature Reviews Molecular Cell Biology, 10*(3), 178–191.

Hagler, D. J., & Goda, Y. (2001). Properties of synchronous and asynchronous release during pulse train depression in cultured hippocampal neurons. *Journal of Neurophysiology, 85*(6), 2324–2334.

Haller, M., Heinemann, C., Chow, R. H., Heidelberger, R., & Neher, E. (1998). Comparison of Secretory responses as measured by membrane capacitance and by amperometry. *Biophysical Journal, 74*(4), 2100–2113.

Herrick, D. Z., Kuo, W., Huang, H., Schwieters, C. D., Ellena, J. F., & Cafiso, D. S. (2009). Solution and membrane-bound conformations of the tandem C2A and C2B domains of synaptotagmin 1: Evidence for bilayer bridging. *Journal of Molecular Biology, 390*(5), 913–923.

Herrick, D. Z., Sterbling, S., Rasch, K. A., Hinderliter, A., & Cafiso, D. S. (2006). Position of synaptotagmin I at the membrane interface: Cooperative interactions of tandem C2 domains. *Biochemistry, 45*(32), 9668–9674.

Hoppins, S., Lackner, L., & Nunnari, J. (2007). The machines that divide and fuse mitochondria. *Annual Review of Biochemistry, 76*, 751–780.

Hui, E., Bai, J., & Chapman, E. R. (2006). Ca^{2+}-triggered simultaneous membrane penetration of the tandem C2-domains of synaptotagmin I. *Biophysical Journal, 91*(5), 1767–1777.

Hui, E., Johnson, C. P., Yao, J., Dunning, F. M., & Chapman, E. R. (2009). Synaptotagmin-mediated bending of the target membrane is a critical step in Ca^{2+}-regulated fusion. *Cell, 138*(4), 709–721.

Jahn, R., Lang, T., & Sudhof, T. C. (2003). Membrane fusion. *Cell, 112*(4), 519–533.

Jahn, R., & Scheller, R. H. (2006). SNAREs—Engines for membrane fusion. *Nature Reviews Molecular Cell Biology, 7*(9), 631–643.

Kesavan, J., Borisovska, M., & Bruns, D. (2007). v-SNARE actions during Ca(2+)-triggered exocytosis. *Cell, 131*(2), 351–363.

Kiessling, V., Domanska, M. K., & Tamm, L. K. (2010). Single SNARE-mediated vesicle fusion observed *in vitro* by polarized TIRFM. *Biophysical Journal, 99*(12), 4047–4055.

Lai, A. L., Huang, H., Herrick, D. Z., Epp, N., & Cafiso, D. S. (2011). Synaptotagmin 1 and SNAREs form a complex that is structurally heterogeneous. *Journal of Molecular Biology, 405*(3), 696–706.

Lee, M., Orci, L., Hamamoto, S., Futai, E., Ravazzola, M., & Schekman, R. (2005). Sar1p N-terminal helix initiates membrane curvature and completes the fission of a COPII vesicle. *Cell, 122*(4), 605–617.

Littleton, J. T., Stern, M., Perin, M., & Bellen, H. J. (1994). Calcium dependence of neurotransmitter release and rate of spontaneous vesicle fusions are altered in Drosophila synaptotagmin mutants. *Proceedings of the National Academy of Sciences USA, 91*(23), 10888–10892.

Liu, T., Tucker, W. C., Bhalla, A., Chapman, E. R., & Weisshaar, J. C. (2005). SNARE-driven, 25-millisecond vesicle fusion *in vitro*. *Biophysical Journal, 89*(4), 2458–2472.

Llano, I., Gonzalez, J., Caputo, C., Lai, F. A., Blayney, L. M., Tan, Y. P., et al. (2000). Presynaptic calcium stores underlie large-amplitude miniature IPSCs and spontaneous calcium transients. *Nature Neuroscience, 3*(12), 1256–1265.

Lynch, K., Gerona, R., Larsen, E., Marcia, R., Mitchell, J., & Martin, T. (2007). Synaptotagmin C2A Loop 2 Mediates Ca^{2+}-dependent SNARE interactions essential for Ca^{2+}-triggered vesicle exocytosis. *Molecular Biology of the Cell, 18*(12), 4957–4968.

Martens, S. (2010). Role of C2 domain proteins during synaptic vesicle exocytosis. *Biochemical Society Transactions, 38*(Pt 1), 213–216.

Martens, S., Kozlov, M., & McMahon, H. (2007). How synaptotagmin promotes membrane fusion. *Science, 316*(5828), 1205–1208.

Martens, S., & McMahon, H. (2008). Mechanisms of membrane fusion: Disparate players and common principles. *Nature Reviews Molecular Cell Biology, 9*(7), 543–556.

Matthew, W., Tsavaler, L., & Reichardt, L. (1981). Identification of a synaptic vesicle-specific membrane protein with a wide distribution in neuronal and neurosecretory tissue. *Journal of Cell Biology, 91*(1), 257–269.

Maximov, A., Lao, Y., Li, H., Chen, X., Rizo, J., Sørensen, J. B., et al. (2008). Genetic analysis of synaptotagmin-7 function in synaptic vesicle exocytosis. *Proceedings of the National Academy of Sciences of the United States of America, 105*(10), 3986–3991.

Maximov, A., & Sudhof, T. C. (2005). Autonomous function of synaptotagmin 1 in triggering synchronous release independent of asynchronous release. *Neuron, 48*(4), 547–554.

McMahon, H. T., Kozlov, M. M., & Martens, S. (2010). Membrane curvature in synaptic vesicle fusion and beyond. *Cell, 140*(5), 601–605.

McMahon, H. T., Ushkaryov, Y. A., Edelmann, L., Link, E., Binz, T., Niemann, H., et al. (1993). Cellubrevin is a ubiquitous tetanus-toxin substrate homologous to a putative synaptic vesicle fusion protein. *Nature, 364*(6435), 346–349.

Mohrmann, R., de Wit, H., Verhage, M., Neher, E., & Sørensen, J. B. (2010). Fast vesicle fusion in living cells requires at least three SNARE complexes. *Science, 330*(6003), 502–505.

Montecucco, C., Schiavo, G., & Pantano, S. (2005). SNARE complexes and neuroexocytosis: How many, how close? *Trends in Biochemical Sciences, 30*(7), 367–372.

Neher, E., & Sakaba, T. (2008). Multiple roles of calcium ions in the regulation of neurotransmitter release. *Neuron, 59*(6), 861–872.

Oren-Suissa, M., & Podbilewicz, B. (2007). Cell fusion during development. *Trends in Cell Biology, 17*(11), 537–546.

Pang, Z. P., Sun, J., Rizo, J., Maximov, A., & Sudhof, T. C. (2006). Genetic analysis of synaptotagmin 2 in spontaneous and Ca^{2+}-triggered neurotransmitter release. *EMBO Journal, 25*(10), 2039–2050.

Perin, M. S., Fried, V. A., Mignery, G. A., Jahn, R., & Sudhof, T. C. (1990). Phospholipid binding by a synaptic vesicle protein homologous to the regulatory region of protein kinase C. *Nature, 345* (6272), 260–263.

Perin, M. S., Johnston, P. A., Ozcelik, T., Jahn, R., Francke, U., & Sudhof, T. C. (1991). Structural and functional conservation of synaptotagmin (p65) in Drosophila and humans. *Journal of Biological Chemistry, 266*(1), 615–622.

Peter, B., Kent, H., Mills, I., Vallis, Y., Butler, P., Evans, P., et al. (2004). BAR domains as sensors of membrane curvature: The amphiphysin BAR structure. *Science, 303*(5657), 495–499.

Pobbati, A., Stein, A., & Fasshauer, D. (2006). N- to C-terminal SNARE complex assembly promotes rapid membrane fusion. *Science, 313*(5787), 673–676.

Rickman, C., Jimenez, J. L., Graham, M. E., Archer, D. A., Soloviev, M., Burgoyne, R. D., et al. (2006). Conserved prefusion protein assembly in regulated exocytosis. *Molecular Biology of the Cell, 17*(1), 283–294.

Rizo, J., & Rosenmund, C. (2008). Synaptic vesicle fusion. *Nature Structural & Molecular Biology, 15*(n7), 665–674.

Rizo, J., & Sudhof, T. C. (1998). C2-domains, structure and function of a universal Ca^{2+}-binding domain. *Journal of Biological Chemistry, 273*(26), 15879–15882.

Schiavo, G., Benfenati, F., Poulain, B., Rossetto, O., Polverino de Laureto, P., DasGupta, B., et al. (1992). Tetanus and botulinum-B neurotoxins block neurotransmitter release by proteolytic cleavage of synaptobrevin. *Nature, 359*(6398), 832–835.

Schoch, S., Deak, F., Konigstorfer, A., Mozhayeva, M., Sara, Y., Sudhof, T., et al. (2001). SNARE function analyzed in synaptobrevin/VAMP knockout mice. *Science, 294*(5544), 1117–1122.

Schonn, J. -S., Maximov, A., Lao, Y., Südhof, T. C., & Sørensen, J. B. (2008). Synaptotagmin-1 and -7 are functionally overlapping Ca^{2+} sensors for exocytosis in adrenal chromaffin cells. *Proceedings of the National Academy of Sciences of the United States of America, 105*(10), 3998–4003.

Shorter, J., & Warren, G. (2002). Golgi architecture and inheritance. *Annual Review of Cell and Developmental Biology, 18*, 379–420.

Skehel, J. J., & Wiley, D. C. (2000). Receptor binding and membrane fusion in virus entry: The influenza hemagglutinin. *Annual Review of Biochemistry, 69*, 531–569.

Sollner, T., Whiteheart, S. W., Brunner, M., Erdjument-Bromage, H., Geromanos, S., Tempst, P et al. (1993). SNAP receptors implicated in vesicle targeting and fusion. *Nature, 362*(6418), 318–324.

Sorensen, J. B. (2004). Formation, stabilisation and fusion of the readily releasable pool of secretory vesicles. *Pflugers Archiv, 448*(4), 347–362.

Sorensen, J. B., Wiederhold, K., Muller, E., Milosevic, I., Nagy, G., de Groot, B., et al. (2006). Sequential N- to C-terminal SNARE complex assembly drives priming and fusion of secretory vesicles. *EMBO Journal, 25*(5), 955–966.

Stein, A., Weber, G., Wahl, M. C., & Jahn, R. (2009). Helical extension of the neuronal SNARE complex into the membrane. *Nature, 460*(7254), 525–528.

Sudhof, T. C. (2004). The synaptic vesicle cycle. *Annual Review of Neuroscience, 27*, 509–547.

Südhof, T. C., & Rothman, J. E. (2009). Membrane fusion: Grappling with SNARE and SM proteins. *Science, 323*(5913), 474–477.

Sun, J., Pang, Z., Qin, D., Fahim, A., Adachi, R., & Sudhof, T. (2007). A dual-Ca^{2+}-sensor model for neurotransmitter release in a central synapse. *Nature, 450*(7170), 676–682.

Sutton, R. B., Davletov, B. A., Berghuis, A. M., Sudhof, T. C., & Sprang, S. R. (1995). Structure of the first C2 domain of synaptotagmin I: A novel Ca^{2+}/phospholipid-binding fold. *Cell, 80*(6), 929–938.

Sutton, R. B., Fasshauer, D., Jahn, R., & Brunger, A. T. (1998). Crystal structure of a SNARE complex involved in synaptic exocytosis at 2.4 A resolution. *Nature, 395*(6700), 347–353.

Takamori, S., Holt, M., Stenius, K., Lemke, E. A., Gronborg, M., Riedel, D., et al. (2006). Molecular anatomy of a trafficking organelle. *Cell, 127*(4), 831–846.

Tang, J., Maximov, A., Shin, O. H., Dai, H., Rizo, J., & Sudhof, T. C. (2006). A complexin/synaptotagmin 1 switch controls fast synaptic vesicle exocytosis. *Cell, 126*(6), 1175–1187.

van den Bogaart, G., Holt, M. G., Bunt, G., Riedel, D., Wouters, F. S., & Jahn, R. (2010). One SNARE complex is sufficient for membrane fusion. *Nature Structural & Molecular Biology, 17* (3), 358–364.

Verhage, M., de Vries, K., Roshol, H., Burbach, J., Gispen, W., & Sudhof, T. (1997). DOC2 proteins in rat brain: Complementary distribution and proposed function as vesicular adapter proteins in early stages of secretion. *Neuron, 18*(3), 453–461.

Verhage, M., Maia, A. S., Plomp, J. J., Brussaard, A. B., Heeroma, J. H., Vermeer, H., et al. (2000). Synaptic assembly of the brain in the absence of neurotransmitter secretion. *Science, 287*(5454), 864–869.

Verhage, M., & Sørensen, J. B. (2008). Vesicle docking in regulated exocytosis. *Traffic, 9*(9), 1414–1424.

Voets, T., Moser, T., Lund, P. E., Chow, R. H., Geppert, M., Sudhof, T. C., et al. (2001). Intracellular calcium dependence of large dense-core vesicle exocytosis in the absence of synaptotagmin I. *Proceedings of the National Academy of Sciences U S A, 98*(20), 11680–11685.

Vrljic, M., Strop, P., Ernst, J. A., Sutton, R. B., Chu, S., & Brunger, A. T. (2010). Molecular mechanism of the synaptotagmin-SNARE interaction in Ca^{2+}-triggered vesicle fusion. *Nature Structural & Molecular Biology, 17*(3), 325–331.

Walter, A. M., Wiederhold, K., Bruns, D., Fasshauer, D., & Sørensen, J. B. (2010). Synaptobrevin N-terminally bound to syntaxin–SNAP-25 defines the primed vesicle state in regulated exocytosis. *The Journal of Cell Biology, 188*(3), 401–413.

Washbourne, P., Thompson, P. M., Carta, M., Costa, E. T., Mathews, J. R., Lopez-Bendito, G., et al. (2002). Genetic ablation of the t-SNARE SNAP-25 distinguishes mechanisms of neuroexocytosis. *Nature Neuroscience, 5*(1), 19–26.

Weber, T., Zemelman, B., McNew, J., Westermann, B., Gmachl, M., Parlati, F., et al. (1998). SNAREpins: Minimal machinery for membrane fusion. *Cell, 92*(6), 759–772.

Xu, J., Mashimo, T., & Sudhof, T. C. (2007). Synaptotagmin-1, -2, and -9: Ca(2+) sensors for fast release that specify distinct presynaptic properties in subsets of neurons. *Neuron, 54*(4), 567–581.

Xu, J., Pang, Z. P., Shin, O. -H., & Sudhof, T. C. (2009). Synaptotagmin-1 functions as a Ca^{2+} sensor for spontaneous release. *Nature Neuroscience, 12*(6), 759–766.

Zerial, M., & McBride, H. (2001). Rab proteins as membrane organizers. *Nature Reviews Molecular Cell Biology, 2*(2), 107–117.

Zhang, X., Rizo, J., & Sudhof, T. C. (1998). Mechanism of phospholipid binding by the C2A-domain of synaptotagmin I. *Biochemistry, 37*(36), 12395–12403.

CHAPTER 7

Chasing the Trails of SNAREs and Lipids Along the Membrane Fusion Pathway

Tae-Young Yoon[1], Dae-Hyuk Kweon[2] and Yeon-Kyun Shin[3]
[1]Department of Physics and KAIST Institute for the BioCentury, KAIST, Daejeon, South Korea
[2]School of Life Science and Biotechnology and Center for Human Interface Nanotechnology, Sungkyunkwan University, Suwon, South Korea
[3]Department of Biochemistry, Biophysics, and Molecular Biology, Iowa State University, Ames, Iowa, USA

I. OVERVIEW

All exocytotic pathways in eukaryotic cells require vesicle fusion, which is thought to be mediated by SNAREs. Cognate vesicle (v-) and target membrane (t-)SNAREs form a four-stranded parallel coiled-coil that bridges two membranes, facilitating fusion. The structural and biophysical studies have demonstrated that the SNARE complex is an efficient fusogen. However, an orchestrated array of help from auxiliary factors such as SM proteins, synaptotagmin, and complexin appears to be necessary for fast neuroexocytosis that demands high-power thrust. SNARE-dependent membrane fusion experiences several

Current Topics in Membranes, Volume 68 0065-230X/10 $35.00

DOI: 10.1016/B978-0-12-385891-7.00001-5

dynamic events such as docking, hemifusion, and fusion pore opening. There is compelling evidence that the hemifusion state is a bona fide intermediate along the fusion pathway. Although electrophysiological studies support that fusion pore is protein-lined, it is possible that the fusion pore is of lipidic nature. Some natural polyphenols are potent inhibitors of SNARE-dependent fusion. Myricetin stops SNARE complex formation in the half way by intercalating into the middle of the helix bundle. These small molecules have potential to complement "BOTOX therapies" in cosmetics and medicine. Recently, an optical microscopy technique to detect a single vesicle fusion event has proven powerful in investigating the mechanism of SNARE-mediated membrane fusion. This new approach can identify the short-lived intermediates that are typically hidden in ensemble measurements. With the single vesicle fusion assays, it is now possible to dissect the effects of regulatory proteins on individual steps along the fusion pathway.

II. INTRODUCTION

Vesicle fusion is a ubiquitously required step for secretory pathways in eukaryotic cells including neurotransmitter release at the synapse, insulin secretion from pancreatic β-cells, and trafficking of newly made proteins. Intracellular vesicle fusion is mediated by complex machinery, but SNAREs are thought to be the core fusogen. Viral fusion proteins are self-sufficient fusogens that are activated by either receptor binding or low pH conditions. In contrast, SNAREs become fusogenic only when two separate cognate vesicle (v-) and target membrane (t-) SNAREs meet and form the SNARE complex in *trans* between two membranes (Fig. 1A). SNARE proteins are highly conserved from yeast to human. Among those, the SNARE family, involved in synaptic vesicle fusion, two t-SNARE proteins syntaxin 1A and SNAP-25 and one v-SNARE synaptobrevin 2 (also called VAMP2) are the favorite prototype system studied extensively. Syntaxin 1A and synaptobrevin 2 have membrane-anchor domains while SNAP-25 is attached to the plasma membrane via post-translationally added palmitoyl chains. All SNARE proteins contain coiled-coil motifs, called SNARE motifs, which mediate the interaction between cognate v- and t-SNARE pairs: One each from syntaxin 1A and synaptobrevin 2 and two from SNAP-25 form a four-helix coiled coil.

With this brief introduction of SNAREs, we divide our chapter into three sections. In Section III, we walk you through the SNARE-dependent fusion pathway with the biophysicist point of view, paying particular attention to structures, energies, and their changes. In Section IV, we describe efforts to discover potent SNARE blockers that can be potentially used for therapeutic purposes as well as for mechanistic studies. In the final Section V, we overview the single vesicle fusion techniques, which has emerged as a powerful new approach to dissect microsteps in SNARE-dependent membrane fusion.

III. SNARE-DEPENDENT MEMBRANE FUSION

A. The SNARE Structure

As the structure of the bromelin-cleaved flu HA solved by Wiley and coworkers in 1981 (Wilson, Skehel, & Wiley, 1981) had been a landmark for viral membrane fusion, the revelation of the SNARE core structures in 1998 (Poirier et al., 1998; Sutton, Fasshauer, Jahn, & Brunger, 1998) was an important turning point for SNARE-dependent membrane fusion. The first SNARE core structure was determined by Shin, Bennett, and coworkers using spin labeling EPR (inset in Fig. 1) (Poirier et al., 1998). In their work, pairs of nitroxide spin labels were

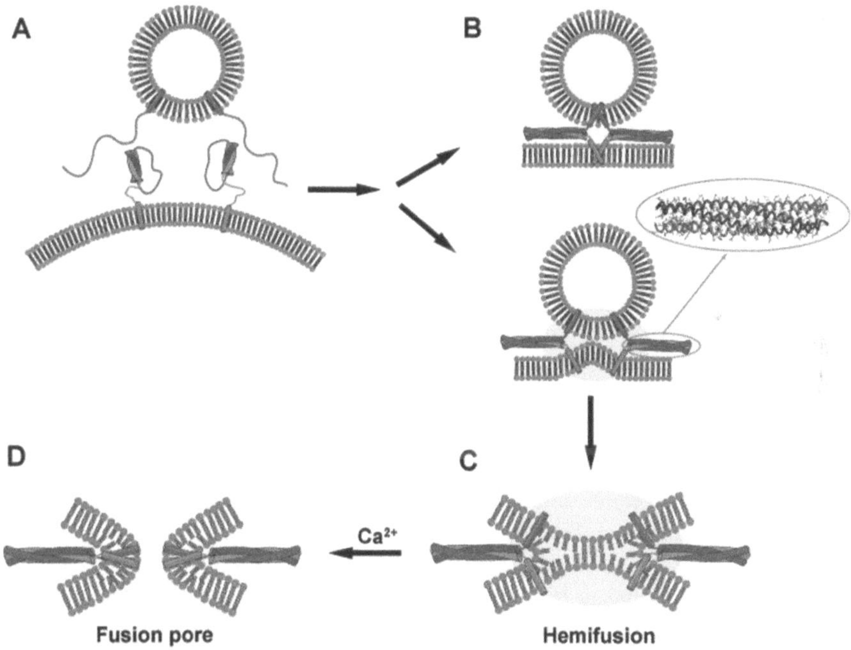

FIGURE 1 A proposed SNARE-mediated fusion pathway. (A) Individual SNAREs are on separate membranes. Target membrane (t-)SNAREs syntaxin 1A in green and SNAP-25 in blue and purple are attached to the plasma membrane and vesicle (v-)SNARE synaptobrevin is on the vesicle. Syntaxin 1A and SNAP-25 form a binary complex that serves as a receptor for synaptobrevin. (B) Vesicle docks to the plasma membrane via SNARE complex formation. TMDs are either clustered (upper panel) together or form a fence to create a patch of pure lipids in the middle (lower panel). Hypothetically, the SNARE complex or SNARE–synaptotagmin 1 complex drives formation of lipid protrusion, which reduces the energy barrier to form the lipid stalk. (C) The hemifusion intermediate in which the proximal leaflets are mixed but the distal leaflets remain intact. (D) Fusion pore. The cooperative action of synaptotagmins, complexins and Ca^{2+} on the SNARE complex drives the opening of the fusion pore. (See Color Insert.)

placed site-specifically at strategic positions and distances between the nitroxides were determined by analyzing the interaction between nitroxide pairs (Rabenstein & Shin, 1995). Such EPR-determined long-range distances were used to determine the placement, orientation and registry of the helical components in a coiled coil. The best fitting structural model reveals that SNAREs form an all parallel four-stranded coiled coil. The fact that the coiled coil is in parallel configuration demonstrates that the membrane attachment points are close to one another, which later turns out to be a general feature shared by nearly all fusion proteins (Skehel & Wiley, 1998). With this structure in mind, one can envision that formation of the SNARE complex reduces the gap between the two membranes, which would facilitate fusion. In the same year, a high-resolution X-ray structure was determined by Jahn, Brunger, and coworkers (Sutton et al., 1998) and this structure basically confirmed all the essential structural features seen in the EPR-based structure. In fact, the two structures overlap within 1.2 Å in backbone. Later, a few other crystal structures have been determined (Antonin, Fasshauer, Becker, Jahn, & Schneider, 2002; Strop, Kaiser, Vrljic, & Brunger, 2008) and it is confirmed that the parallel four-stranded coiled coil is conserved for other SNARE systems.

Recently, an X-ray structure of the neuronal SNARE complex including its membrane-proximal linker and transmembrane domains (TMDs) has been reported. This structure shows that the coiled coil in the SNARE core extends all the way through TMDs (Stein, Weber, Wahl, & Jahn, 2009). Thus, it appears that the SNARE core is structurally and consequently, energetically coupled to TMDs via the helical linker region. This structure is likely to represent the final stage of SNARE assembly (i.e., *cis* complex). Based on this structure, Stein et al. (2009) propose that SNARE core formation is directly coupled to the merger of two membranes and argues that SNARE complex is a robust fusion catalyst. On the other hand, EPR studies show that the linker regions, when separately anchored to membranes in *trans*, may be partially or fully unstructured but remain emerged in membranes (Kim, Kweon, & Shin, 2002; Kweon, Kim, & Shin, 2003a, 2003b). The insertion of the linker regions into membranes would promote a tight coupling between the SNARE core and TMDs, which would similarly facilitate the apposition of two membranes. Thus, these two studies argue that the SNARE complex is structurally competent to mediate membrane fusion.

B. Sequential SNARE Assembly: A Zipper Model

It is now widely accepted that SNARE core formation proceeds from the membrane distal N-terminal region toward the membrane proximal C-terminal region (zipper model) (Fiebig, Rice, Pollock, & Brunger, 1999). In support of this model, mutational analysis of SNAP-25 and VAMP2 in chromaffin cells (Sorensen et al., 2006; Walter, Wiederhold, Bruns, Fasshauer, & Sorensen, 2010) indicated that SNAREs are checked at partially assembled state prior to the calcium influx. Recently, using genetically dye-labeled SNAREs, Takahashi et al. (2010) showed that assembly of the SNARE complex varied from region to region across the insulin-secreting cell. Regions with preassembled SNAREs showed fast exocytosis in response to Ca^{2+}, while in the region where SNARE assembly was triggered by Ca^{2+} exocytosis was slower (Takahashi et al.). These results indicate that a N-terminally partially zipped state represents a primed or prepared state for the Ca^{2+} arrival, which is followed by calcium-triggered residual SNARE assembly at the membrane proximal C-terminal.

C. Is the SNAREpin a Self-Propelling Fusogen for Neuroexocytosis?

Synaptic vesicle fusion, which is required for the release of neurotransmitters, has fascinated us for many years because of its unusually fast time scales (submicroseconds) and the tight regulation by Ca^{2+}. There are several key players other than SNAREs such as synaptotagmin 1, complexin, and Munc18, which interact with SNAREs in an orchestrated fashion to induce this remarkable event. Dissection of functions of individual proteins and their interplay with SNAREs should hold the key to the understanding of mechanism.

One of the most pressing questions regarding SNARE-dependent membrane fusion is: Are SNAREs alone, without the help of other proteins, energetically powerful enough to thrust complete fusion in a few 100 μs? It is postulated that a clue to this question may be found in the unusual stability of the SNARE core. The SNARE core is highly thermostable with a melting temperature greater than 90 °C and it is SDS-resistant, too (Fasshauer, Otto, Eliason, Jahn, & Brunger, 1997). Thus, one might argue that the energy released from SNARE-assembly should be sufficient to drive the fast fusion. Indeed, several *in vitro* reconstitution experiments using the supported bilayer showed that membrane fusion by only SNAREs can happen in a few microseconds time scales (Domanska, Kiessling, Stein, Fasshauer, & Tamm, 2009; Karatekin et al., 2010; Liu, Tucker, Bhalla, Chapman, & Weisshaar, 2005). In contrast, the fusion between two SNARE-carrying proteoliposomes is much slower (a few hundred microseconds) (Yoon et al., 2008), raising concerns whether the supported bilayer serves as a good model membrane to study SNARE-dependent membrane fusion.

Regardless of the validity, under the assumption that SNAREs are self-sufficient and self-propelling fusogen for the fast fusion, two scenarios of the tight regulation by Ca^{2+} are possible. The first scenario is that the system is checked at an intermediate state along the pathway until the arrival of Ca^{2+} signal. One mechanistic model proposes that complexins, a family of small soluble neuronal proteins, clamp the completion of SNARE complex formation and the complexin clamp is lifted by the action of synaptotagmin 1 upon Ca^{2+} arrival (Giraudo, Eng, Melia, & Rothman, 2006; Schaub, Lu, Doneske, Shin, & McNew, 2006; Tang et al., 2006). Consistent with this model complexin is shown to have the capacity to replace the membrane proximal part of synaptobrevin in the SNARE core (Giraudo et al., 2008; B. Lu, Song, & Shin, 2010). There is also a report suggesting that complexin can check the fusion at the hemifusion state (Schaub et al., 2006).

The second scenario with the same "mighty SNARE" idea is that t-SNARE is an intrinsically poor receptor for v-SNARE and it needs to be primed or prepared by regulators. The basis for this idea have stemmed from the observation that a small peptide derived from the C-terminal region of v-SNARE can accelerate the fusion significantly (Pobbati, Stein, & Fasshauer, 2006). Jahn and coworkers believe that the binding of this peptide to t-SNARE primes it for the fast nucleation of the SNARE complex. One could envision that complexin and synaptotagmin 1, which are both known to bind t-SNARE, play similar priming roles.

An entirely different scenario assumes that SNAREs need to be assisted by other fusion regulators to drive the fusion in such a fast time scale. The supporting evidence for this scenario can be found in the results from a single vesicle fusion assay (Lee et al., 2010). This method makes it possible to analyze the individual fusion steps separately (see Section V). The results show that, complexin and synaptotagmin 1–Ca^{2+} are both potent stimulators of the later steps after docking, supporting the idea that the regulators are to assist SNAREs energetically to stimulate the fusion.

Another relevant question is: Is one SNARE complex sufficiently strong to fuel complete membrane fusion? Although there is a report proposing that one SNARE complex is sufficient to do the job (van den Bogaart et al., 2010) there is evidence that more than one complex cooperate for successful fast fusion. The actual number varies significantly depending on the methods and samples used for the particular measurements. But a general consensus is that at least three complexes participate in the process (Mohrmann, de Wit, Verhage, Neher, & Sorensen, 2010).

D. Do SNAREs Corral the Lipids?

For protein-free membranes, it is theoretically shown that fusion can happen if a lipid protrusion is created from one membrane toward the other (Chernomordik & Kozlov, 2008; Efrat, Chernomordik, & Kozlov, 2007).

Such a lipid dimple can reduce the activation energy barrier for the stalk intermediate significantly. Although experimentally not tested, it is possible that the biological fusion may occur via a similar mechanism. To achieve this, the fusion proteins should form a hypothetical fence surrounding the pure lipid lawn in the middle. The concerted motion of the proteins would then create a dimple in the pure lipid patch in the middle (lower panel in Fig. 1B). According to a recent model, synaptotagmin 1 might play a primary role in bending the target membrane in calcium-dependent manners (Hui, Johnson, Yao, Dunning, & Chapman, 2009). One implication of this model is that synaptotagmin 1 is not just a regulator of the timing or extent of membrane fusion, but an essential part of fusion machinery itself. This radical new idea needs further experimental verification.

Such a lipid-centric model however appears to be inconsistent with the experimental data showing SNAREs have tendency to aggregate in large numbers near the fusion site (Barg, Knowles, Chen, Midorikawa, & Almers, 2010; Han, Wang, Bai, Chapman, & Jackson, 2004; Sieber et al., 2007). In particular, t-SNARE syntaxin 1A in the plasma membrane exists as large clusters consisting of 20–70 molecules (Barg et al., 2010; Sieber et al., 2007). One could argue that these patches may disperse to create a pure lipid patch in the middle. But there is no experimental evidence that supports such a theory. Further, an EPR experiment (X. Lu, Zhang, & Shin, 2008) and electrophysiological experiments with the mutants of syntaxin 1A (Han et al., 2004) showed that the TMD of t-SNARE may form a homo oligomer of three to eight monomers. Thus, it is likely that the very early stage of the fusion complex is the cluster of a few SNARE complexes (upper panel in Fig. 1B).

Starting the fusion complex with an oligomeric state has a geometric advantage because the SNARE complexes in the membrane gap are arranged outward from the center not to interfere directly with the apposition of two membranes. We speculate that the fluctuation of the TMD oligomer leads to the mixing of outer leaflets to create the hemifusion state.

E. Hemifusion

In 1993, White and coworkers demonstrated that GPI-anchored flu HA lacking the TMD arrest HA-induced fusion at the hemifusion state at which lipids mix freely while the contents remain still separated (Kemble, Danieli, & White, 1994) (see Fig. 1C). The results have two important meanings. First, this is the first demonstration of the hemifusion state, which had been seen only in the protein-free model membrane systems. Second, the TMD plays an important role in driving the fusion reaction beyond hemifusion. In cells, Chernomordik and coworkers observed that lysolipids favoring positive-curvature blocked the secretion

(Chernomordik et al., 1993), implying that the negatively curved membrane such as the hemifusion state might exists along the exocytotic pathway.

Inspired by these two seminal experiments, Shin and coworkers designed an *in vitro* experiment to observe the hypothetical hemifusion state in SNARE-dependent membrane fusion (Xu, Zhang, Su, McNew, & Shin, 2005). They first determined the transmembrane structure of the membrane anchor region of v-SNARE sso1p, which is an yeast counterpart of syntaxin 1A. Then, they prepared a truncation mutant in which the half of the TMD that spans the distal leaflet of the bilayer is deleted. They found that this mutant quenched the fusion reaction at the hemifusion state. Further, that the hemifusion state is an on-pathway intermediate, which is verified by the single vesicle fusion assay, which tracks the dynamic changes in a single fusion event. Later, more *in vivo* as well as *in vitro* data have helped to establish that the hemifusion state is indeed a bona fide intermediate for SNARE-mediated membrane fusion (Abdulreda, Bhalla, Chapman, & Moy, 2008; Jun & Wickner, 2007; Wong, Koppel, Cowan, & Wessel, 2007; Yoon, Okumus, Zhang, Shin, & Ha, 2006; Zampighi et al., 2006).

For example, it was shown that the hemifusion state is a stable intermediate of exocytosis in neuronal cells *in vivo* and two membranes may be hemifused before Ca^{2+} influx (Zampighi et al., 2006). Wong et al. (2007) suggested a stable hemifusion state before calcium influx between large cortical granules and the plasma membrane of sea urchin eggs. By using fluorescence recovery after photobleaching (FRAP), they showed that granules could exchange lipids, but not proteins, indicating that membranes are resting in a hemifusion state. However, there are contradicting results against the hemifusion intermediate model (Fernandez-Busnadiego et al., 2010). Fernandez-Busnadiego et al. used cryo-electron tomography to investigate the structure of the presynapse, and they were able to observe direct contact between two membranes only when the vesicles were being exocytosed, not before the calcium influx, indicating that two membranes are perhaps not hemifused before calcium influx.

F. Fusion Pore

Over the past several years, a pressing issue for the fusion pore has been what lines the fusion pore (Chernomordik & Kozlov, 2003; Jackson & Chapman, 2006; White & Castle, 2005) (see Fig. 1D). Provided that the hemifusion state is the bona fide fusion intermediate, it is hard to imagine that the fusion pore is not lined with the lipid molecules, where the head groups face the pore direction and the acyl chains point away from the pore. However, traditionally electrophysiologists favored the protein-lined pore (Jackson & Chapman, 2006). They envision that the initial fusion pore is formed by the docking of the two hemi-proteins in two membranes due to SNARE assembly. The interruption of the flow of neurotransmitters due to the mutations in TMD supported such a protein-centric model, although an alternative explanation is possible for these results.

It is certain that fusion regulators such as complexins and synaptotagmin 1 influence the timing and the duration of fusion pore formation as well as the size of the pore. Understanding the functions of fusion regulators on fusion pore formation has been slow due to the lack of a reliable *in vitro* fusion assay to detect the content mixing. Progress toward to this goal has been made recently (Diao et al., 2010).

G. Lipids as Active Regulators for the SNARE Function

Traditionally, effects of lipids on membrane fusion have been mainly attributed to their molecular shapes that favor a particular monolayer curvature (Chernomordik & Kozlov, 2005; Chernomordik, Zimmerberg, & Kozlov, 2006). Extreme examples are lysolipids: Lysolipids have a larger head than the tail, resembling an inverse cone, which fits better to the positively curved monolayer than to the negatively curved monolayer. Thus, in the presence of lysolipids the transition to the hemifusion state, which accompanies creation of a negatively curved surface, would be inhibited. On the other hand, PE, which has a larger tail than the head due to its small head group, favors the negatively curved monolayer and therefore, it would stimulate hemifusion.

Cholesterol draws our particular attention because of its large abundance in synaptic membranes; it constitutes as much as 40% of all lipids in synaptic vesicles (Takamori et al., 2006). There is evidence that cholesterol accelerates membrane fusion in cells. Cyclodextrans are generally used to extract cholesterol from cell membranes (Churchward, Rogasevskaia, Hofgen, Bau, & Coorssen, 2005). But extraction of cholesterol from the selected membrane and if possible, by desired amount are very difficult with this approach.

Alternatively, *in vitro* reconstitution fusion assays provide great opportunities to explore the role of cholesterol in SNARE-dependent membrane fusion (Chang et al., 2009). One can add or subtract desired amounts of cholesterol from selected membranes. In the proteoliposome fusion assay, it was found that cholesterol indeed stimulate lipid mixing. However, cholesterol was much more effective when it was present in the v-SNARE side than when it was in the t-SNARE side (Tong, Borbat, Freed, & Shin, 2009). Such an asymmetric effect cannot be explained only by the curvature argument because if cholesterol effect is exclusively through its curvature preference we should have observed similar effects from both sides. One possible explanation for this observation is that cholesterol modifies the protein structure to make it more favorable for fusion. If the hypothetical dimple formation at either one of the membrane were rate-determining, cholesterol might have an asymmetric effect as well.

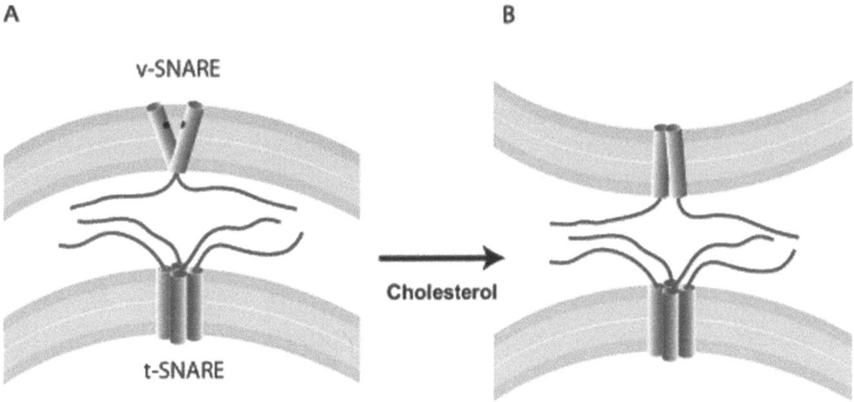

FIGURE 2 The scissors mechanism for the cholesterol action on SNAREs. (A) In the absence of cholesterol the v-SNARE TMD dimer has an open scissors shape with the pivot point in the outer leaflet of the vesicle membrane. This triangle-shaped TMD fits better to the membrane concaved toward the target membrane, which would be unfavorable for the merger of two membranes. (B) In the presence of cholesterol the v-SNARE TMD becomes a parallel dimer, which help reduce the concave curvature in the membrane, making the bilayer favorable for fusion. (See Color Insert.)

However, cholesterol has the intrinsic negative curvature and therefore, it would be inhibitory instead of stimulatory because the dimple is positively curved.

EPR is an excellent method to investigate the structure of the TMD. Shin, Freed, and coworkers used EPR to determine structures of the v-SNARE TMD in the presence and in the absence of cholesterol, respectively (Tong et al., 2009) (Fig. 2). They found that the v-SNARE TMD exists as dimers. In the absence of cholesterol the two helices form an open scissors-type dimer with the pivot point in the N-terminal region and they diverge in the C-terminal region. In contrast, in the presence of cholesterol the two helices get close in the C-terminal region and form a closed scissors-type dimer. The open scissors-type conformation would be unfavorable because it causes the bilayer to have negatively curved toward the target membrane.

This work shows that lipids such as cholesterol can be involved in membrane fusion not just simply as a passive participant with a certain curvature preference but as an active player, which can modify the protein structure and regulate the protein activity.

IV. SMALL MOLECULES AS INTERROGATORS FOR SNARE-MEDIATED MEMBRANE FUSION

It is widely accepted that the nucleation of the SNARE complex at the membrane distal N-terminal region corresponds to vesicle priming, whereas complex formation at the membrane-proximal C-terminal end drives membrane fusion. C-terminal complex formation may be assisted by auxiliary proteins such as synaptotagmins and complexins. Recently, Yang et al. (2010) asked if such sequential SNARE zippering could be blocked by small molecules. Finding small molecule SNARE blockers appears to be relatively tangible because coiled coils are known to be excellent targets for small molecules. For example, a few molecules were identified to inhibit formation of the six-helical bundle of HIV gp41 that is considered a key structural intermediate for membrane fusion required for vial entry (Balogh, Wu, Zhou, & Gochin, 2009; Cai & Gochin, 2007; Frey et al., 2006). A recombinant leucine zipper, in which a core-forming residue was changed to alanine, was shown to bind ring-structured small compounds such as benzene, cyclohexane, and toluene (Doerr, Case, Pelczer, & McLendon, 2004; Gonzalez, Plecs, & Alber, 1996). Furthermore, it was also shown that a designed four-helix bundle protein accepts anesthetics in a core hydrophobic layer (Zhang & Johansson, 2003, 2005). The existence of natural compounds that block SNARE complex formation has been recently verified (Jung et al., 2009), although the chemical properties of the candidate compounds were unknown.

Among natural compounds, polyphenols deserve special attention because they are abundant in natural products, have ring structures, and are known to be versatile in protein binding. Some polyphenols in diets can cross the blood–brain barrier (Ehrnhoefer et al., 2006; Kocisko et al., 2003), suggesting that they can target SNARE proteins in central nervous system once identified as SNARE blockers. Yang et al. (2010) showed that certain small hydrophobic molecules (SHM) enabled a layer-by-layer control of SNARE zippering by intercalating into various points of the SNARE complex. As an initial step, 39 polyphenolic compounds representing 12 subgroups were screened for an inhibitory activity against SNARE-driven proteoliposome fusion. Ten polyphenol compounds showed significant inhibitory effect on membrane fusion, and the degree of SNARE complex formation, assessed by Western blotting, was reduced for all 10 compounds. By virtue of the simplicity of the *in vitro* fusion assay, this observation suggests that inhibition of membrane fusion by SHM was likely a direct consequence of inhibition of SNARE zippering.

Yang et al. performed further experiments using the three most potent compounds: delphinidin, cyanidin, and myricetin, and found out that these three SHMs intercalate into the inner layer of the SNARE core complex during helical bundle formation. Delphinidin and cyanidin bind to the N-terminal part of SNARE motif, inhibiting N-terminal nucleation of SNARE complex formation. On the other hand, myricetin has two binding sites: one in the middle of the SNARE motif and the other in the N-terminal part. This second binding site has an important functional consequence: It is observed that myricetin stops SNARE-mediated membrane fusion at the hemifusion state. Neurotransmission and SNARE complex formation in neuronal PC12 cells also support the results from the *in vitro* study.

On the basis of these observations, it is proposed that the dynamic SNARE zippering process could be stopped at the point of interest by wedging an appropriate SHM into the zipper (Fig. 3). Delphinidin and cyanidin wedges into the SNARE zipper at the far N-terminus (P20∼V42 of VAMP2), thereby preventing N-terminal nucleation of SNARE complex formation. Meanwhile, myricetin stops subsequent zippering toward the membrane-proximal part by wedging into the middle region of the SNARE zipper. The detailed Ala-scanning study by Yang et al. shows that the four-stranded coiled coil up to residue M46 with the frayed C-terminal halves may represent the *trans* SNARE complex blocked by myricetin.

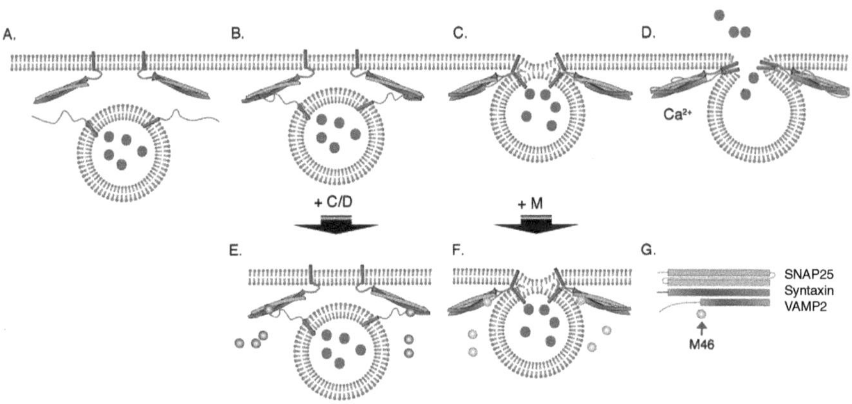

FIGURE 3 Small molecule SNARE inhibitors. Several small molecules stop SNARE zippering by wedging the complex. After stopping membrane fusion by this wedge-like action of small molecules membrane structure was analyzed to correlate the SNARE conformation to membrane fusion stage. (A) Predocking state of vesicle. (B) The N-terminal partial complex is formed as observed by 4 °C preincubation. This partial complex docks secretary vesicles near the plasma membrane. (C) In one of the previously popular SNARE-driven membrane fusion models, the assembled four-helical bundle with TMDs still at separate membranes waits for a calcium trigger. In this structure, hemifusion is assumed to be achieved by full zippering of the SNARE motifs, and inner leaflet mixing is coupled to TMD interaction. (D) Calcium promotes full SNARE zippering extending to the TMDs, which opens fusion pore. (E) Delphinidin/cyanidin wedges into the SNARE zipper at the far N-terminus (P20∼V42 of VAMP2), thereby preventing N-terminal nucleation of SNARE complex formation and even outer leaflet mixing. (F) Myricetin wedges into the SNARE zipper midway (N49∼L63 of VAMP2), hampering residual C-terminal zippering and inner leaflet mixing. Hemifusion assay raises the possibility that this halfway zippering is sufficient to induce hemifusion. Results also raise the possibility that the N-terminal partial complex shown in (B) may be present only at 4 °C but this partial complex seems to be enough for hemifusion at physiological temperatures. (G) In the *trans*-complex, VAMP2 coiling extends up to M46 and the residual C-terminal region is still frayed. (See Color Insert.)

Clostridium botulinum neurotoxins (BoNTs) are one of the most potent natural inhibitors of neurotransmitter release. BoNTs elicit neuron-specific flaccid paralysis by specifically cleaving neuronal SNARE proteins. After realization of the therapeutic potentials of these otherwise fatal toxins, the controlled use of BoNT/A was approved by the FDA for the treatment of several hypersecretion-related neurological diseases such as strabismus, blepharospam, hemificial spasm, and cervical dystonia (Dolly, Lawrence, Meng, Wang, & Ovsepian, 2009). Recently, the cosmetic use of these toxins for treating glabellar facial lines and auxiliary hyperhidrosis has gained immense popularity. Based on the ongoing successes of BoNT for the treatment of a variety of hypersecretion-related neurological disorders, a therapeutic strategy to block SNARE function for the treatment of diseases of central nervous system appears to have great potential, too. Several hideous mental diseases such as schizophrenia, epilepsy, and autism are believed to be linked to the hypersecretion of neurotransmitters. The use of BoNT for this purpose, however, faces insurmountable challenges in safely delivering the toxin across the blood–brain barrier. Small molecule SNARE blocker may provide a solution for targeting the exocytosis machinery in the central nervous system.

V. SINGLE MOLECULE (OR PARTICLE) FLUORESCENCE TECHNIQUES TO STUDY SNARE-MEDIATED MEMBRANE FUSION

The past decade has witnessed the remarkable development of new biophysical/biochemical tools called single molecule techniques, which allow one to observe and/or interrogate individual molecules (Weiss, 1999). The single molecule techniques have proven to be especially useful when the biological sample of interest is either highly heterogeneous or unsynchronized in its dynamics. For instance, when the same molecules show different levels of signal, that is, a heterogeneous sample, an ensemble measurement only measures the mean value of the signal that is grossly averaged over all molecules. A single molecule measurement would reveal the whole spectrum of such signals, providing the detailed information that would be almost impossible to obtain from the simple mean value. Also, when the molecules are going back and forth between several discrete states and this state transition is not synchronized even between neighboring molecules, that is, a unsynchronized sample, an ensemble measurement again measures the mean value whereas a single molecule measurement would allow for direct observation of the transition dynamics of individual molecules.

Currently, the single molecule techniques can be generally classified into two categories: fluorescence and force spectroscopy (readers are referred to recent excellent reviews on both types of techniques) (Greenleaf, Woodside, & Block,

2007; Joo, Balci, Ishitsuka, Buranachai, & Ha, 2008; Moffitt, Chemla, Smith, & Bustamante, 2008). In this article, we will be much more focused on the single molecule fluorescence techniques. This is because our biological process of interest is membrane fusion driven by the SNARE proteins and the already established *in vitro* assay of SNARE-mediated fusion uses fluorescence signals (Arac et al., 2006; Stein, Radhakrishnan, Riedel, Fasshauer, & Jahn, 2007; Tucker, Weber, & Chapman, 2004; Weber et al., 1998). Thus, adoption of the single molecule "fluorescence" techniques directly introduces single molecule resolution to the same *in vitro* assay (as discussed below). Comparison between the ensemble fusion and the new single molecule data provides novel insights toward the molecular mechanism of membrane fusion and its regulation by fusion effectors.

The *in vitro* assay of SNARE-mediated fusion, originally invented by the Rothman group, consists of two groups of vesicles, each of which harbors one of two types of SNAREs (v- and t-SNAREs) that constitute a cognate pair for making SNARE complex (Fig. 4a) (Weber et al., 1998). This assay reproduces membrane fusion mediated by SNARE proteins in a well-defined *in vitro* setting, providing a way one studies the molecular mechanism of SNARE and their regulators while controlling protein sequences and molecular environments. This assay has indeed critically contributed to elucidation of the molecular mechanism of exocytosis. Yet, there is a large room for another improvement because a fusion reaction contains more than 10^{10} vesicles, and this vesicle solution is a representative of a highly heterogeneous and unsynchronized sample. Individual fusion events can be trapped in different intermediate states, and 10^{10} vesicle pairs start their fusion reaction in a stochastic manner, so their progression along the fusion pathway should be extremely unsynchronized.

Application of single molecule techniques should make it possible to identify where individual fusion events are placed along the fusion pathway and measure the kinetics of transition to following states. At the simplest level, a fusion event can be decomposed into the docking and actual fusion steps (Fig. 4). These two steps may bear direct implications for the real synaptic vesicle fusion. Synaptic vesicles are thought to be first primed (or prepared) and then fusing with the plasma membrane triggered by influx of Ca^{2+} ions (Fig. 4a). Therefore, if one is to dissect the effects on the priming step, it is essential to observe the docking event separately and measure changes in its kinetics. Likewise, to study the effects on the fusion step, one should be able to exclusively measure the kinetic rate of the actual fusion step and its changes. Therefore, the single molecule techniques, which allow for the separate measurement of the docking and fusion steps in individual fusion events, could be an ideal technique for studying SNARE-mediated membrane fusion. Hereafter, we will call such assay the single vesicle fusion assay.

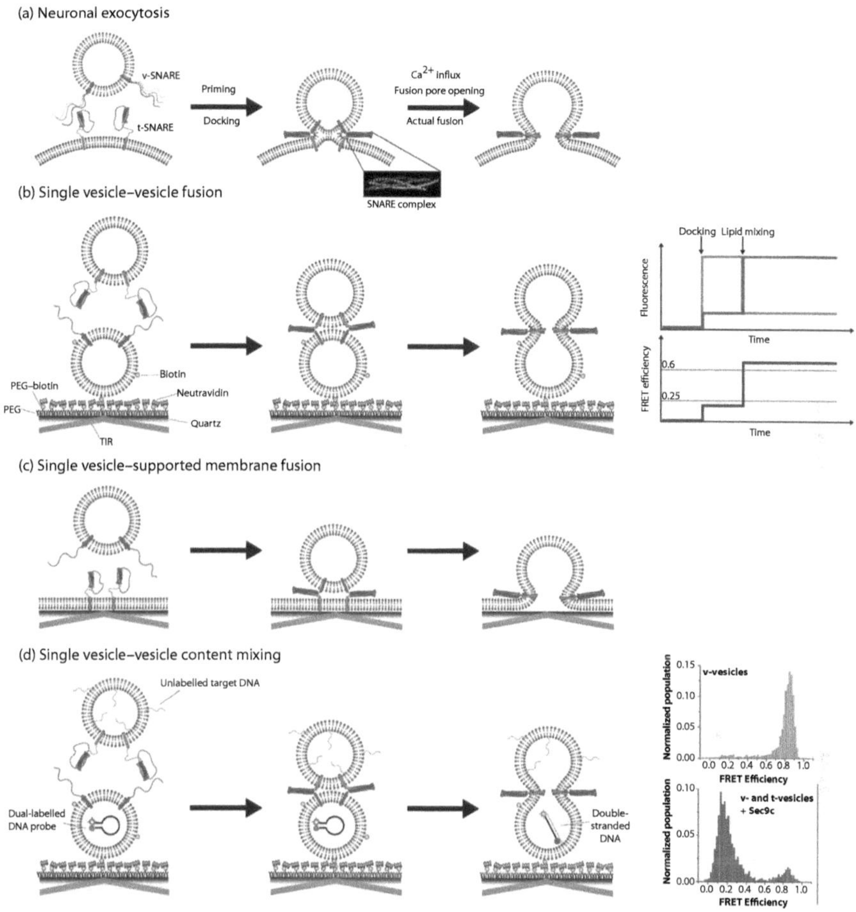

FIGURE 4 Single-vesicle fusion study of neuronal exocytosis. (a) Simplified schematic for neuronal exocytosis. A synaptic vesicle fusion process can be separated into the docking and actual fusion steps. (b) To study the neuronal exocytosis, a single vesicle–vesicle fusion assay detects the docking and actual fusion between two single proteoliposomes. The surface attached vesicles contain the acceptor dyes. Upon actual fusion, these acceptor dyes are mixed with the donor dyes in the microinjected vesicles for fusion reaction, which results in an increase of the FRET efficiency (measured over two single vesicles) above 0.7. (c) The surface immobilized vesicles can be replaced by a supported membrane containing synaptic fusion protein. The microinjected vesicles contain fluorescent dyes, which diffuse into the supported membrane subsequent to the actual fusion between a microinjected vesicle and supported membrane. (d) To monitor content mixing, fluorescently labeled DNA probes can be encapsulated in the surface immobilized vesicles. For instance, the molecular beacon inside the vesicle shows transition to a lower FRET efficiency, when there is content mixing between two single vesicles and the target DNA of the molecular beacon migrates into the surface-immobilized vesicle or vice versa. Adapted from Diao et al. (2010). Reprinted by permission from Macmillan Publishers Ltd. (See Color Insert.)

One natural question here is how to introduce the single molecule techniques to the *in vitro* assay of SNARE-mediated fusion. This *in vitro* assay uses two types of vesicles, one with v-SNAREs (referred to as v-vesicles) and the other with t-SNAREs (t-vesicles), and label these vesicles with fluorescent dyes. The original design by the Rothman group labeled the v-vesicle membrane with two types of fluorescent dyes (dyes are attached to the headgroup of PE lipid), NBD and rhodamine, which constitute a pair of FRET donor and acceptor (Weber et al., 1998). When fusion proceeds, the v-vesicle membrane is progressively diluted by unlabeled t-vesicle membranes. The FRET efficiency then goes down and the donor fluorescence intensity, which was initially quenched by FRET to the acceptor, grows in its magnitude. To adopt the single molecule fluorescence techniques, the Ha group has changed the scheme for fluorescence labeling that the t- and v-vesicle membranes are now labeled with the FRET donor and acceptor dyes, respectively (Fig. 4b) (Yoon et al., 2006). The lipophilic cyanine dyes, DiI and DiD, are chosen as the FRET donor and acceptor dyes because these dyes show similar fluorescence characteristics to those of Cy3 and Cy5, which make the most widely used dye pair for single molecule FRET. With this labeling, close contact between the v- and t-vesicle membranes give low, but nonzero, FRET efficiencies (less than 0.25) mainly from the contacting areas. Upon fusion, two membranes completely mix with each other, resulting in high FRET efficiencies larger than 0.6–0.7, which can be clearly distinguished from the signals generated by close contact (note that such distinction is only possible in the single vesicle fusion assay) (Yoon et al., 2006). The signal for fusion progression under this new scheme is an increase in the FRET efficiency, not a decrease, which is usually considered a safe design that is free from many false positives such as photobleaching of acceptor dyes.

To make it possible to track individual fusion events, the first step should be immobilization of one type of vesicles. This immobilization is essential for prolonged observation because a free floating vesicle stays within the excitation volume of conventional optical microscopy for no more than 100 ms, which is a very limited time window for doing meaningful manipulations and observations. To immobilize vesicles on a glass (or quartz) surface, a conventional imaging surface of optical microscope, a surface passivation procedure is absolutely required otherwise the strong hydrophilicity of hydroxyl group of glass surface would strongly attract and finally rupture vesicles to form a supported membrane, instead of point-contacting surface-immobilized vesicles. The well-established protocol of surface passivation in the single molecule field is to coat polyethylene glycol (PEG) polymers on the glass surface, with a self-assembled monolayer of aminosilane as the linker between them. Almost 99 mol% of PEG molecules have methyl group at their terminals, thereby providing hydrophobic surface properties (Fig. 4b). About 1 mol% of PEG molecules have the biotin group, on which neutravidin, a variant of streptavidin, can be layered. By including ~0.1 mol% biotinylated lipids (biotin is attached to the lipid headgroup) in the lipid composition, the product vesicles can be immobilized on the surface through simple incubation. The Stamou group showed that if the vesicle diameter is smaller than 100 nm, the immobilized vesicles indeed make point contact with the surface (Bendix, Pedersen, & Stamou, 2009).

Another viable option for such immobilization is to form supported membrane through vesicle rupture (Fig. 4c) (Sackmann, 1996). In this case, the complicated procedure like PEGylation is not required because the hydrophilicity of the glass surface draws vesicles toward it without requiring additional treatments. However, the marginal gap between the supported membrane and its solid support, measured to be ~1 nm, is narrow for cytosolic parts of many membrane proteins. The consequent hydrodynamic dragging makes many membrane proteins practically immobile in the supported membrane, which is usually detrimental to their function. To circumvent this technical issue, the Tamm group makes the supported membrane through two sequential procedures (Domanska et al., 2009). First, one lipid monolayer is deposited through Langmuir–Blodgett transfer. When vesicle solution containing membrane proteins is added, the lipid monolayer that show its hydrophobic tails to the vesicle solution attracts vesicles and those adsorbed vesicles rearrange themselves, making the other leaflet to complete the bilayer structure. It has been reported that the t-SNARE proteins are as mobile as lipid molecules in this two-step bilayer system.

After one type of membrane has been immobilized, either in the form of vesicles or supported membrane, the other group of vesicles is introduced for fusion reaction usually through microfluidic buffer exchange. In the case of single vesicle–vesicle fusion used by the Yoon group, the v-vesicles are labeled with acceptor DiD and immobilized on the surface, while the t-vesicles are labeled with donor DiI and microinjected for fusion reaction (Fig. 4b) (Lee et al., 2010; Yoon et al., 2006, 2008). The concentration of the injected t-vesicles is kept low, typically less than tens of pM [vesicle], which is to ensure that the distance between two docking events of t-vesicles is larger than the diffraction limit, a requirement for observing individual fusion events. The moment of t-vesicle docking is then detected as a sudden spike of the donor fluorescence signal. Multiple docking events are readily detectable since such multiple docking would give more than one donor spikes. Finally, when the two vesicles are fully fused, the FRET efficiency passes the threshold of 0.6–0.7. In the case of the supported membrane–vesicle fusion assay developed by the Tamm, Weisshaar, and Rothman groups (Domanska et al., 2009; Karatekin et al., 2010; Wang, Smith, Chapman, & Weisshaar, 2009), the t-vesicle membranes are surface deposited and the v-vesicles are injected for reaction, and only the v-vesicles are fluorescently labeled (Fig. 4c). The docking event can be identified as the abrupt increase in the fluorescence signal as was the case for the vesicle–vesicle assay. This supported membrane assay however measures the start of actual fusion step, not the completion of it, which is reported as spreading of the fluorescent dyes to the supported membrane through two dimensional Brownian diffusion. Therefore, both single vesicle–vesicle and single vesicle-supported systems allow for the detection of the moments of docking and actual fusion.

The detection of these two moments is crucial because the kinetic rates for docking and fusion can be directly extracted from such data. Collection of the docking moments after injection of vesicles directly gives the relative kinetic rate for docking. For example, when the number of docking events is doubled within the same time window, it is straightforward to conclude that the kinetic rate for docking has been doubled. We note however that one should be cautious in making this statement because there are always nonspecific binding events in the given time window. Assessment of the fusion kinetics requires measurement of the time gap between the docking and the completion of actual fusion. After these time gaps collected for many individual events, the exponential fitting of their distribution gives the kinetic rate for the actual fusion step. As discussed above, changes in the docking kinetics could be relevant to modulation of the priming step. Acceleration or retardation of the fusion kinetics may be recapitulating the modulation of the late steps of neurotransmitter release.

It should be noted that an important technical issue has been raised in measuring the fusion kinetics. *In vitro* assays have mainly used lipid mixing as the measure of fusion, but the correlation between the lipid mixing and content mixing could be much lower than thought. The Boxer group quantitatively reported that they could induce the lipid mixing for 60% of their vesicles while only 5% actually showed content mixing (Chan, van Lengerich, & Boxer, 2009). This observation implies that the outer and inner leaflets of two fusing vesicles can be completely mixed with each other, but their inner vesicle lumens remain separated. The signals for complete lipid mixing (high vesicle–vesicle FRET or complete spread out of fluorescent dyes) cannot be put equivalent to content mixing in many cases. The fusion kinetics will be best measured when one can monitor the moment of content mixing in individual fusion events.

Recently, the Shin and Ha groups moved the position of fluorescent labels from vesicle membranes to vesicle lumens (Diao et al., 2010). In their experimental design, a pair of FRET dyes, Cy3 and Cy5, is placed at the end of the stem region (5 base pairs) of a molecular beacon, in which this stem region is connected by 20 base pair single-stranded (ss) loop region (Fig. 4d). This probe is encapsulated in the surface-immobilized vesicles. The microinjected vesicles have the target DNA that can hybridize with the molecular beacon and convert the loop region into a double-stranded (ds) helix. Because the persistence length of the dsDNA is increased by more than one order of magnitude (typically up to 50 nm), the molecular beacon is converted into a linear, dsDNA after reacting with the target DNA. Consequently, when there is content mixing that entails the reaction between the molecular beacon and target DNA, the inter-dye distance between Cy3 and Cy5 is substantially increased (from the diameter of dsDNA to 25 base pair separation in a dsDNA), which should result in a very low FRET efficiency between two dyes. The Shin and Ha groups measured such changes in the FRET efficiency and indeed observed a high-to-low FRET transition in single vesicle–vesicle fusion event induced by yeast SNAREs. This experiment seems to be important because it demonstrates that the content mixing, also, could be measured at the individual fusion event level. Furthermore, the well-developed DNA engineering enables one to control the size of molecular beacon and target DNA at the resolution of one base pair. So different sizes of DNA structure could be used as the content mixing probe, which could lead to novel experiments where one can measure the size of the fusion pore with the sub-nanometer accuracy.

Is there any issue that has been previously obscure but resolved by the observation through the single vesicle fusion assay? One of the most fundamental observations would be that at the vesicle concentrations used for *in vitro* assays, docking becomes the rate-limiting step especially when the fusion process is driven by SNAREs only (Yoon et al., 2006). The single-vesicle assay clearly shows that the typical waiting time to see actual fusion complete is around a few seconds, but the average waiting time for docking is at least two orders of magnitude longer. This observation implies that the slowly increasing curve, a typical result of an ensemble fusion assay, reports on the docking step in many cases, and the relatively fast fusion kinetics and its modulation would be largely hidden behind the rate-limiting docking kinetics.

In addition, the Shin and Yoon groups have been using the single vesicle–vesicle assay to study the regulators of neuronal SNARE-mediated fusion (Lee et al., 2010; Yoon et al., 2008). For example, a soluble fusion regulator, named complexin, is added in the neuronal SNARE-mediated fusion reaction. It is observed that in the presence of complexin, the docking kinetics is retarded while the fusion kinetics is accelerated. Such fusion step-specific effect of complexin seems to result from different interaction modes of complexin with individual SNARE proteins and SNARE complexes (Yoon et al., 2008). The function of the putative Ca^{2+} switch of neuronal exocytosis, synaptotagmin 1, has also been studied. Previous bulk fusion studies mainly used soluble version of synaptotagmin 1, keeping only its tandem C2 domains while truncating out the TMD and linker region to it. The Shin and Yoon groups have reconstituted synaptotagmin 1 in the form of membrane proteins, and using the high sensitivity of the single vesicle assay, they have observed the strong fusion stimulation by membrane-anchored synaptotagmin 1, which accelerates neuronal SNARE-mediated fusion to hundreds of milliseconds (Lee et al., 2010).

The development of the single vesicle fusion assay is being vigorously pursued by several groups, and the technique seems to be maturing very fast that the single vesicle fusion assay has already begun to produce previously hidden insights toward SNARE-mediated membrane fusion. The single vesicle assay has clear advantages over the conventional ensemble fusion assay in a sense that it enables separate measurement of the docking and fusion kinetics and their respective modulation by fusion regulators. The high sensitivity to fluorescence signals and time resolution up to milliseconds scale provide very different windows for doing experiments, especially suited for studying the mechanism of neuronal exocytosis that occurs on the millisecond or submillisecond time scale in the central nervous system. It is highly likely that the coming years will witness increasing use of the single vesicle fusion assay, and the new information from this assay will greatly help reshape our view on the molecular mechanism of membrane fusion.

References

Abdulreda, M. H., Bhalla, A., Chapman, E. R., & Moy, V. T. (2008). Atomic force microscope spectroscopy reveals a hemifusion intermediate during soluble *N*-ethylmaleimide-sensitive factor-attachment protein receptors-mediated membrane fusion. *Biophysical Journal, 94*, 648–655.

Antonin, W., Fasshauer, D., Becker, S., Jahn, R., & Schneider, T. R. (2002). Crystal structure of the endosomal SNARE complex reveals common structural principles of all SNAREs. *Nature Structural Biology, 9*, 107–111.

Arac, D., Chen, X., Khant, H. A., Ubach, J., Ludtke, S. J., Kikkawa, M., et al. (2006). Close membrane-membrane proximity induced by Ca^{2+}-dependent multivalent binding of synaptotagmin-1 to phospholipids. *Nature Structural & Molecular Biology, 13*, 209–217.

Balogh, E., Wu, D., Zhou, G., & Gochin, M. (2009). NMR second site screening for structure determination of ligands bound in the hydrophobic pocket of HIV-1 gp41. *Journal of the American Chemical Society, 131*, 2821–2823.

Barg, S., Knowles, M. K., Chen, X., Midorikawa, M., & Almers, W. (2010). Syntaxin clusters assemble reversibly at sites of secretory granules in live cells. *Proceedings of the National Academy of Sciences of the United States of America, 107*, 20804–20809.

Bendix, P. M., Pedersen, M. S., & Stamou, D. (2009). Quantification of nano-scale intermembrane contact areas by using fluorescence resonance energy transfer. *Proceedings of the National Academy of Sciences of the United States of America, 106*, 12341–12346.

Cai, L., & Gochin, M. (2007). A novel fluorescence intensity screening assay identifies new low-molecular-weight inhibitors of the gp41 coiled-coil domain of human immunodeficiency virus type 1. *Antimicrobial Agents and Chemotherapy, 51*, 2388–2395.

Chan, Y. H., van Lengerich, B., & Boxer, S. G. (2009). Effects of linker sequences on vesicle fusion mediated by lipid-anchored DNA oligonucleotides. *Proceedings of the National Academy of Sciences of the United States of America, 106*, 979–984.

Chang, J., Kim, S. A., Lu, X., Su, Z., Kim, S. K., & Shin, Y. K. (2009). Fusion step-specific influence of cholesterol on SNARE-mediated membrane fusion. *Biophysical Journal, 96*, 1839–1846.

Chernomordik, L. V., & Kozlov, M. M. (2003). Protein-lipid interplay in fusion and fission of biological membranes. *Annual Review of Biochemistry, 72*, 175–207.

Chernomordik, L. V., & Kozlov, M. M. (2005). Membrane hemifusion: Crossing a chasm in two leaps. *Cell, 123*, 375–382.

Chernomordik, L. V., & Kozlov, M. M. (2008). Mechanics of membrane fusion. *Nature Structural & Molecular Biology, 15*, 675–683.

Chernomordik, L. V., Vogel, S. S., Sokoloff, A., Onaran, H. O., Leikina, E. A., & Zimmerberg, J. (1993). Lysolipids reversibly inhibit Ca(2+), GTP- and pH-dependent fusion of biological membranes. *FEBS Letters, 318*, 71–76.

Chernomordik, L. V., Zimmerberg, J., & Kozlov, M. M. (2006). Membranes of the world unite!. *Journal of Cell Biology, 175*, 201–207.
Churchward, M. A., Rogasevskaia, T., Hofgen, J., Bau, J., & Coorssen, J. R. (2005). Cholesterol facilitates the native mechanism of Ca^{2+}-triggered membrane fusion. *Journal of Cell Science, 118*, 4833–4848.
Diao, J., Su, Z., Ishitsuka, Y., Lu, B., Lee, K. S., Lai, Y., et al. (2010). A single-vesicle content mixing assay for SNARE-mediated membrane fusion. *Nature Communications, 1*, 1–6.
Doerr, A. J., Case, M. A., Pelczer, I., & McLendon, G. L. (2004). Design of a functional protein for molecular recognition: Specificity of ligand binding in a metal-assembled protein cavity probed by 19f NMR. *Journal of the American Chemical Society, 126*, 4192–4198.
Dolly, J. O., Lawrence, G. W., Meng, J., Wang, J., & Ovsepian, S. V. (2009). Neuro-exocytosis: Botulinum toxins as inhibitory probes and versatile therapeutics. *Current Opinion in Pharmacology, 9*, 326–335.
Domanska, M. K., Kiessling, V., Stein, A., Fasshauer, D., & Tamm, L. K. (2009). Single vesicle millisecond fusion kinetics reveals number of SNARE complexes optimal for fast SNARE-mediated membrane fusion. *The Journal of Biological Chemistry, 284*, 32158–32166.
Efrat, A., Chernomordik, L. V., & Kozlov, M. M. (2007). Point-like protrusion as a prestalk intermediate in membrane fusion pathway. *Biophysical Journal, 92*, L61–L63.
Ehrnhoefer, D. E., Duennwald, M., Markovic, P., Wacker, J. L., Engemann, S., Roark, M., et al. (2006). Green tea (-)-epigallocatechin-gallate modulates early events in huntingtin misfolding and reduces toxicity in Huntington's disease models. *Human Molecular Genetics, 15*, 2743–2751.
Fasshauer, D., Otto, H., Eliason, W. K., Jahn, R., & Brunger, A. T. (1997). Structural changes are associated with soluble *N*-ethylmaleimide-sensitive fusion protein attachment protein receptor complex formation. *The Journal of Biological Chemistry, 272*, 28036–28041.
Fernandez-Busnadiego, R., Zuber, B., Maurer, U. E., Cyrklaff, M., Baumeister, W., & Lucic, V. (2010). Quantitative analysis of the native presynaptic cytomatrix by cryoelectron tomography. *Journal of Cell Biology, 188*, 145–156.
Fiebig, K. M., Rice, L. M., Pollock, E., & Brunger, A. T. (1999). Folding intermediates of SNARE complex assembly. *Nature Structural Biology, 6*, 117–123.
Frey, G., Rits-Volloch, S., Zhang, X. Q., Schooley, R. T., Chen, B., & Harrison, S. C. (2006). Small molecules that bind the inner core of gp41 and inhibit HIV envelope-mediated fusion. *Proceedings of the National Academy of Sciences of the United States of America, 103*, 13938–13943.
Giraudo, C. G., Eng, W. S., Melia, T. J., & Rothman, J. E. (2006). A clamping mechanism involved in SNARE-dependent exocytosis. *Science, 313*, 676–680.
Giraudo, C. G., Garcia-Diaz, A., Eng, W. S., Yamamoto, A., Melia, T. J., & Rothman, J. E. (2008). Distinct domains of complexins bind SNARE complexes and clamp fusion in vitro. *The Journal of Biological Chemistry, 283*, 21211–21219.
Gonzalez Jr., L., Plecs, J. J., & Alber, T. (1996). An engineered allosteric switch in leucine-zipper oligomerization. *Nature Structural Biology, 3*, 510–515.
Greenleaf, W. J., Woodside, M. T., & Block, S. M. (2007). High-resolution, single-molecule measurements of biomolecular motion. *Annual Review of Biophysics and Biomolecular Structure, 36*, 171–190.
Han, X., Wang, C. T., Bai, J., Chapman, E. R., & Jackson, M. B. (2004). Transmembrane segments of syntaxin line the fusion pore of Ca^{2+}-triggered exocytosis. *Science, 304*, 289–292.
Hui, E., Johnson, C. P., Yao, J., Dunning, F. M., & Chapman, E. R. (2009). Synaptotagmin-mediated bending of the target membrane is a critical step in Ca^{2+}-regulated fusion. *Cell, 138*, 709–721.
Jackson, M. B., & Chapman, E. R. (2006). Fusion pores and fusion machines in Ca^{2+}-triggered exocytosis. *Annual Review of Biophysics and Biomolecular Structure, 35*, 135–160.

Joo, C., Balci, H., Ishitsuka, Y., Buranachai, C., & Ha, T. (2008). Advances in single-molecule fluorescence methods for molecular biology. *Annual Review of Biochemistry, 77*, 51–76.

Jun, Y., & Wickner, W. (2007). Assays of vacuole fusion resolve the stages of docking, lipid mixing, and content mixing. *Proceedings of the National Academy of Sciences of the United States of America, 104*, 13010–13015.

Jung, C. H., Yang, Y. S., Kim, J. S., Shin, Y. K., Hwang, J. S., Son, E. D., et al. (2009). Inhibition of SNARE-driven neuroexocytosis by plant extracts. *Biotechnology Letters, 31*, 361–369.

Karatekin, E., Di Giovanni, J., Iborra, C., Coleman, J., O'Shaughnessy, B., Seagar, M., et al. (2010). A fast, single-vesicle fusion assay mimics physiological SNARE requirements. *Proceedings of the National Academy of Sciences of the United States of America, 107*, 3517–3521.

Kemble, G. W., Danieli, T., & White, J. M. (1994). Lipid-anchored influenza hemagglutinin promotes hemifusion, not complete fusion. *Cell, 76*, 383–391.

Kim, C. S., Kweon, D. H., & Shin, Y. K. (2002). Membrane topologies of neuronal SNARE folding intermediates. *Biochemistry, 41*, 10928–10933.

Kocisko, D. A., Baron, G. S., Rubenstein, R., Chen, J., Kuizon, S., & Caughey, B. (2003). New inhibitors of scrapie-associated prion protein formation in a library of 2000 drugs and natural products. *The Journal of Virology, 77*, 10288–10294.

Kweon, D. H., Kim, C. S., & Shin, Y. K. (2003a). Insertion of the membrane-proximal region of the neuronal SNARE coiled coil into the membrane. *Journal of Biological Chemistry, 278*, 12367–12373.

Kweon, D. H., Kim, C. S., & Shin, Y. K. (2003b). Regulation of neuronal SNARE assembly by the membrane. *Nature Structural Biology, 10*, 440–447.

Lee, H. K., Yang, Y., Su, Z., Hyeon, C., Lee, T. S., Lee, H. W., et al. (2010). Dynamic Ca^{2+}-dependent stimulation of vesicle fusion by membrane-anchored synaptotagmin 1. *Science, 328*, 760–763.

Liu, T., Tucker, W. C., Bhalla, A., Chapman, E. R., & Weisshaar, J. C. (2005). SNARE-driven, 25-millisecond vesicle fusion in vitro. *Biophysical Journal, 89*, 2458–2472.

Lu, B., Song, S., & Shin, Y. K. (2010). Accessory alpha-helix of complexin I can displace VAMP2 locally in the complexin-SNARE quaternary complex. *Journal of Molecular Biology, 396*, 602–609.

Lu, X., Zhang, Y., & Shin, Y. K. (2008). Supramolecular SNARE assembly precedes hemifusion in SNARE-mediated membrane fusion. *Nature Structural & Molecular Biology, 15*, 700–706.

Moffitt, J. R., Chemla, Y. R., Smith, S. B., & Bustamante, C. (2008). Recent advances in optical tweezers. *Annual Review of Biochemistry, 77*, 205–228.

Mohrmann, R., de Wit, H., Verhage, M., Neher, E., & Sorensen, J. B. (2010). Fast vesicle fusion in living cells requires at least three SNARE complexes. *Science, 330*, 502–505.

Pobbati, A. V., Stein, A., & Fasshauer, D. (2006). N- to C-terminal SNARE complex assembly promotes rapid membrane fusion. *Science, 313*, 673–676.

Poirier, M. A., Xiao, W., Macosko, J. C., Chan, C., Shin, Y. K., & Bennett, M. K. (1998). The synaptic SNARE complex is a parallel four-stranded helical bundle. *Nature Structural Biology, 5*, 765–769.

Rabenstein, M. D., & Shin, Y. K. (1995). Determination of the distance between two spin labels attached to a macromolecule. *Proceedings of the National Academy of Sciences of the United States of America, 92*, 8239–8243.

Sackmann, E. (1996). Supported membranes: Scientific and practical applications. *Science, 271*, 43–48.

Schaub, J. R., Lu, X., Doneske, B., Shin, Y. K., & McNew, J. A. (2006). Hemifusion arrest by complexin is relieved by Ca^{2+}-synaptotagmin I. *Nature Structural & Molecular Biology, 13*, 748–750.

Sieber, J. J., Willig, K. I., Kutzner, C., Gerding-Reimers, C., Harke, B., Donnert, G., et al. (2007). Anatomy and dynamics of a supramolecular membrane protein cluster. *Science, 317*, 1072–1076.

Skehel, J. J., & Wiley, D. C. (1998). Coiled coils in both intracellular vesicle and viral membrane fusion. *Cell, 95*, 871–874.

Sorensen, J. B., Wiederhold, K., Muller, E. M., Milosevic, I., Nagy, G., de Groot, B. L., et al. (2006). Sequential N- to C-terminal SNARE complex assembly drives priming and fusion of secretory vesicles. *EMBO Journal, 25*, 955–966.

Stein, A., Radhakrishnan, A., Riedel, D., Fasshauer, D., & Jahn, R. (2007). Synaptotagmin activates membrane fusion through a Ca(2+)-dependent trans interaction with phospholipids. *Nature Structural & Molecular Biology, 14*, 904–911.

Stein, A., Weber, G., Wahl, M. C., & Jahn, R. (2009). Helical extension of the neuronal SNARE complex into the membrane. *Nature, 460*, 525–528.

Strop, P., Kaiser, S. E., Vrljic, M., & Brunger, A. T. (2008). The structure of the yeast plasma membrane SNARE complex reveals destabilizing water-filled cavities. *The Journal of Biological Chemistry, 283*, 1113–1119.

Sutton, R. B., Fasshauer, D., Jahn, R., & Brunger, A. T. (1998). Crystal structure of a SNARE complex involved in synaptic exocytosis at 2.4 A resolution. *Nature, 395*, 347–353.

Takahashi, N., Hatakeyama, H., Okado, H., Noguchi, J., Ohno, M., & Kasai, H. (2010). SNARE conformational changes that prepare vesicles for exocytosis. *Cell Metabolism, 12*, 19–29.

Takamori, S., Holt, M., Stenius, K., Lemke, E. A., Gronborg, M., Riedel, D., et al. (2006). Molecular anatomy of a trafficking organelle. *Cell, 127*, 831–846.

Tang, J., Maximov, A., Shin, O. H., Dai, H., Rizo, J., & Sudhof, T. C. (2006). A complexin/synaptotagmin 1 switch controls fast synaptic vesicle exocytosis. *Cell, 126*, 1175–1187.

Tong, J., Borbat, P. P., Freed, J. H., & Shin, Y. K. (2009). A scissors mechanism for stimulation of SNARE-mediated lipid mixing by cholesterol. *Proceedings of the National Academy of Sciences of the United States of America, 106*, 5141–5146.

Tucker, W. C., Weber, T., & Chapman, E. R. (2004). Reconstitution of Ca^{2+}-regulated membrane fusion by synaptotagmin and SNAREs. *Science, 304*, 435–438.

van den Bogaart, G., Holt, M. G., Bunt, G., Riedel, D., Wouters, F. S., & Jahn, R. (2010). One SNARE complex is sufficient for membrane fusion. *Nature Structural & Molecular Biology, 17*, 358–364.

Walter, A. M., Wiederhold, K., Bruns, D., Fasshauer, D., & Sorensen, J. B. (2010). Synaptobrevin N-terminally bound to syntaxin-SNAP-25 defines the primed vesicle state in regulated exocytosis. *Journal of Cell Biology, 188*, 401–413.

Wang, T., Smith, E. A., Chapman, E. R., & Weisshaar, J. C. (2009). Lipid mixing and content release in single-vesicle, SNARE-driven fusion assay with 1-5 ms resolution. *Biophysical Journal, 96*, 4122–4131.

Weber, T., Zemelman, B. V., McNew, J. A., Westermann, B., Gmachl, M., Parlati, F., et al. (1998). SNAREpins: Minimal machinery for membrane fusion. *Cell, 92*, 759–772.

Weiss, S. (1999). Fluorescence spectroscopy of single biomolecules. *Science, 283*, 1676–1683.

White, J. M., & Castle, J. D. (2005). Searching for the silver lining. *Nature Structural & Molecular Biology, 12*, 382–384.

Wilson, I. A., Skehel, J. J., & Wiley, D. C. (1981). Structure of the haemagglutinin membrane glycoprotein of influenza virus at 3 A resolution. *Nature, 289*, 366–373.

Wong, J. L., Koppel, D. E., Cowan, A. E., & Wessel, G. M. (2007). Membrane hemifusion is a stable intermediate of exocytosis. *Developmental Cell, 12*, 653–659.

Xu, Y., Zhang, F., Su, Z., McNew, J. A., & Shin, Y. K. (2005). Hemifusion in SNARE-mediated membrane fusion. *Nature Structural & Molecular Biology, 12*, 417–422.

Yang, Y., Shin, J. Y., Oh, J. -M., Jung, C. H., Hwang, Y., Kim, S., et al. (2010). Dissection of SNARE-driven membrane fusion and neuroexocytosis by wedging small hydrophobic molecules into the SNARE zipper. *Proceedings of the National Academy of Sciences of the United States of America, 107*, 22145–22150.

Yoon, T. Y., Lu, X., Diao, J., Lee, S. M., Ha, T., & Shin, Y. K. (2008). Complexin and Ca^{2+} stimulate SNARE-mediated membrane fusion. *Nature Structural & Molecular Biology, 15*, 707–713.

Yoon, T. Y., Okumus, B., Zhang, F., Shin, Y. K., & Ha, T. (2006). Multiple intermediates in SNARE-induced membrane fusion. *Proceedings of the National Academy of Sciences of the United States of America, 103*, 19731–19736.

Zampighi, G. A., Zampighi, L. M., Fain, N., Lanzavecchia, S., Simon, S. A., & Wright, E. M. (2006). Conical electron tomography of a chemical synapse: Vesicles docked to the active zone are hemi-fused. *Biophysical Journal, 91*, 2910–2918.

Zhang, T., & Johansson, J. S. (2003). An isothermal titration calorimetry study on the binding of four volatile general anesthetics to the hydrophobic core of a four-alpha-helix bundle protein. *Biophysical Journal, 85*, 3279–3285.

Zhang, T., & Johansson, J. S. (2005). A calorimetric study on the binding of six general anesthetics to the hydrophobic core of a model protein. *Biophysical Chemistry, 113*, 169–174.

CHAPTER 8

Inferring Structures of Kinetic Intermediates in Ca^{2+}-Triggered Exocytosis

Meyer B. Jackson
Department of Neuroscience, University of Wisconsin School of Medicine and Public Health, Madison, Wisconsin, USA

I. OVERVIEW

Fusion of biological membranes proceeds through a series of intermediates that are generated by structural transitions in proteins and deformation of lipid bilayers. In membrane fusion during Ca^{2+}-triggered exocytosis the key proteins have been identified and subjected to structural analysis. These proteins have been used to develop hypothetical mechanisms for exocytosis, often invoking explicit structures for key kinetic intermediates. The structures of these intermediates posit unique relations between proteins and lipids. These models of fusion intermediates have helped guide the design of experiments to elucidate the mechanism of exocytosis. This chapter evaluates efforts to probe the

Current Topics in Membranes, Volume 68 0065-230X/10 $35.00

DOI: 10.1016/B978-0-12-385891-7.00008-8

structures of kinetic intermediates of exocytosis, and reviews the experiments that have tested competing models.

II. INTRODUCTION

The fusion of two membranes requires the distortion and controlled disruption of lipid bilayers. After being brought very close together, two lipid bilayer membranes are strained to the point where they can relax to a new bilayer structure in which the previously separate membranes have become contiguous. Elucidating the mechanism of membrane fusion is in essence an effort to characterize the distortions and to search for structures that can serve as kinetic intermediates. These intermediates of membrane fusion have proven elusive. A considerable amount of theoretical work has explored the membrane fusion energy landscape, but experimental approaches have provided only limited glimpses of the fleeting intermediates, and limited tests for the theories. This chapter reviews work on Ca^{2+}-triggered exocytosis with the goal of gaining insight into the mechanism of fusion. Here, we will examine the structures of putative intermediates and evaluate the evidence for and against them.

The search for structures of intermediates in exocytosis began with the classical work in electron microscopy. Pictures of samples frozen a few milliseconds after triggering exocytosis revealed so-called omega figures in which the still round vesicle membrane was joined with the plasma membrane. These studies found that the pores connecting the vesicle lumen to the extracellular space were never less than 20 nm in diameter (Heuser & Reese, 1981). Because the membranes were already fused and the size of the pore much greater than molecular dimensions, these omega figures cannot be regarded as true fusion intermediates. They were too late. As much as we would like to see the intermediates of membrane fusion in an electron microscope, there have been no significant advances along these lines since the original work demonstrating membrane fusion. Thus, while the pictures of bona fide fusion intermediates have yet to be taken with the electron microscope, these early studies defined the problem and helped frame the question. What do the membranes look like as they first contact one another, and what structures appear en route to the omega figures seen in the electron microscope?

The studies of exocytosis with the electron microscope created the hope that technical advances, perhaps with another structural technique such as super-resolution microscopy, will bring these structures into view. However, the nature of kinetic intermediates makes them especially difficult to probe by direct methods. They generally have a high energy and a short lifetime. Whether the major problems of exocytosis will be solved by directly detecting the kinetic intermediates or by indirect measurements based on kinetic analysis remains an open question. The present discussion will have its greatest value in the design of kinetic experiments to test competing models. On the other hand, a major advance in direct observation may render the speculations offered here obsolete and irrelevant.

III. HYPOTHETICAL INTERMEDIATES

The major intermediate structures invoked in studies of exocytosis are depicted in Fig. 1. They include a fusion dimple, fusion stalk, hemifusion diaphragm, gap junction-like channel (closed and open), and lipidic fusion pore. Fusion dimples form as two bilayers are drawn together in a highly localized contact. One or both of bilayers can protrude in a dimple while the bilayers remain intact. Membrane bending will increase their energy. The positive curvature of the opposing monolayers[1] will increase the exposure of the hydrocarbon core to water, and provide limited surfaces that could serve to nucleate a contact between the hydrocarbon cores of each monolayer. Such a contact could become a fusion stalk (Kozlov, Leikin, Chernomordik, Markin, & Chizmadzhev, 1989). A fusion stalk joins the two proximal monolayers of each bilayer into one contiguous monolayer. In this nonbilayer structure the very high curvature in the connecting region of the merged monolayer will result in high bending energy. Hydrocarbon chains will tilt off the normal to the monolayer surface. Small high-energy cavities or "voids" can form where the hydrocarbon chains cannot pack very well. Filling these voids with small hydrophobic

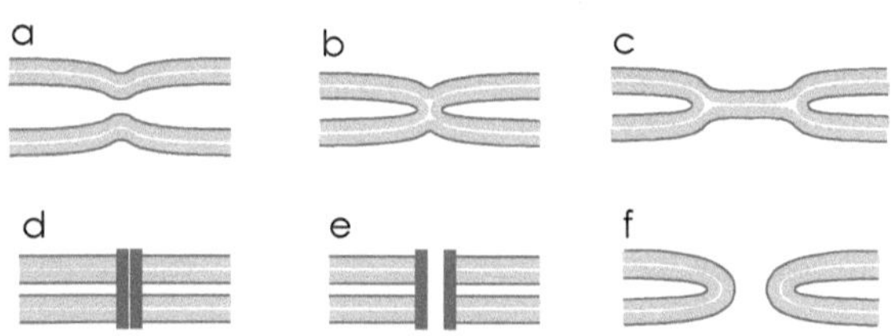

FIGURE 1 Structures invoked in models of exocytosis. (a) *Fusion dimple*: Protrusions of two bilayers toward one another in a very early stage of fusion. (b) *Fusion stalk*: The two proximal monolayers of each bilayer merge in a very narrow region. (c) *Hemifusion diaphragm*: The stalk contact between the proximal monolayers enlarges to form a bilayer. (d and e) *Gap junction-like fusion pore*: Proteins span both plasma and vesicle membranes to form a channel. When the channel is open it connects the vesicle lumen with the extracellular space. (f) *The lipidic fusion pore*: At the earliest instant that two bilayers become one and a narrow lipidic pore forms. This is the earliest embodiment of the omega figures seen in the electron microscope.

[1]Positive monolayer curvature is defined as curvature in which the headgroups spread out relative to the hydrocarbon changes. Negative monolayer curvature spreads out the hydrocarbons relative to the headgroups.

molecules will reduce their energies considerably (Siegel, 1999). The hemifusion diaphragm will form from a fusion stalk as the curved region of the merged monolayer within the stalk spreads out to leave the two distal monolayers in direct contact in a new bilayer. The proximal monolayers are contiguous as they are in the stalk and form a widening circular zone of high curvature. The widening of this circle imposes tension on the bilayer forming the central diaphragm, and this tension could drive the formation of a lipid lined pore (Kozlov et al., 1989).

Gap junctions formed between pairs of cells consist of proteinaceous channels that span two plasma membranes and connect the cytoplasm of the two cells. The proteins that drive exocytosis could form a similar structure, initially appearing as a closed channel and then undergoing a conformational transition to form an open channel. Models invoking this type of structure draw from the extensive literature on ion channel gating in membrane excitability.

Dimples, stalks, hemifusion diaphragms, and lipidic fusion pores contain stretches of lipid monolayer that are highly curved. The membrane curvature in these structures is energetically unfavorable, and the role of membrane bending energetics in the stability of these structures will be discussed below. The gap junction-like channel can form without significant bending of membranes. However, the formation of such a multiprotein complex could pull the membranes together and dimple them. Thus, a gap junction-like channel could be a hybrid if it incorporates features of a fusion dimple.

Fig. 1f depicts the earliest stage where the two bilayers become one, a lipidic fusion pore. Indeed, the omega figures seen in the electron microscope could be regarded as an expanded lipidic fusion pore, but the earliest such pore formed during fusion must be much narrower. The lipidic fusion pore is an essential state; any model of membrane fusion must create such a structure. It is tempting to view it as the endpoint of membrane fusion. However, further expansion of the lipidic fusion pore cannot be taken for granted. This metastable structure can close off in a reversal of fusion (Chanturiya, Chernomordik, & Zimmerberg, 1997; Razinkov, Melikyan, Epand, Epand, & Cohen, 1998). Theoretical work suggests that lipidic fusion pores lie in a minimum in the energy surface (Jackson, 2009; Katsov, Muller, & Schick, 2004). Work will then be required to drive pore expansion and bring the newly joined bilayers into a flat planar state (Chizmadzhev, Cohen, Shcherbakov, & Zimmerberg, 1995; Cohen & Melikyan, 2004).

IV. LIPID CONTACT

The hypothetical intermediates depicted in Fig. 1 make predictions about pathways for flux of the fluid membrane and the aqueous solution between compartments. In the stalk and hemifusion diaphragm only lipid in the proximal leaflets can flow between the two membranes. There will be no aqueous solution flow and no distal bilayer flow. In the lipidic fusion pore aqueous content can flow as well as lipid in both the proximal and distal leaflets. Thus, experimental measurements of aqueous flux and lipid flux, and especially measurements of lipid flux that resolve flow of each leaflet, offer tests for the different intermediates. The present and following sections assess experimental work on these two issues.

Lipid flux can be monitored using lipophilic fluorescent dyes, and chemical destruction of the outer leaflet fluorophores of a vesicle leaves label only on the inner leaflet. In liposomes reconstituted with neuronal SNARE proteins[2], a comparison of fusion kinetics of fully labeled versus inner labeled liposomes showed that the proximal leaflets mixed before the distal leaflets (Lu, Zhang, McNew, & Shin, 2005). In liposomes reconstituted with yeast SNAREs, inner leaflet lipid mixing closely followed total lipid mixing when the protein-to-lipid ratio was high, but when the ratio was reduced the inner leaflet mixing efficiency fell below the total mixing efficiency. Thus, vesicles containing fewer SNAREs will form hemifused states that do not continue on to full fusion (Xu, Zhang, Su, McNew, & Shin, 2005). Total internal reflection fluorescence microscopy of SNARE-mediated fusion of single liposomes with planar bilayers showed that fusion proceeded in two distinct temporal phases (Liu, Wang, Chapman, & Weisshaar, 2008). These experiments make a strong case for the appearance of hemifusion intermediates during SNARE-mediated fusion of artificial lipid vesicles. We must presume that these hemifusion intermediates arose by expanding stalks, but the stalk cannot be detected in these assays because it is too narrow to allow a significant amount of dye flux.

In the SNARE-mediated fusion of yeast vacuoles SNARE pairing occurred very rapidly, followed next by lipid mixing and finally by aqueous content mixing. The content mixing was much slower than the lipid mixing (Jun & Wickner, 2007). Although the mixing of the inner and outer leaflets was not resolved, this study indicates that a hemifusion intermediate allows lipids to mix well before the formation of an aqueous pore. However, the content mixing assay used in this study depends on passage of large protein molecules through the pore. If a pore structure such as shown in Fig. 1f forms and is too small to allow passage of protein, then the delay in content mixing may actually reflect the time needed for pore expansion. In a fusion assay using cells expressing "flipped" SNAREs, a significant amount of lipid mixing occurred without content mixing (Giraudo et al., 2005); in this study the hemifusion state was a dead end that failed to proceed to complete fusion.

[2]SNAREs (soluble *N*-ethylmaleimide sensitive factor attachment protein receptors) are discussed in more detail in Section VIII.B.1. Because SNARE proteins are essential for Ca^{2+}-triggered exocytosis, work with these proteins in model systems is relevant to the present chapter.

Capacitance recording from mast cells showed that a vesicle's area becomes smaller during transient kiss-and-run fusion events, indicating vesicle membrane flux (Monck, Alvarez de Toledo, & Fernandez, 1990). In exocytosis in PC12 cells loss of a membrane dye from vesicles was reported for kiss-and-run events (Taraska & Almers, 2004). However, such flux was not seen in similar capacitance recordings during transient kiss-and-run events in nerve terminals (Klyachko & Jackson, 2002). In general, observations of lipid flux suggest that a transient lipidic pore forms in concert with the access of vesicle content to the extracellular fluid. These results are consistent with the formation of a lipidic fusion pore that can close off after pore formation, but these studies do not establish lipidic pore formation as concomitant with aqueous contact, and so it remains unclear whether the lipidic fusion pore forms from a hemifusion diaphragm or a gap junction-like channel. An electron microscopic study reported synaptic vesicles in intimate contact with the plasma membrane, and these contacts resembled stalks and hemifusion diaphragms (Zampighi et al., 2006). This study was conducted on heavily fixed brain tissue, and it will be interesting to see if such figures turn up in samples prepared under milder conditions with improved membrane preservation.

V. FLUID CONTACT

Fluid contact can be monitored with exceptional sensitivity by either amperometry, which measures the flux of an oxidizable substance from a single vesicle (Chow & von Rüden, 1995; Wightman et al., 1991), or complex impedance, which measures the ionic conductance of the fusion pore (Debus & Lindau, 2000). The conductance measurement offers a crude method of estimating the pore diameter: one assumes a cylindrical geometry and calculates the conductance as the pore area divided by saline resistivity and length

$$\gamma = \frac{A}{\rho l} \tag{1}$$

Assuming a length l approximately equal to the thickness of two lipid bilayers and a saline resistivity of 100 Ω-cm allows one to calculate the area, A. The pore radius can then be determined by assuming the pore is circular. For ion channels this approximation is within a factor of 2 or 3 for larger channels; the conductance is generally lower than the value obtained from Eq. 1, so the radius determined in this way can be regarded as a lower bound (Jackson, 2006). This calculation is appropriate for a gap junction-like channel, which is known to have a roughly cylindrical shape. Lipidic fusion pores are not cylindrical (Fig. 1f). This effectively shortens their length so that the conductance for a given area will be greater.

Fusion pore conductances in Ca^{2+}-triggered exocytosis vary over a very wide range, from 19 to 330 pS (Klyachko & Jackson, 2002; Lindau & Almers, 1995;

Lindau & Alvarez de Toledo, 2003). The wide range of diameters implied by this diversity of structures is difficult to reconcile with a lipidic structure, where the dimensions should reflect the structure of the metastable state that minimizes the elastic bending energy (Jackson, 2009). Although the dimensions of an energy-minimized fusion pore may vary with lipid composition, the finding of a threefold difference between the estimated fusion pore diameters for large dense-core vesicles and small microvesicles in the same nerve terminal (Klyachko & Jackson, 2002) makes this an unlikely explanation for the wide disparity in conductances. Differences in the protein composition of a gap junction-like channel, or a difference in the number of SNARE complexes (Han, Wang, T., Bai, Chapman, & Jackson, 2004; Z. Zhang, Zhang, & Jackson, 2010), offer a more plausible explanation.

For the 19 pS conductance of microvesicles in nerve terminals, a radius of 0.3 Å was estimated (Klyachko & Jackson, 2002). By comparison, the conductances of lipidic fusion pores formed during fusion of lipids (Chanturiya et al., 1997) or during viral fusion (Razinkov et al., 1998), are on the order of hundreds of pS. The diameters of these pores from Eq. 1 are consistent with theoretical estimates based on minimizing membrane bending energy (Jackson, 2009) and quite a bit larger than the smaller pores of microvesicles in nerve terminals. This analysis thus favors a gap junction-like fusion pore, at least for the small pores of microvesicle and synaptic vesicle exocytosis (Klyachko & Jackson, 2002; He, Wu, Mohan, & Wu, 2006).

VI. HYDROCARBON–WATER CONTACT

Although not depicted in Fig. 1, the disruption of a lipid bilayer could entail the exposure of the hydrocarbon tails to water. This will be strongly opposed by the hydrophobic interaction. The advantage of the stalk mechanism is that membranes can fuse while incurring hardly any hydrocarbon–water exposure (Chernomordik & Kozlov, 2003; Kozlov et al., 1989). There is no experimental work assessing exposure of the lipid tails to water during any form of membrane fusion. However, the pore lining residues of ion channels are quite hydrophobic (Doyle et al., 1998).

In studies of membrane fusion the feasibility of such contacts rests on theoretical estimates of the energy. To calculate the energy one estimates the area of a putative interface and multiplies by an empirical parameter determined from experimental measurements. Measurements of interfacial surface tension yield a value of roughly 50 erg/cm^2, and calculations using this factor yielded high energies in excess of 100 kT for hydrophobic pores through membranes (Chernomordik & Kozlov, 2003). However, from measurements of free energies of transfer of molecules between water and nonpolar solvents one obtains a value of roughly 25 cal/(mol Å^2) (Tanford, 1980), which works out to 17.5 erg/cm^2.

Thus, the thermodynamic measurements give a number that is about one-third of the interfacial surface tension value. This serious discrepancy was pointed out by Tanford (1979), and lacks a satisfactory explanation. The smaller thermodynamic factor was recently used in calculations of the energy cost of exposing lipid hydrocarbon tails to water during the expansion of a gap junction-like fusion pore formed from SNARE protein membrane anchors. The resulting energies for fusion intermediates were significant but surmountable (Jackson, 2010).

VII. MEMBRANE BENDING

A. Theory

The high membrane curvature of some of the putative intermediates in Fig. 1 will increase the energy, thus presenting a kinetic barrier to processes that depend on membrane bending. The energetics of membrane bending is an important part of the energetic landscape of fusion. It is difficult to study membrane elasticity experimentally on such microscopic scales, but theoretical methods exist for evaluating the cost of creating such structures. This analysis starts with a fundamental expression relating energy to the curvature of a surface (Boal, 2002; Helfrich, 1973).

$$E_b = \tfrac{1}{2}\kappa_m \int_S \left(\frac{1}{R_1} + \frac{1}{R_2} - C_0\right)^2 dS + \kappa_g \int_S \frac{1}{R_1 R_2} dS \qquad (2)$$

This expression treats the membrane as an elastic object in which strains in the form of bending meet with a linear restoring force. This continuum model ignores molecular detail. R_1 and R_2 are the two principle radii of curvature. The first integral represents the contribution of mean curvature and the second represents the contribution of Gaussian curvature; κ_m and κ_g are the moduli or flexural rigidities that relate these curvatures to energy. C_0 is the spontaneous curvature of the membrane. This equation provides a general strategy for evaluating hypothetical structures with curved membrane. A chosen shape can be evaluated by integrating over the surface, S. A shape with a lower energy computed from Eq. 2 will be favored. Subject to specific geometric constraints, a surface will assume a shape that minimizes Eq. 2.

Eq. 2 has been used to estimate the energies of some of the fusion intermediates depicted in Fig. 1 (Chizmadzhev et al., 1995; Jackson, 2009; Kozlov & Chernomordik, 1998; Kozlov et al., 1989; Kozlovsky & Kozlov, 2002; Markin & Albanesi, 2002; Siegel, 1993, 1999). Early studies focused on mean curvature and assumed that the curved regions within stalks, hemifusion diaphragms, and lipidic pores have a toroidal structure formed as a surface of revolution of a circular contour rotated around a vertical axis (Kozlov et al., 1989). A toroidal fusion pore will have a membrane bending energy of ~75 kT (Chizmadzhev et al., 1995), and such a high energy raises the question of how it can form.

The incorporation of additional features into theoretical models has led to major revisions in the estimates of the relevant energies. Hydrocarbon chain tilt plays an important role in reducing the energy of hemifusion diaphragms and stalks (Kozlovsky & Kozlov, 2002; Kuzmin, Zimmerberg, Chizmadzhev, & Cohen, 2001). Furthermore, in lipidic fusion pores each monolayer of the bilayer must be treated as a separate surface with the energy calculated as a sum of two mean curvature integrals of the form of Eq. 2. Finally, a toroid is not the shape with the lowest energy (Markin & Albanesi, 2002). A surface of revolution with zero mean curvature can be generated with a hyperbolic cosine function (a catenary) (Kreyszig, 1991). A catenoid surface generated in this way will incur no increase in bending energy from mean curvature (Kozlov & Chernomordik, 1998; Siegel, 2006, 2008). However, in a lipidic fusion pore the two monolayers are coupled elastic surfaces and they cannot be both catenoids with a fixed spacing. When the elastic bending energy was minimized for an arbitrary contour, nonzero energies were achieved that were much lower than that of a toroid, but still quite substantial (Jackson, 2009). Fig. 2a illustrates a nontoroidal lipidic fusion pore with a shape that minimizes the mean curvature. This continuum elasticity approach generates fusion pores as metastable states (Fig. 2b). The energy surface formed by varying pore length and diameter has a minimum for specified dimensions and the radius

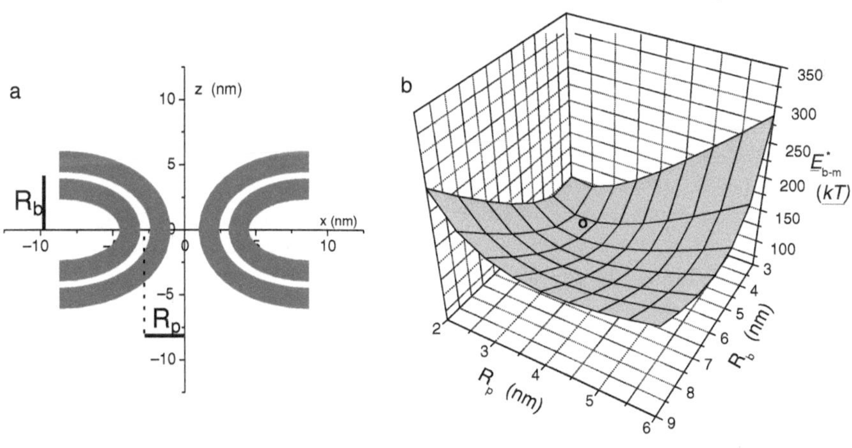

FIGURE 2 Lipidic fusion pore with minimized membrane bending energy (from Jackson, 2009). (a) The shape of a fusion pore determined by minimization of the elastic bending energy (the mean curvature of Eq. 2). (b) The energy surface of a lipidic fusion pore with energy plotted as a function of pore radius, R_p, and pore length, R_b (see figure (a)). These calculations were performed for a bilayer with a distance of 2.1 nm between the neutral surfaces of the two monolayers (corresponding to a C-18 lipid). $C_0 = -0.1\ nm^{-1}$ for both monolayers. The **O** marks the energy minimum at $R_p = 2.75$ nm and $R_b = 4.2$ nm.

is in reasonable agreement with experimental estimates of pore size derived from conductance measurements.

An important issue in the usefulness of this approach is the validity of Eq. 2 at very high curvature. A molecular theory for lipid monolayer flexibility suggested that energy is proportional to the square of the mean curvature over an appreciable range (Szliefer et al., 1990). A study of energy minimized models of lipidic fusion pores at the atomistic, course-grained, and continuum levels yielded shapes that were quite similar (Yoo, Jackson, & Cui, in preparation). Thus, the contours obtained from continuum models may provide a reasonable view of the shape of a lipidic fusion pore.

The discussion thus far only addressed the energy associated with the mean curvature. The Gaussian curvature (in the second integral in Eq. 2) can be neglected in many situations because it has the interesting mathematical property of being invariant to changes in shape that do not involve separating or combining surfaces (Kreyszig, 1991). However, the formation of stalks and hemifusion diaphragms combine two monolayers into one, and the formation of a lipidic fusion pore combines four into two. The formation of these structures thus changes the Gaussian curvature by 4π or 8π, respectively (Siegel & Kozlov, 2004). The contribution of Gaussian curvature does not depend on the detailed shape so the shapes determined by minimization of the mean curvature are valid regardless of the contribution of Gaussian curvature. The modulus of mean curvature, κ_m, has been measured for many membranes but the modulus of Gaussian curvature κ_g is much more difficult to determine (Marsh, 2006). One estimate has lead to the conclusion that combining two monolayers into one would require an energy in excess of 100 kT; thus, formation of a lipidic fusion pore would require 200 kT (Siegel, 2008). More experimental work on Gaussian curvature is clearly needed. This need is underscored by theoretical work on lipid bilayers that incorporate molecular detail. A mean-field theory of lipid bilayers indicated that the energy of formation of a lipidic fusion pore is much lower (Katsov et al., 2004). Closing the large gap between these very different approaches may help illuminate the roles of both mean curvature and Gaussian curvature in the membrane fusion energy landscape.

B. Experimental Tests of Membrane Bending Energy

Most experimental efforts to assess the role of membrane bending in exocytosis have employed cone-shaped lipids that pack better into curved lipid bilayers. These lipids produce a change in C_0. In general, making C_0 more negative promotes fusion, and manipulations of C_0 in viral fusion supported a mechanism in which fusion proceeds through a hemifusion intermediate, progressing to an initial fusion pore that is lipidic (Chernomordik & Kozlov,

2003, 2008). However, negative C_0 stabilizes both hemifusion diaphragms and lipidic fusion pores. So the acceleration of exocytosis does not provide a critical test for models of exocytosis. The alterations in fusion rate by perturbing C_0 with cone-shaped or inverted-cone-shaped lipids such as lysophosphatidylcholine (LPC) and arachidonic acid indicate that membrane bending occurs, but do not specify a kinetic step (Amatore et al., 2006; Lam, Tryoen-Toth, Tsai, Vitale, & Stuenkel, 2008). The same argument applies to the result that the addition of cholesterol and other lipids with a negative spontaneous curvature enhance the exocytosis of cortical granules (Churchward et al., 2008).

Kinetic steps in exocytosis can be resolved by amperometry recording from single cells (Chow & von Rüden, 1995; C. T. Wang, Bai, Chang, Chapman, & Jackson, 2006). With this technique the exocytosis of a single vesicle appears as a spike of rapid release, preceded by a foot that reports a small trickle of norepinephrine flux through the initial fusion pore (Chow, von Rüden, & Neher, 1992; C. T. Wang et al., 2001). In contrast to the results with cortical granules (Churchward et al., 2008), experiments in chromaffin cells showed that cholesterol failed to alter the rate of fusion events but did alter the stability of prespike feet (N. Wang et al., 2010).

More detailed analysis of amperometry data provided a critical test of the role of membrane bending. With this technique the spike area provides a measure of the amount of norepinephrine in a vesicle, and larger vesicles contain more norepinephrine. So a large amperometric spike indicates the fusion of a large vesicle (Gong, Hafez, Alvarez de Toledo, & Lindau, 2003). The duration of the prespike foot, denoted as τ, indicates the stability of the initial fusion pore. Amperometric recordings indicated that larger vesicles formed more stable initial fusion pores (Z. Zhang & Jackson, 2010). Since fusion pore lifetime is related to the inverse of the rate with which initial fusion pores progress to the next kinetic state, and rates vary exponentially with energy, the quantity $\ln(1/\tau)$ provides a measure related to the energy of the kinetic transition. The number of norepinephrine molecules in a vesicle is related to vesicle volume, and the cube root of the area of an amperometric event gives a linear dimension of the vesicle. A plot of $\ln(1/\tau)$ versus $Q^{1/3}$ was well fitted by a line (Fig. 3a), indicating that membrane curvature makes an energetic contribution to fusion pore kinetics.

A theoretical analysis of the curvature of a vesicle that has formed a lipidic fusion pore yielded the following relation

$$\ln(1/\tau) = A + \frac{B(R_p + R_h)^2}{R_v}\left(\frac{1}{R_r} - 3C_0 + R_r C_0^2\right) \qquad (3)$$

A and B are geometry independent factors, R_p is the pore radius, R_h and R_r have values close to R_p, and R_v is the vesicle radius (Fig. 3b; Z. Zhang & Jackson, 2010). This equation recapitulated the experimentally observed relation

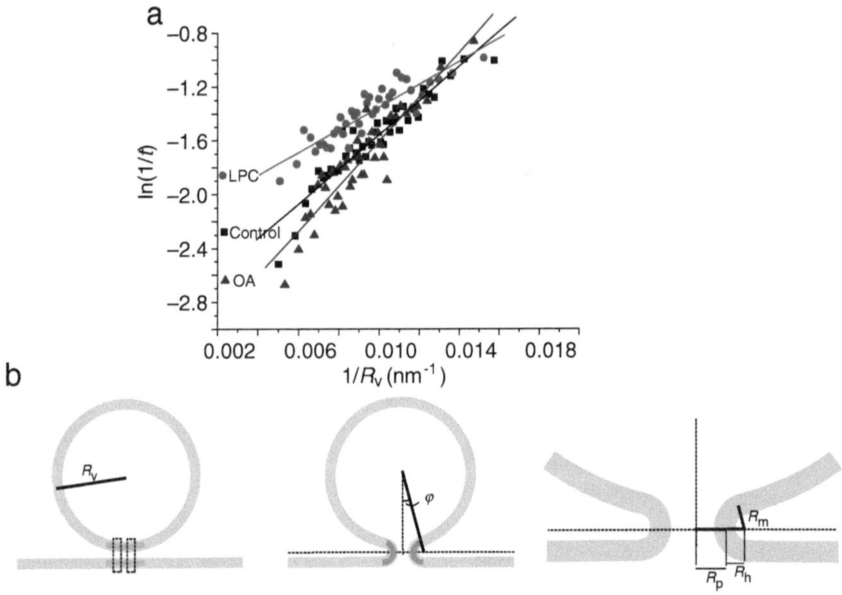

FIGURE 3 (a) Linear relationship between $\ln(1/\tau)$ and $1/R_v$ for exocytosis of norepinephrine from chromaffin cells measured with amperometry (from Z. Zhang & Jackson, 2010). Addition to the extracellular solution of lysophosphatidylcholine (LPC, with positive C_0) and oleic acid (OA, with negative C_0) altered the slopes, as predicted by Eq. 3. (b) A structural transition for which Eq. 3 provides the change in membrane bending energy. (See Color Insert.)

between vesicle size and fusion pore stability. Eq. 3 indicates that the slope of a $\ln(1/\tau)$ versus $Q^{1/3}$ plot depends on spontaneous curvature. When C_0 was altered by the addition of cone shaped lipids, the slopes of these plots did indeed change in the expected direction (Fig. 3a). These results on the stability of the initial fusion pore indicate that significant amounts of membrane bending occur only after the initial fusion pore has appeared. The initial fusion pore is more stable when a vesicle is large because a transition from the initial fusion pore state entails forming a lipidic fusion pore containing more curved membrane. This study thus places the structure depicted in Fig. 1f after the state that gives rise to the prespike foot seen in amperometry recording. This prespike foot state thus has very little curved membrane and is more likely to be a gap junction-like channel (Fig. 1e).

VIII. ROLE OF PROTEINS

Proteins are essential for the rapid and specific fusion of biological membranes. Although the number of proteins associated with synaptic vesicles is quite large (Takamori et al., 2006), the key players most clearly identified with the process of Ca^{2+}-triggered exocytosis are the synaptic SNARE proteins and synaptotagmin. We have quite a bit of structural information about these proteins but the challenge remains to identify these structures with fusion intermediates and explain how these proteins generate structures such as those depicted in Fig. 1.

A. Membrane Bending Proteins

Proteins have been proposed to deform membranes to promote fusion (Monck & Fernandez, 1994), and it is known that proteins that bind to membranes can shape them in a number of ways (Zimmerberg & Kozlov, 2006). Synaptotagmin I, an integral vesicle membrane protein that serves as the Ca^{2+} sensor for exocytosis (Augustine, 2001; Chapman, 2002; Koh & Bellen, 2003), penetrates membranes (Chapman & Davis, 1998), and induces curvature (Hui, Johnson, Yao, Dunning, & Chapman, 2009; Martens, Kozlov, & McMahon, 2007). Synaptotagmin I binds curved membranes more tightly than flat membranes, and this is another manifestation of the same basic phenomenon. This preference for curved membranes suggests that a synaptotagmin–membrane complex constitutes part of a fusion intermediate with curved membrane. Because synaptotagmin I induces positive curvature in the monolayer to which it binds, and because it only has access to the membrane leaflets facing the cytoplasm, membrane dimples (Fig. 1a) are plausible candidates for this complex. Another possibility is a hybrid dimple/gap junction state in which membranes curve around a proteinaceous fusion pore. Such a structure is consistent with the manner in which phosphatidylserine (a lipid required for synaptotagmin I membrane binding) stabilizes open fusion pores (Z. Zhang, Hui, Chapman, & Jackson, 2009). Furthermore, a kinetic analysis revealed two distinct kinetic steps accelerated by Ca^{2+}/synaptotagmin, one of which leads to fusion pore opening (C. T. Wang et al., 2006; Z. Zhang, Hui, Chapman, & Jackson, 2010). This early kinetic step is likely to entail membrane dimpling.

By contrast, lipidic fusion pores have the opposite sign of mean curvature to that preferred by synaptotagmin I. Thus, the synaptotagmin–membrane interaction is more likely to destabilize fusion pores rather than stabilize them. To stabilize a lipidic fusion pore synaptotagmin I would have to do more than simply induce negative curvature. Within a lipidic pore each element of membrane surface is saddle-like; the two principal curvatures are opposite in sign. With respect to Fig. 1f the curvature in the plane of the page (the meridianal curvature) is negative and the curvature in the plane of the membrane, perpendicular to the plane of the page (the parallel curvature) is positive. Meridianal curvature is greater than parallel curvature in a lipidic fusion pore (Jackson,

2009; Razinkov et al., 1998) so the mean curvature will be negative, opposite to that preferred by synaptotagmin. Finally, it is unlikely that synaptotagmin has a preference for bilayers with saddle-like curvature because it induces the formation of cylindrical tubules (Hui et al., 2009; Martens et al., 2007), with zero saddle-like curvature.

If one considers only the sign of mean curvature, the bending of membranes by synaptotagmin would place this interaction in dimples rather than stalks, hemifusion diaphragms, or lipidic fusion pores. However, this view may be an oversimplification. In order to understand how proteins drive membrane bending during fusion we will need to understand their membrane bending activity in greater detail. Synaptotagmin could stabilize the positive parallel curvature in the neck of a fusion pore, even though the net mean curvature is negative. In this regard it may be significant that an arginine-rich viral protein transduction domain can enter cells but a lysine-rich version cannot (Mitchell, Kim, Steinman, Fathman, & Rothbard, 2000). This has been attributed to the unique ability of arginine to form bidentate hydrogen bonds with phospholipid head groups and induce saddle-like curvature (Mishra, Gordon, Yang, Coridan, & Wong, 2008). Synaptotagmin has regions rich in arginine and lysine, and a strategic arrangement of these residues could enable proteins not only to bend membranes but shape them into forms that are ideal intermediates for fusion. *In vitro* studies of peptide membrane interactions combined with functional studies of the kinetics of exocytosis offer an approach to test this idea and probe the structures of fusion intermediates in greater detail.

B. Structural Roles of Proteins

1. SNAREs

The SNARE proteins constitute a large family with a wide range of functions in various forms of membrane trafficking. The synaptic SNAREs, synaptobrevin (VAMP), SNAP-25, and syntaxin, are essential for Ca^{2+}-triggered exocytosis (Jahn & Scheller, 2006; Schiavo, Matteoli, & Montecucco, 2000). Homologous domains within these proteins of $\sim$70 amino acids join together to form a complex consisting of 4 parallel α-helices. The crystal structure of this complex by Sutton, Fassauer, Jahn, and Brunger (1998) has had a major influence on thinking in this field. This structure is an essential element in most mechanisms that researchers have proposed for exocytosis, and has been invoked many times in models of membrane fusion. However, these models hinge on some critical unresolved issues. First of all, if fusion is mediated by more than one SNARE complex (i.e., a complex of SNARE complexes), then we need to know the number of SNARE complexes that come together during the fusion of one vesicle. Second, we need to know the relation between the degree of SNARE complex assembly and the progress of fusion.

Biochemical studies have shown that SNARE complexes form high molecular weight oligomers (Hayashi, Yamasaki, Nauenburg, Binz, & Niemann, 1995; Hayashi et al., 1994), which is consistent with cooperative work in cells. Furthermore, domain swapping has been reported between SNARE complexes (Kweon et al., 2002). An early estimate of the number of SNARE complexes required for fusion was based on the supralinear dependence on concentration of inhibition by a synaptobrevin fragment that prevents SNARE complex formation. This study yielded a value of 3 (Y. Hua & Scheller, 2001), which has been confirmed by a closely related experiment involving competition between functional and nonfunctional forms of SNAP-25 (Mohrmann, de Wit, Verhage, Neher, & Sorensen, 2010). The actions of clostridial toxins, which cleave synaptic SNARE proteins, suggest that the number could be much higher (Montecucco, Schiavo, & Pantano, 2005). These estimates based on the steepness of the dependence on SNARE concentration should be taken as lower bounds to the actual number of participating SNARE complexes.

Liposome fusion assays have also been used to estimate the number of SNARE complexes that drive a membrane fusion event. One study suggested that only one SNARE complex can perform this task (van den Bogaart et al., 2010), but these authors did not evaluate the rate of fusion as a function of the number of SNAREs per vesicle. Other studies in reconstituted systems have suggested that several are needed (5–11, Karatekin et al., 2010; 6–8, Domanska, Kiessling, Stein, Fasshauer, & Tamm, 2009). The actual number may depend on conditions; variations in the number of proteins per liposome lead to different results (Domanska, Kiessling, & Tamm, 2010). If multiple SNARE complexes are necessary for fusion then the most plausible arrangement would be with the 4-helix bundles parallel to the plane of the fusing membranes and radiating outward from a central focus like the spokes of a wheel. These proteins have membrane anchors (discussed below) and placing these motifs at the center of the focus would concentrate forces generated by the SNARE complexes to maximize their capacity to deform membranes (Jackson, 2010) (Fig. 4a). Indeed, electron microscopy has shown that SNARE complexes isolated from brain form star-like aggregates (Hohl et al., 1998; Rickman, Hu, Carroll, & Davletov, 2005).

The SNARE complex is extremely stable, so its formation will release a great deal of energy (Fasshauer, 2003). It is natural to expect that this energy provides a driving force for the deformation of membranes. FRET measurements have indicated that SNARE proteins form associations in cells, and that the spatial relations between these proteins change upon stimulation (An & Almers, 2004; Xia, Zhou, Lin, & Liu, 2001). The association of SNARE motifs *in vitro* during SNARE complex assembly proceeds in two stages (Fasshauer, Antonin, Subramaniam, & Jahn, 2002; Pobbati, Stein, & Fasshauer, 2006), with the membrane-distal sections associating first and the membrane proximal sections

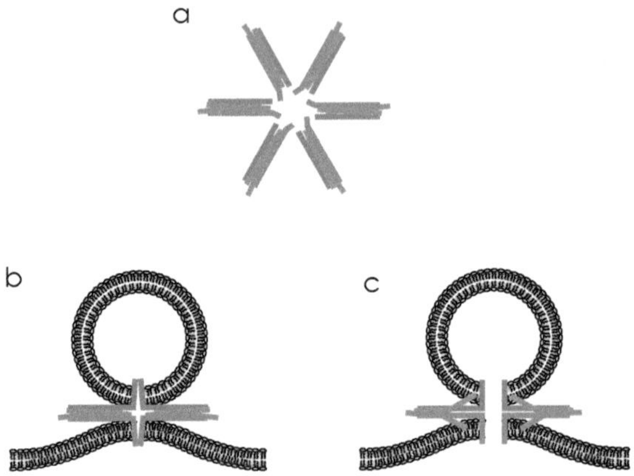

FIGURE 4 Fusion intermediates building on the SNARE proteins. (a) A star-like configuration of SNARE complexes, presumably surrounding a fusion pore. (b) A closed gap junction-like fusion pore formed by the SNARE membrane anchors. (c) An open gap junction-like fusion pore. *Note*: The dimpling of the plasma membrane toward the vesicle makes this a hybrid structure. (See Color Insert.)

associating later. A crystal structure has been reported that includes the membrane anchors as contiguous helical extensions of the SNARE motifs of the 4-helix bundle (Stein, Weber, Wahl, & Jahn, 2009). This structure can only form in a lipidic fusion pore because the SNARE protein membrane anchors of synaptobrevin (a vesicle protein) and syntaxin (a plasma membrane protein), reside in separate membranes prior to fusion. Thus, finding these membrane anchors together in intimate contact can only mean that this structure occurs after the merger of the two membranes. At the present time, there is no way of knowing if the original SNARE complex structure without the membrane anchors (Sutton et al., 1998) is simply a truncated version of the final postfusion structure (Stein et al., 2009).

The sequential assembly of the SNARE complex from the membrane distal N-terminal end toward the membrane proximal C-terminus has been proposed to drive fusion (Hanson, Roth, Morisaki, Jahn, & Heuser, 1997). In this zipping hypothesis each stage of assembly should correspond to a state of deformation of the vesicle and plasma membranes. Some data has supported the progressive zipping of the SNARE complex during fusion. One of the earliest studies was based on the actions of different clostridial toxins in blocking exocytosis (S. Y. Hua & Charlton, 1999). A use-dependence of the action of one of these toxins showed that a specific cleavage site in synaptobrevin was protected prior to the triggering of fusion. This indicated that the SNARE complex was at least partially preassembled prior to the triggering event. Mutagenesis studies of SNAP-25 (Sorensen et al., 2006) and synaptobrevin (Walter, Wiederhold,

Bruns, Fasshauer, & Sorensen, 2010) showed that destabilization of SNARE complexes near the membrane proximal C-terminus slowed the triggering of exocytosis whereas destabilization toward the N-terminus reduced an earlier priming step or had no effect at all. Thus, these results provided further support for SNARE complex zipping during fusion. By contrast, a number of complex destabilizing mutations in all three SNARE proteins were found to stabilize the initial fusion pore; these mutants were scattered throughout the SNARE complex (Han & Jackson, 2006). These results were interpreted with a model in which the SNARE complex is partially disassembled when the fusion pore first opens (Fig. 4c). Furthermore, mutations in the same layer but in different SNARE proteins differed markedly in their effects, and this may indicate that the structural relationships between these proteins during fusion may not match the crystal structure. The difficulty in reconciling much of the detailed kinetic work on exocytosis with zipping models calls to question not only the zipping hypothesis but also whether the SNARE complex crystal structure (Sutton et al., 1998), used in the interpretation of all the above-cited kinetic experiments, represents a structure that actually forms in cells. The SNARE proteins may form additional structures in which the effects of mutations on stability are difficult to predict. That SNARE complex transitions influence specific kinetic steps of exocytosis can be inferred from the effects of mutations in the synaptobrevin linker region that connects the membrane anchor with the SNARE motif (Kesavan, Borisovska, & Bruns, 2007). But the nature of these transitions remains unclear.

The membrane anchor of syntaxin spans the plasma membrane and the membrane anchor of synaptobrevin spans the vesicle membrane. Together, these proteins provide the structural equipment for a gap junction-like channel (Fig. 1e). Thus, the SNARE protein membrane anchors can form either a closed (Fig. 4b) or open (Fig. 4c) pore, and drawing the membranes together could dimple the plasma membrane to create a dimple/gap junction-like hybrid structure. This hypothesis has been tested for syntaxin by expressing membrane anchor mutations in PC12 cells and using amperometry to measure the flux of norepinephrine through the initial fusion pore, as well as by measuring the fusion pore conductance (Han et al., 2004). At a few unique locations the introduction of amino acids with large side chains such as tryptophan reduced the flux and conductance in parallel, while reducing the size of these side chains increased norepinephrine flux. Introducing positive charges reduced flux and introducing negative charges increased flux (Han & Jackson, 2005). Only residues falling on one face of an α-helical wheel had this effect. Residues within the membrane anchor, but on different faces, as well as residues in the SNARE motifs of all three SNARE proteins (Han & Jackson, 2006), failed to alter fusion pore flux. However, one mutation outside of the membrane anchor, a nine amino acids truncation of the C-terminus of SNAP-25, reduced the amplitude of norepinephrine flux through a fusion pore (Fang et al., 2008). This result was interpreted in terms of a fusion pore lined by both lipid and protein.

Based on a measurement of fusion pore conductance, the pore radius was estimated under the assumption of a simple cylindrical pore geometry (Eq. 1). The number of α-helices needed to line a cylinder of this size was estimated to be 5–8 (Han et al., 2004; Z. Zhang, Zhang, et al., 2010). These experiments thus provides another estimate of the number of SNARE complexes that cooperate in the triggering of fusion, as discussed above. The reduction in pore flux seen with C-terminal truncation (Fang et al., 2008) could indicate that the lower fusion pore flux and slower rates of exocytosis results from the formation of fusion pores from fewer SNARE complexes.

2. Synaptotagmin

Synaptotagmin 1 is thought to trigger exocytosis through its Ca^{2+}-dependent interactions with PS-containing liposomes and SNARE proteins (Chapman, 2002). The fact that synaptotagmin 1, its isoforms, and its mutants, influence the stability of initial fusion pores (Bai, Wang, Richards, Jackson, & Chapman, 2004; C. T. Wang et al., 2001, 2006; Z. Zhang, Hui, et al., 2010) suggests that it may interact directly with this fusion intermediate. This presumably reflects the Ca^{2+}-triggered interactions with SNAREs and PS, but dissecting the specific roles of these interactions has proven difficult and the issue is controversial (Bai et al., 2004; Lynch et al., 2007; Pang, Shin, Meyer, Rosenmund, & Sudhof, 2006; Z. Zhang, Hui, et al., 2010).

Given the hypothetical gap junction-like fusion pore formed by SNARE membrane anchors discussed above, one would hope to build synaptotagmin into these models by placing it at specific sites on the SNARE proteins. A mutagenesis study indicated that synaptotagmin 1 binds to the membrane proximal half of SNAP-25 (X. Zhang, Kim-Miller, Fukuda, Kowalchyk, & Martin, 2002). There are many synaptotagmin crystal and NMR structures (Fernandez et al., 2001; Sutton, Davletov, Berghuis, Sudhof, & Sprang, 1995; Ubach, Zhang, Shao, Sudhof, & Rizo, 1998) but no structures of synaptotagmin–SNARE complexes. Although crystals formed from synaptotagmin in the presence of SNAREs differ from crystals formed in the absence of SNAREs (Vrljic et al., 2010), SNAREs were absent from the actual crystals so the implications of this study for the sites of interaction between synaptotagmin and SNARE proteins are unclear.

IX. CONCLUSION

Research on exocytosis supports the existence of many of the hypothetical membrane fusion intermediates depicted in Fig. 1. It is interesting that evidence supports the formation of a hemifusion intermediate for fusion of liposomes by

SNARE proteins but during exocytosis in cells a hemifusion intermediate has not been directly established. Instead, exocytosis appears to employ a gap junction-like fusion pore with Ca^{2+}-triggered exocytosis in cells. Thus, SNARE proteins may be capable of fusing membranes by more than one mechanism. As long as we remain unable observe membrane fusion in microscopic detail and capture membranes in the act, we will have to resort to inferences based on functional studies and kinetics. The present contribution has illustrated some of the successes of this approach and offered a few ideas about where such studies might lead.

References

Amatore, C., Arbault, S., Bouret, Y., Guille, M., Lemaitre, F., & Verchier, Y. (2006). Regulation of exocytosis in chromaffin cells by trans-insertion of lysophosphatidylcholine and arachidonic acid into the outer leaflet of the cell membrane. *Chembiochem, 7*(12), 1998–2003.

An, S. J., & Almers, W. (2004). Tracking SNARE complex formation in live endocrine cells. *Science, 306*(5698), 1042–1046. doi: 306/5698/1042.

Augustine, G. J. (2001). How does calcium trigger neurotransmitter release? *Current Opinion in Neurobiology, 11*(3), 320–326.

Bai, J., Wang, C. T., Richards, D. A., Jackson, M. B., & Chapman, E. R. (2004). Fusion pore dynamics are regulated by synaptotagmin*t-SNARE interactions. *Neuron, 41*(6), 929–942.

Boal, D. (2002). *Mechanics of the cell.* Cambridge University Press, Cambridge.

Chanturiya, A., Chernomordik, L. V., & Zimmerberg, J. (1997). Flickering fusion pores comparable with initial exocytotic pores occur in protein-free phospholipid bilayers. *Proceedings of the National Academy of Sciences USA, 94*(26), 14423–14428.

Chapman, E. R. (2002). Synaptotagmin: A Ca(2+) sensor that triggers exocytosis? *Nature Reviews Molecular Cell Biology, 3*, 498–508.

Chapman, E. R., & Davis, A. (1998). Direct interaction of a Ca^{2+}-binding loop of synaptotagmin with lipid bilayers. *Journal of Biological Chemistry, 273*, 13995–14001.

Chernomordik, L. V., & Kozlov, M. M. (2003). Protein-lipid interplay in fusion and fission of biological membranes. *Annual Review of Biochemistry, 72*, 175–207.

Chernomordik, L. V., & Kozlov, M. M. (2008). Mechanics of membrane fusion. *Nature Structural & Molecular Biology, 15*(7), 675–683.

Chizmadzhev, Y. A., Cohen, F. S., Shcherbakov, A., & Zimmerberg, J. (1995). Membrane mechanics can account for fusion pore dilation in stages. *Biophysical Journal, 69*(6), 2489–2500.

Chow, R. H., & von Rüden, L. (1995). Electrochemical detection of secretion from single cells. *In* B. Sakmann & E. Neher, eds., Single-channel recording. pp. 245–275. Plenum Press, New York.

Chow, R. H., von Rüden, L., & Neher, E. (1992). Delay in vesicle fusion revealed by electrochemical monitoring of single secretory events in adrenal chromaffin cells. *Nature, 356*, 60–63.

Churchward, M. A., Rogasevskaia, T., Brandman, D. M., Khosravani, H., Nava, P., & Atkinson, J. K., et al. (2008). Specific lipids supply critical negative spontaneous curvature—An essential component of native Ca^{2+}-triggered membrane fusion. *Biophysical Journal, 94*(10), 3976–3986. doi: S0006-3495(08)70398-9.

Cohen, F. S., & Melikyan, G. B. (2004). The energetics of membrane fusion from binding, through hemifusion, pore formation, and pore enlargement. *Journal of Membrane Biology, 199*(1), 1–14.

Debus, K., & Lindau, M. (2000). Resolution of patch capacitance recordings and of fusion pore conductance in small vesicles. *Biophysical Journal, 78*, 2983–2997.

Domanska, M. K., Kiessling, V., Stein, A., Fasshauer, D., & Tamm, L. K. (2009). Single vesicle millisecond fusion kinetics reveals number of SNARE complexes optimal for fast SNARE-mediated membrane fusion. *Journal of Biological Chemistry, 284*(46), 32158–32166. doi: M109.047381.

Domanska, M. K., Kiessling, V., & Tamm, L. K. (2010). Docking and fast fusion of synaptobrevin vesicles depends on the lipid compositions of the vesicle and the acceptor SNARE complex-containing target membrane. *Biophysical Journal, 99*(9), 2936–2946. doi: S0006-3495(10)01112-4.

Doyle, D. A., Cabral, J. M., Pfuetzner, R. A., Kuo, A., Gulbis, J. M., & Cohen, S. L., et al. (1998). The structure of the potassium channel: Molecular basis of K^+ conduction and selectivity. *Science, 280*, 69–77.

Fang, Q., Berberian, K., Gong, L. W., Hafez, I., Sorensen, J. B., & Lindau, M. (2008). The role of the C terminus of the SNARE protein SNAP-25 in fusion pore opening and a model for fusion pore mechanics. *Proceedings of the National Academy of Sciences of the United States of America, 105*(40), 15388–15392. doi: 0805377105.

Fasshauer, D. (2003). Structural insights into the SNARE mechanism. *Biochimica et Biophysica Acta, 1641*(2–3), 87–97.

Fasshauer, D., Antonin, W., Subramaniam, V., & Jahn, R. (2002). SNARE assembly and disassembly exhibit a pronounced hysteresis. *Nature Structural Biology, 9*(2), 144–151. doi: 10.1038/nsb750.

Fernandez, I., Arac, D., Ubach, J., Gerber, S. H., Shin, O., & Gao, Y., et al. (2001). Three-dimensional structure of the synaptotagmin 1 C2B-domain: Synaptotagmin 1 as a phospholipid binding machine. *Neuron, 32*(6), 1057–1069. doi: S0896-6273(01)00548-7.

Giraudo, C. G., Hu, C., You, D., Slovic, A. M., Mosharov, E. V., & Sulzer, D., et al. (2005). SNAREs can promote complete fusion and hemifusion as alternative outcomes. *Journal of Cell Biology, 170*(2), 249–260.

Gong, L. W., Hafez, I., Alvarez de Toledo, G., & Lindau, M. (2003). Secretory vesicles membrane area is regulated in tandem with quantal size in chromaffin cells. *Journal of Neuroscience, 23*(21), 7917–7921.

Han, X., & Jackson, M. B. (2005). Electrostatic interactions between the syntaxin membrane anchor and neurotransmitter passing through the fusion pore. *Biophysical Journal, 88*(3), L20–L22.

Han, X., & Jackson, M. B. (2006). Structural transitions in the synaptic SNARE complex during Ca^{2+}-triggered exocytosis. *Journal of Cell Biology, 172*(2), 281–293.

Han, X., Wang, C. T., Bai, J., Chapman, E. R., & Jackson, M. B. (2004). Transmembrane segments of syntaxin line the fusion pore of Ca^{2+}-triggered exocytosis. *Science, 304*(5668), 289–292.

Hanson, P. I., Roth, R., Morisaki, H., Jahn, R., & Heuser, J. E. (1997). Structure and conformational changes in NSF and its membrane receptor complexes visualized by quick-freeze/deep-etch electron microscopy. *Cell, 90*(3), 523–535. doi: S0092-8674(00)80512-7.

Hayashi, T., McMahon, H., Yamasaki, S., Binz, T., Hata, Y., & Sudhof, T. C., et al. (1994). Synaptic vesicle membrane fusion complex: Action of clostridial neurotoxins on assembly. *EMBO Journal, 13*(21), 5051–5061.

Hayashi, T., Yamasaki, S., Nauenburg, S., Binz, T., & Niemann, H. (1995). Disassembly of the reconstituted synaptic vesicle membrane fusion complex in vitro. *EMBO Journal, 14*(10), 2317–2325.

He, L., Wu, X.-S., Mohan, R., & Wu, L.-G. (2006). Two modes of fusion pore opening revealed by cell-attached recordings at a synapse. *Nature, 444*(7115), 102–105. doi: 10.1038/nature05250.

Helfrich, W. (1973). Elastic properties of lipid bilayers: Theory and possible experiments. *Zeitschrift für Naturforschung [C], 28*(11), 693–703.

Heuser, J. E., & Reese, T. S. (1981). Structural changes after transmitter release at the frog neuromuscular junction. *Journal of Cell Biology, 88*, 564–580.

Hohl, T. M., Parlati, F., Wimmer, C., Rothman, J. E., Sollner, T. H., & Engelhardt, H. (1998). Arrangement of subunits in 20 S particles consisting of NSF. *SNAPs, and SNARE complexes. Molecular Cell, 2*(5), 539–548. doi: S1097-2765(00)80153-7.

Hua, S. Y., & Charlton, M. P. (1999). Activity-dependent changes in partial VAMP complexes during neurotransmitter release. *Nature Neuroscience, 2*(12), 1078–1083. doi: 10.1038/16005.

Hua, Y., & Scheller, R. H. (2001). Three SNARE complexes cooperate to mediate membrane fusion. *Proceedings of the National Academy of Sciences, 98*, 8065–8070.

Hui, E., Johnson, C. P., Yao, J., Dunning, F. M., & Chapman, E. R. (2009). Synaptotagmin-mediated bending of the target membrane is a critical step in Ca(2+)-regulated fusion. *Cell, 138*(4), 709–721. doi: S0092-8674(09)00654-0.

Jackson, M. B. (2006). Molecular and cellular biophysics. Cambridge University Press, Cambridge.

Jackson, M. B. (2009). Minimum membrane bending energies of fusion pores. *Journal of Membrane Biology, 231*, 101–115. doi: 10.1007/s00232-009-9209-x.

Jackson, M. B. (2010). SNARE complex zipping as a driving force in the dilation of proteinaceous fusion pores. *Journal of Membrane Biology, 235*(2), 89–100. doi: 10.1007/s00232-010-9258-1.

Jahn, R., & Scheller, R. H. (2006). SNAREs—Engines for membrane fusion. *Nature Reviews Molecular Cell Biology, 7*(9), 631–643.

Jun, Y., & Wickner, W. (2007). Assays of vacuole fusion resolve the stages of docking, lipid mixing, and content mixing. *Proceedings of the National Academy of Sciences of the United States of America, 104*(32), 13010–13015.

Karatekin, E., Di Giovanni, J., Iborra, C., Coleman, J., O'Shaughnessy, B., & Seagar, M., et al. (2010). A fast, single-vesicle fusion assay mimics physiological SNARE requirements. *Proceedings of the National Academy of Sciences of the United States of America, 107*(8), 3517–3521. doi: 0914723107.

Katsov, K., Muller, M., & Schick, M. (2004). Field theoretic study of bilayer membrane fusion. *I. Hemifusion mechanism. Biophysical Journal, 87*(5), 3277–3290. doi: 10.1529/biophysj.103.038943.

Kesavan, J., Borisovska, M., & Bruns, D. (2007). v-SNARE actions during Ca(2+)-triggered exocytosis. *Cell, 131*(2), 351–363.

Klyachko, V. A., & Jackson, M. B. (2002). Capacitance steps and fusion pores of small and large-dense-core vesicles in nerve terminals. *Nature, 418*(6893), 89–92. doi: 10.1038/nature00852.

Koh, T. W., & Bellen, H. J. (2003). Synaptotagmin I, a Ca2+ sensor for neurotransmitter release. *Trends in Neurosciences, 26*(8), 413–422. doi: S0166223603001954.

Kozlov, M. M., & Chernomordik, L. V. (1998). A mechanism of protein-mediated fusion: Coupling between refolding of the influenza hemagglutinin and lipid rearrangements. *Biophysical Journal, 75*(3), 1384–1396. doi: S0006-3495(98)74056-1.

Kozlov, M. M., Leikin, S. L., Chernomordik, L. V., Markin, V. S., & Chizmadzhev, Y. A. (1989). Stalk mechanism of vesicle fusion. *Intermixing of aqueous contents. European Biophysics Journal, 17*(3), 121–129.

Kozlovsky, Y., & Kozlov, M. M. (2002). Stalk model of membrane fusion: Solution of energy crisis. *Biophysical Journal, 82*(2), 882–895.

Kreyszig, E. (1991). *Differential geometry.* Dover Publications, New York.

Kuzmin, P. I., Zimmerberg, J., Chizmadzhev, Y. A., & Cohen, F. S. (2001). A quantitative model for membrane fusion based on low-energy intermediates. *Proceedings of the National Academy of Sciences of the United States of America, 98*(13), 7235–7240.

Kweon, D. H., Chen, Y., Zhang, F., Poirier, M., Kim, C. S., & Shin, Y. K. (2002). Probing domain swapping for the neuronal SNARE complex with electron paramagnetic resonance. *Biochemistry, 41*(17), 5449–5452.

Lam, A. D., Tryoen-Toth, P., Tsai, B., Vitale, N., & Stuenkel, E. L. (2008). SNARE-catalyzed fusion events are regulated by syntaxin1A-lipid interactions. *Molecular Biology of the Cell, 19*(2), 485–497.

Lindau, M., & Almers, W. (1995). Structure and function of fusion pores in exocytosis and ectoplasmic membrane fusion. *Current Opinion in Cell Biology, 7*(4), 509–517.

Lindau, M., & Alvarez de Toledo, G. (2003). The fusion pore. *Biochimica et Biophysica Acta, 1641* (2–3), 167–173.

Liu, T., Wang, T., Chapman, E. R., & Weisshaar, J. C. (2008). Productive hemifusion intermediates in fast vesicle fusion driven by neuronal SNAREs. *Biophysical Journal, 94*(4), 1303–1314.

Lu, X., Zhang, F., McNew, J. A., & Shin, Y. K. (2005). Membrane fusion induced by neuronal SNAREs transits through hemifusion. *Journal of Biological Chemistry, 280*(34), 30538–30541. doi: M506862200.

Lynch, K. L., Gerona, R. R., Larsen, E. C., Marcia, R. F., Mitchell, J. C., & Martin, T. F. (2007). Synaptotagmin C2A loop 2 mediates Ca^{2+}-dependent SNARE interactions essential for Ca^{2+}-triggered vesicle exocytosis. *Molecular Biology of the Cell, 18*(12), 4957–4968 doi: E07-04-0368.

Markin, V. S., & Albanesi, J. P. (2002). Membrane fusion: Stalk model revisited. *Biophysical Journal, 82*(2), 693–712.

Marsh, D. (2006). Elastic curvature constants of lipid monolayers and bilayers. *Chemistry and Physics of Lipids, 144*(2), 146–159. doi: S0009-3084(06)00147-2.

Martens, S., Kozlov, M. M., & McMahon, H. T. (2007). How synaptotagmin promotes membrane fusion. *Science, 316*(5828), 1205–1208.

Mishra, A., Gordon, V. D., Yang, L., Coridan, R., & Wong, G. C. (2008). HIV TAT forms pores in membranes by inducing saddle-splay curvature: Potential role of bidentate hydrogen bonding. *Angewandte Chemie International Edition English, 47*(16), 2986–2989. doi: 10.1002/anie. 200704444.

Mitchell, D. J., Kim, D. T., Steinman, L., Fathman, C. G., & Rothbard, J. B. (2000). Polyarginine enters cells more efficiently than other polycationic homopolymers. *Journal of Peptide Research, 56*(5), 318–325.

Mohrmann, R., de Wit, H., Verhage, M., Neher, E., & Sorensen, J. B. (2010). Fast vesicle fusion in living cells requires at least three SNARE complexes. *Science, 330*(6003), 502–505. doi: science.1193134.

Monck, J. R., Alvarez de Toledo, G., & Fernandez, J. M. (1990). Tension in secretory granule membranes causes extensive membrane transfer through the exocytotic fusion pore. *Proceedings of the National Academy of Sciences of the United States of America, 87*(20), 7804–7808.

Monck, J. R., & Fernandez, J. M. (1994). The exocytotic fusion pore and neurotransmitter release. *Neuron, 12*(4), 707–716.

Montecucco, C., Schiavo, G., & Pantano, S. (2005). SNARE complexes and neuroexocytosis: How many, how close? *Trends in Biochemical Sciences, 30*(7), 367–372. doi: S0968-0004(05)00147-7.

Pang, Z. P., Shin, O. H., Meyer, A. C., Rosenmund, C., & Sudhof, T. C. (2006). A gain-of-function mutation in synaptotagmin-1 reveals a critical role of Ca^{2+}-dependent soluble *N*-ethylmaleimide-sensitive factor attachment protein receptor complex binding in synaptic exocytosis. *Journal of Neuroscience, 26*(48), 12556–12565 doi: 26/48/12556.

Pobbati, A. V., Stein, A., & Fasshauer, D. (2006). N- to C-terminal SNARE complex assembly promotes rapid membrane fusion. *Science, 313*(5787), 673–676. doi: 313/5787/673.

Razinkov, V. I., Melikyan, G. B., Epand, R. M., Epand, R. F., & Cohen, F. S. (1998). Effects of spontaneous bilayer curvature on influenza virus-mediated fusion pores. *Journal of General Physiology, 112*(4), 409–422.

Rickman, C., Hu, K., Carroll, J., & Davletov, B. (2005). Self-assembly of SNARE fusion proteins into star-shaped oligomers. *Biochemical Journal, 388*(Pt 1), 75–79. doi: BJ20041818.

Schiavo, G., Matteoli, M., & Montecucco, C. (2000). Neurotoxins affecting neuroexocytosis. *Physiological Reviews, 80*(2), 717–766.

Siegel, D. P. (1993). Energetics of intermediates in membrane fusion: Comparison of stalk and inverted micellar intermediate mechanisms. *Biophysical Journal, 65*(5), 2124–2140.
Siegel, D. P. (1999). The modified stalk mechanism of lamellar/inverted phase transitions and its implications for membrane fusion. *Biophysical Journal, 76*(1 Pt 1), 291–313. doi: S0006-3495 (99)77197-3.
Siegel, D. P. (2006). Determining the ratio of the Gaussian curvature and bending elastic moduli of phospholipids from Q(II) phase unit cell dimensions. *Biophysical Journal, 91*(2), 608–618. doi: S0006-3495(06)71759-3.
Siegel, D. P. (2008). The Gaussian curvature elastic energy of intermediates in membrane fusion. *Biophysical Journal, 95*(11), 5200–5215. doi: S0006-3495(08)78946-X.
Siegel, D. P., & Kozlov, M. M. (2004). The Gaussian curvature elastic modulus of *N*-monomethylated dioleoylphosphatidylethanolamine: Relevance to membrane fusion and lipid phase behavior. *Biophysical Journal, 87*(1), 366–374. doi: 10.1529/biophysj.104.040782.
Sorensen, J. B., Wiederhold, K., Muller, E. M., Milosevic, I., Nagy, G., & de Groot, B. L., et al. (2006). Sequential N- to C-terminal SNARE complex assembly drives priming and fusion of secretory vesicles. *EMBO Journal, 25*(5), 955–966.
Stein, A., Weber, G., Wahl, M. C., & Jahn, R. (2009). Helical extension of the neuronal SNARE complex into the membrane. *Nature, 460*(7254), 525–528. doi: nature08156.
Sutton, R. B., Davletov, B. A., Berghuis, A. M., Sudhof, T. C., & Sprang, S. R. (1995). Structure of the first C2 domain of synaptotagmin I: A novel Ca^{2+}/phospholipid-binding fold. *Cell, 80*(6), 929–938.
Sutton, R. B., Fassauer, D., Jahn, R., & Brunger, A. T. (1998). Crystal structure of a SNARE complex involved in synaptic exocytosis at 2.4 Å resolution. *Nature, 395*, 347–353.
Szliefer, I., Kramer, D., Ben-Shaul, A., Gelbart, W. M., & Safran, S. A. (1990). Molecular theory of curvature elasticity in surfactant films. *Journal of Chemical Physics, 92*, 6800–6817.
Takamori, S., Holt, M., Stenius, K., Lemke, E. A., Gronborg, M., & Riedel, D., et al. (2006). Molecular anatomy of a trafficking organelle. *Cell, 127*(4), 831–846. doi: S0092-8674(06) 01400-0.
Tanford, C. (1979). Interfacial free energy and the hydrophobic effect. *Proceedings of the National Academy of Sciences of the United States of America, 76*(9), 4175–4176.
Tanford, C. (1980). *The hydrophobic effect*. John Wiley and Sons, New York.
Taraska, J. W., & Almers, W. (2004). Bilayers merge even when exocytosis is transient. *Proceedings of the National Academy of Sciences of the United States of America, 101*(23), 8780–8785.
Ubach, J., Zhang, X., Shao, X., Sudhof, T. C., & Rizo, J. (1998). Ca^{2+} binding to synaptotagmin: How many Ca^{2+} ions bind to the tip of a C2-domain? *EMBO Journal, 17*, 3921–3930.
van den Bogaart, G., Holt, M. G., Bunt, G., Riedel, D., Wouters, F. S., & Jahn, R. (2010). One SNARE complex is sufficient for membrane fusion. *Nature Structural & Molecular Biology, 17* (3), 358–364. doi: nsmb.1748.
Vrljic, M., Strop, P., Ernst, J. A., Sutton, R. B., Chu, S., & Brunger, A. T. (2010). Molecular mechanism of the synaptotagmin-SNARE interaction in Ca^{2+}-triggered vesicle fusion. *Nature Structural & Molecular Biology, 17*(3), 325–331. doi: nsmb.1764.
Walter, A. M., Wiederhold, K., Bruns, D., Fasshauer, D., & Sorensen, J. B. (2010). Synaptobrevin N-terminally bound to syntaxin-SNAP-25 defines the primed vesicle state in regulated exocytosis. *Journal of Cell Biology, 188*(3), 401–413. doi: jcb.200907018.
Wang, C. T., Bai, J., Chang, P. Y., Chapman, E. R., & Jackson, M. B. (2006). Synaptotagmin-Ca^{2+} triggers two sequential steps in regulated exocytosis in rat PC12 cells: Fusion pore opening and fusion pore dilation. *Journal of Physiology, 570*(Pt 2), 295–307.
Wang, C. T., Grishanin, R., Earles, C. A., Chang, P.-Y., Martin, T. F. J., & Chapman, E. R., et al. (2001). Synaptotagmin modulation of fusion pore kinetics in regulated exocytosis of dense-core vesicles. *Science, 294*, 1111–1115.

Wang, N., Kwan, C., Gong, X., de Chaves, E. P., Tse, A., & Tse, F. W. (2010). Influence of cholesterol on catecholamine release from the fusion pore of large dense core chromaffin granules. *Journal of Neuroscience, 30*(11), 3904–3911.

Wightman, R. M., Jankowski, J. A., Kennedy, R. T., Kawagoe, K. T., Schroeder, T. J., & Leszczyszyn, D. J., et al. (1991). Temporally resolved catecholamine spikes correspond to single vesicle release from individual chromaffin cells. *Proceedings of the National Academy of Sciences of the United States of America, 88*, 10754–10758.

Xia, Z., Zhou, Q., Lin, J., & Liu, Y. (2001). Stable SNARE complex prior to evoked synaptic vesicle fusion revealed by fluorescence resonance energy transfer. *Journal of Biological Chemistry, 276* (3), 1766–1771. doi: 10.1074/jbc.M008741200.

Xu, Y., Zhang, F., Su, Z., McNew, J. A., & Shin, Y. K. (2005). Hemifusion in SNARE-mediated membrane fusion. *Nature Structural & Molecular Biology, 12*(5), 417–422.

Zampighi, G. A., Zampighi, L. M., Fain, N., Lanzavecchia, S., Simon, S. A., & Wright, E. M. (2006). Conical electron tomography of a chemical synapse: Vesicles docked to the active zone are hemi-fused. *Biophysical Journal, 91*(8), 2910–2918. doi: S0006-3495(06)72005-7.

Zhang, X., Kim-Miller, M. J., Fukuda, M., Kowalchyk, J. A., & Martin, T. F. (2002). Ca^{2+}-dependent synaptotagmin binding to SNAP-25 is essential for Ca2+-triggered exocytosis. *Neuron, 34* (4), 599–611. doi: S0896627302006712.

Zhang, Z., Hui, E., Chapman, E. R., & Jackson, M. B. (2009). Phosphatidylserine regulation of Ca^{2+}-triggered exocytosis and fusion pores in PC12 cells. *Molecular Biology of the Cell, 20*(24), 5086–5095. doi: E09-08-0691.

Zhang, Z., Hui, E., Chapman, E. R., & Jackson, M. B. (2010). Regulation of exocytosis and fusion pores by synaptotagmin-effector interactions. *Molecular Biology of the Cell, 21*(16), 2821–2831 doi: E10-04-0285.

Zhang, Z., & Jackson, M. B. (2010). Membrane bending energy and fusion pore kinetics in Ca(2+)-triggered exocytosis. *Biophysical Journal, 98*(11), 2524–2534 doi: S0006-3495(10) 00323-1.

Zhang, Z., Zhang, Z., & Jackson, M. B. (2010). Synaptotagmin IV modulation of vesicle size and fusion pores in PC12 cells. *Biophysical Journal, 98*(6), 968–978.

Zimmerberg, J., & Kozlov, M. M. (2006). How proteins produce cellular membrane curvature. *Nature Reviews Molecular Cell Biology, 7*(1), 9–19.

PART 4

DEVELOPMENTAL CELL FUSION

CHAPTER 9

Eukaryotic Cell–Cell Fusion Families

Ori Avinoam and Benjamin Podbilewicz
Department of Biology, Technion, Israel Institute of Technology, Haifa, Israel

I. OVERVIEW

The capacity of cells to unite in a regulated fashion, through fusion of their plasma membranes, is essential for reproduction and development in eukaryotes. Yet, the majority of proteins that mediate this process (i.e. fusogens) remain unidentified or uncharacterized. The Fusion Family (FF) and the Syncytins are the first families of cell-cell fusogens. The FF fusion proteins are required in both fusing cell membranes and their structure reveals the formation of flower-like super-complexes. The Syncytins were recently acquired from retroviruses and are similar to the Env glycoprotein of HIV. While there are apparent similarities between these families, their mechanisms are different. These similar and disparate principles may be relevant to other cell fusion events. Thus, FF proteins and Syncytins serve as paradigms for the mechanisms of eukaryotic cell fusion and provide insight to the ongoing quest for the missing fusogens.

Current Topics in Membranes, Volume 68 0065-230X/10 $35.00

DOI: 10.1016/B978-0-12-385891-7.00009-X

II. INTRODUCTION

Omnis cellula e cellula (all cells are derived from cells) conventionally reflects the significance of cell division to cell biology. However, this key concept is rarely considered in the context of multinucleated or hybrid cells, that rise through fusion of two or more cells into one cell. Cell fusion is critical for many developmental processes such as, fertilization, muscle development, bone maintenance, and immune function in humans (Brodbeck & Anderson, 2009; Helming & Gordon, 2009; Pavlath, 2010; Rochlin, Yu, Roy, & Baylies, 2010; Schejter & Baylies, 2010; Sutovsky, 2009; Vignery, 2000, 2005, 2008). It has also been implicated in neuronal function, stem-cell reprogramming, and carcinogenesis (Ambrosi & Rasmussen, 2005; Lu & Kang, 2009; Oren-Suissa, Hall, Treinin, Shemer, & Podbilewicz, 2010; Rodic, Rutenberg, & Terada, 2004). Accordingly, cell–cell fusion is necessary for embryonic and postembryonic development. While it is evident that cell-to-cell fusion occurs in most eukaryotes, the molecular mechanism is not well-understood. A prerequisite of cell fusion is the capacity to mediate specific and nontoxic membrane bilayer fusion. This process is thought to depend on the function of specialized proteins that act on the bilayers directly (e.g., fusogens). The role of fusogenic proteins is predominantly supported by experimental evidence from a plethora of viral membrane glycoproteins that mediate virus–target–cell fusion or cell-to-cell fusion following infection (Backovic & Jardetzky, 2009; Barry, Key, Haddad, & Duncan, 2010; Clancy, Barry, Ciechonska, & Duncan, 2010; Harrison, 2005, 2008; Haywood, 2010; Kielian, 2006; Martens & McMahon, 2008; Sapir, Avinoam, Podbilewicz, & Chernomordik, 2008; Weissenhorn, Hinz, & Gaudin, 2007; White, Delos, Brecher, & Schornberg, 2008). Additional support for the roles of proteins in membrane fusion is found in protein-mediated membrane fusion that occurs between intracellular membranes (Hughson & Reinisch, 2010; Jena, 2009; Scott & Youle, 2010; Westermann, 2010; Wickner & Schekman, 2008). In consequence, the search for the missing cell–cell fusogens in eukaryotes relies on insights from these models in conjunction with a classical forward genetics approach (Bate, 1990; Baylies, Bate, & Ruiz Gomez, 1998; Chen, Grote, Mohler, & Vignery, 2007; Chen & Olson, 2005; Onel & Renkawitz-Pohl, 2009; Oren-Suissa & Podbilewicz, 2007, 2010; Rochlin et al., 2010; Sapir et al., 2008; Schejter & Baylies, 2010). Moreover, many resident *e*ndogenous *r*etro-*v*iruses (ERVs) can be identified in the genomes of eukaryotes. Human ERVs (HERV) make up 8% of the human genome, of which, 18 HERV elements have been shown to encode intact viral envelope glycoproteins (ENVs) that can potentially mediate cell–cell fusion upon expression (Bannert & Kurth, 2004; Benit, Dessen, & Heidmann, 2001; Blaise, de Parseval, Benit, & Heidmann, 2003; Blaise, de Parseval, & Heidmann, 2005; de Parseval & Heidmann, 2005; de Parseval, Lazar, Casella, Benit, & Heidmann,

2003; Villesen, Aagaard, Wiuf, & Pedersen, 2004). This line of evidence lead to the identification of Syncytin as a functional fusogenic glycoprotein derived from a provirus (Blond et al., 1999; Mi et al., 2000). Syncytin is but one example of a retroviral fusogen that has been selectively preserved while the *gag* and *pol* genes accumulated inactivating mutations. Nevertheless, the biological significance of ERV-derived ENVs in the evolutionary acquisition of cell fusion is unclear. With the exception of these virus-derived fusogens, the only other identified family of fusogens is the "fusion family" (FF) from *Caenorhabditis elegans* (Avinoam et al., 2011; Sapir et al., 2007, 2008). Thus, the large majority of fusogens remain unidentified or uncharacterized. Here, we review the chronological steps leading to the identification and characterization of these founding families of eukaryotic cell–cell fusogens.

III. CHOOSING A MODEL SYSTEM: *C. ELEGANS* AS AN ORGANISM TO STUDY CELL–CELL FUSION

Since its introduction in 1963 by Sydney Brenner, *C. elegans* proved to be an outstanding model system to study biological processes. Distinctively, when the detailed cell lineage was completed by John Sulston and collaborators, in the 1980s, it became apparent that *C. elegans* is particularly compelling for the study of cell–cell fusion. For the first time, the morphology and development of each of the 959 somatic nuclei could be studied from cell birth to terminal cell fate through the transparent worm and in the context of the whole animal. Between 1970 and 1988 essentially all syncytial tissues were revealed using serial-section electron microscopy (Sulston & Horvitz, 1977; Sulston, Schierenberg, White, & Thomson, 1983; White, Southgate, Thomson, & Brenner, 1986). By the year 1999, 44 syncytial cells, containing 300 different somatic nuclei that account for approximately 30% of the soma, were reported in the *C. elegans* hermaphrodite alone (Podbilewicz, 2000, 2006; Shemer & Podbilewicz, 2003).

A complete and detailed account of all syncytial tissues was essential for further exploration. In 1994 a spatial and temporal account of embryonic and postembryonic cell-to-cell fusion of epithelial cells was generated (Podbilewicz & White, 1994). Since it is not possible to reliably resolve membrane fusion using Nomarski optics, immunofluorescence of an antigen, that is associated with the apical junctions, was used (MH27; Hall, 1996; Labouesse, 1997). Developing a method to follow the membranes and adding live-temporal-data proved critical to facilitate a forward-genetics approach to seek out the missing fusogens. In 1995, the vital fluorescent membrane probe FM 4–64 was introduced, providing the means for live imaging of cell fusion *in vivo* (Mohler, Simske, Williams-Masson, Hardin, & White, 1998; Vida & Emr, 1995). In parallel, a transgenic worm strain, expressing a reporter construct combining the AJM-1 apical junction protein (MH27 antigen) with GFP, was constructed (AJM-1::GFP). These reporters were coupled to transmission electron microscopy (TEM) to generate an in-depth analysis of fusion in the epithelia of *C. elegans*. It was observed that fusion begins at a single location on the apical cell membrane and spreads both laterally and basally, reminiscent of viral fusion pores (Chernomordik & Kozlov, 2003, 2005; Chernomordik, Zimmerberg, & Kozlov, 2006; Earp, Delos, Park, & White, 2005; Mohler et al., 1998). Although this apical fusion pore could form at any position along the length of the membrane the correlation with the apical junction marker revealed that the site of initiation was either at or directly apical to the cell junction. In contrast, other studies in *C. elegans* embryos and in cells in tissue culture, support the model in which several fusion pores may form simultaneously and then coalesce as they expand (Cassata et al., 2005; Chen et al., 2008; Gattegno et al., 2007). In addition, since an accumulation of vesicles was observed at the site of fusion it was suggested that plasma membrane vesiculation may play a role in the process of pore expansion (Mohler et al., 1998). In 1999, this study was complemented by a detailed description of male-tail and vulva formation that engage several fusion events, necessary for the formation of functional copulation apparatuses. Serial TEM, immunofluorescence (IF), and AJM-1::GFP transgenic worms were used to characterize the sequence of cell–cell fusion events. In the male tail spatial correlation between sites of fusion and apical junction also suggested that fusion initiates at or near the apical junction, however, plasma membrane vesiculation was not observed (Nguyen, Hall, Yang, & Fitch, 1999). Interestingly, this study found sex-specific differences in the timing and sequence of fusion events during tail formation suggesting a sex-specific mode of regulation. Vulva morphogenesis in hermaphrodites also involves a sequence of cell fusions (Kimble, 1981; Sharma-Kishore, White, Southgate, & Podbilewicz, 1999; Sulston & Horvitz, 1977; Sulston & White, 1980). Initially, 5 of the 12 ventral P(n).p cells fuse with the surrounding syncytium (hyp7). Subsequently, three out of six vulva precursor cells (VPCs; P(3,4,8).p) divide and contribute six additional nuclei by fusion to hyp7. Finally, the 22 cells derived from the remaining 3 VPCs (P(5–7).p) migrate and form seven ring-shaped cells. These stacked concentric rings sculpt a tube that connects the uterus with the external world (Kiontke et al., 2007; Kolotuev & Podbilewicz, 2008; Sharma-Kishore et al., 1999). Strikingly, one of the signals that induces the VPCs to adopt the vulval cell fate originates from a cell that is also destined to fuse (anchor cell, AC) (see Section III. C; Newman & Sternberg, 1996; Sharma-Kishore et al., 1999). If the VPCs are not exposed to AC-derived EGF signaling they adopt a different cell fate and fuse to the hypodermis (Sternberg, 2005). A complex pattern of cell fusions, utilized to sculpt a functional tube by forming rings and eliminating cells via fusion, was also observed in other organs in *C. elegans* such as the pharynx (Albertson & Thomson, 1976; Rasmussen, English, Tenlen, & Priess, 2008; Sulston & Horvitz, 1977; Sulston et al., 1983), the male tail (Nguyen et al., 1999), and uterus (Newman & Sternberg, 1996; Newman, White, & Sternberg, 1996; Sapir et al., 2007). Thus, the evolution and function of cell fusion and organ formation are tightly coupled in *C. elegans* and related nematodes (Felix et al., 2000; Kiontke et al., 2007; Kolotuev & Podbilewicz, 2004, 2008; Louvet-Vallee, Kolotuev, Podbilewicz, & Felix, 2003).

The critical role of cell fusion in the structure and function of organs was also observed in humans where some myoblasts, macrophages, and trophoblasts undergo programmed cell fusion fates during the organogenesis or maintenance of the muscles, bones, and placenta. Therefore, cell-to-cell fusion emerges as a process that complements cell recognition, migration, invasion, apoptosis, and attachment in the morphogenesis of tissues. In order to identify the mediators of cell fusion an efficient screen was required to discriminate between loss-of-fusion phenotypes and wild-type, and to recover mutant progeny specifically affected in somatic cell fusion.

A. Genetic Screens for Cell Fusion Defects in C. elegans

Forward genetic screening for fusion defective mutants requires several crucial tools. First, it is necessary to effectively visualize dynamic changes in cell boundaries (Podbilewicz & White, 1994). This can be achieved by using reliable fluorescent membrane markers (Mohler et al., 2002; Shemer & Podbilewicz, 2000). Second, it is important to discriminate between genes encoding proteins, that specifically induce cell fusion and proteins that affect migration, signaling, and attachment or phenotypes associated with variation in cell fate (Mohler et al., 2002; Shemer, Kishore, & Podbilewicz, 2000). In this respect, *C. elegans* is a practical model because cells can be followed from birth to their prefusion state and ultrastructural studies can be performed to determine whether a candidate mutation affected the acquisition of fusion competence or the distinct stages of membrane fusion (Podbilewicz & White, 1994). Additionally, in order to discriminate between the initiation and completion of membrane fusion and to avoid misinterpreting artifacts associated with the dynamic displacement of membrane markers it is necessary to evaluate the level of mixing between fusing cytoplasms (Mohler et al., 1998, 2002). Conventionally, cytoplasmic mixing is evaluated by following the diffusion of fluorescent markers from one cell to the other or by measuring the reaction output generated by the encounter of molecules (e.g., enzyme and substrate) expressed differentially in the two fusing cells (Avinoam et al.,

2011; Hu et al., 2003; Mohler et al., 2002; Podbilewicz et al., 2006; Shemer & Podbilewicz, 2000; Shemer et al., 2004).

B. EFF-1

The possibility to follow cell migration, attachment, and fusion *in vivo* facilitated the endeavor to differentiate between gene functions associated with differentiation and pattering that precede fusion and membrane fusion *per se*. Subsequently, two *e*pithelial *f*usion-*f*ailure (*eff*) mutants were isolated in independent screens and upon failure to complement each other were attributed to the same gene and annotated as *eff-1(oj55)* and *eff-1(hy21)* (Mohler et al., 2002). The publication of the essentially complete *C. elegans* genome in 1998 and a detailed physical map of the genome obtained by Alan Coulson facilitated the mapping and identification of *eff-1* by transformation rescue (Coulson, Sulston, Brenner, & Karn, 1986; The-*C.-elegans*-Sequencing-Consortium, 1998). By 2002, *eff-1* and its promoter were cloned and transgenic animals, expressing an *eff-1* promoter fused to GFP (*eff-1*p::GFP), were generated. The observed green fluorescent cells confirmed that *eff-1* RNA was expressed in most of the known fusing cells and assisted the identification of cells that were unknown to fuse at the time. Most puzzling was the putative expression of *eff-1* in neurons, which was not predicted by previous works (Mohler et al., 2002). In 2004, an elegant assay to test whether *eff-1* is sufficient to fuse cells that normally do not fuse *in vivo* was established. To ubiquitously express *eff-1*, it was placed under the control of an endogenous heat shock inducible promoter (Fire, Harrison, & Dixon, 1990; Jones, Russnak, Kay, & Candido, 1986). This inducible construct was used to generate transgenic animals also carrying the membrane marker AJM-1::GFP, the cytoplasmic marker *eff-1*p::GFP, and the *eff-1* mutant allele *hy21*. Induction by heat shock treatment resulted in fusion of most embryonic cells. Ectopic fusion was also detected in larvae and in adult animals (Shemer et al., 2004). These results taken together suggested that most cells *in vivo* are fusion-competent and that EFF-1 expression in these cells is sufficient to fuse them. In this same study, thin TEM sectioning of wild type and *eff-1* mutant animals revealed that *eff-1* was required both for initiating and expansion of fusion, from an initial pore to complete membrane merger. EFF-1-induced cell fusion *in vivo* was further demonstrated by an independent study (del Campo et al., 2005); del Campo et al. showed that *eff-1* ectopic expression is sufficient to fuse the cells of the early embryo (at the ~100 cell stage) suggesting that all embryonic cells are fusion competent. This study made two additional contributions by showing both the behavior of an EFF-1::GFP translational fusion construct and the effect of several mutations on the behavior of EFF-1. However, the dynamic subcellular localization of EFF-1::GFP does not match observations made by IF staining using anti EFF-1 monoclonal antibodies (Valansi, Smurova, Gonzalez, & Podbilewicz, unpublished results). Furthermore, the EFF-1:: GFP construct harbors two point mutations at highly conserved sites, T176A and N529D, and does not rescue *eff-1(hy21)* animals (OA, unpublished results). In addition, the structure–function relationship of putative motifs in *eff-1* (del Campo et al., 2005) remained elusive as mutations mostly resulted in EFF-1 protein misfolding and retention in the endoplasmic reticulum.

The accumulation of evidence suggesting that EFF-1 is a *bona fide* fusogen necessitated the utilization of research tools designed for the characterization of viral fusogens. Consequently, in 2006, *eff-1* was expressed in heterologous insect cells that do not fuse under normal growth conditions (Podbilewicz et al., 2006). *eff-1* expression at the cell membrane was sufficient to fuse these cells. Furthermore, fusion between cells was shown to depend on expression of *eff-1* in both fusing partners. This was shown by using a combination of cytoplasmic markers in Sf9 insect cells and complemented with classical genetic mosaic analysis *in vivo* (in *C. elegans* embryos, larvae, and adults), which yielded a similar conclusion. Researchers also reported that EFF-1 forms complexes on the membrane and that the extracellular domain stimulates fusion when added to cells expressing the full length transmembrane isoforms (Figure 2) (Podbilewicz et al., 2006). These results provided insights into the mechanism of EFF-1-mediated fusion. Most striking is the evidence showing that *eff-1*-mediated fusion proceeds through a hemifusion intermediate. This was deduced from experiments showing that membrane lipid mixing between fusing partners occurred prior to cytoplasmic mixing. Thus, it seems that while *eff-1* displays a novel homotypic mechanism it converges with viral and intracellular fusogens at the molecular level through the conserved hemifusion intermediate (Podbilewicz et al., 2006). More studies are required to determine the structure–function relationship of EFF-1 and to understand its activity at the molecular level. Furthermore, *eff-1*'s role in the nervous system and the function of its alternative-splicing variants (isoforms) require further examination.

The characterization of cell fusion in a variety of *eff-1* null-alleles (*np29* and *ok1021*) suggested that some cells still fuse in the absence of *eff-1* (del Campo et al., 2005; Mason, Rabinowitz, & Portman, 2008; Mohler et al., 2002; Oren-Suissa et al., 2010; Podbilewicz et al., 2006; Sapir et al., 2007; Shemer et al., 2004). These findings coupled to the annotation of EFF-1's sequence provided the basis for sequence similarity searches to other predicted or known proteins in *C. elegans* that may function as fusogens.

C. AFF-1

Two primary observations supported the hypothesis that other fusogens are encoded in the *C. elegans* genome. First, some of the cells that were known to fuse did not express the *eff-1p*::GFP construct, suggesting that EFF-1 was not acting in these cells. Second, these cells reproducibly fused in worms carrying strong loss of function *eff-1* mutations, suggesting that a different fusogen was acting in these cells. Between 2004 and 2006, attention focused on the finding that EFF-1 was not required for sperm–egg fusion and for AC fusion to the gonadal–utse cell that forms the worm hymen (Choi, 2006; del Campo et al., 2005; Podbilewicz et al., 2006; Shemer et al., 2004). At the time, two putative open reading frames with homology to *eff-1* could be identified by sequence comparisons.

The predicted gene *C26D10.7* shares 90% identity and 95% similarity with *eff-1*. It is located directly 12,596 bp upstream and next to *eff-1*. These results suggest that *C26D10.7* represents a gene duplication event that occurred relatively recently in the evolutionary history of *C. elegans*. It had been speculated that *C26D10.7* is a pseudogene (Mohler et al., 2002; Sapir et al., 2007). However, this is unlikely because variations in the DNA sequence occurred mostly in introns suggesting that C26D10.7 is responsive to selection forces.

The predicted gene C44B7.3, also located on chromosome II, showed only moderate homology to EFF-1 (26% identity and 46% similarity) (Mohler et al., 2002; Sapir et al., 2007). In 2003, Choi and Newman performed forward genetic screens for mutants in which the vulva and uterus failed to connect properly. The screens yielded two mutations that demonstrated complete AC fusion failure and defective connection between the gonad and the vulva. The first mutation identified was in the *N*-ethylmaleimide-*S*ensitive-*F*actor (*nsf-1*; Choi, Richards, Cinar, & Newman, 2006) and the second mutation was mapped to C44B7.3 (Sapir et al., 2007). This mutant specifically failed to complement a deletion in C44B7.3 and had the same phenotype as did RNAi directed against C44B7.3 (Sapir et al., 2007). Therefore, the gene was named *aff-1* (*A*C-*f*usion-*f*ailure). Subsequently, it was determined that AFF-1 is also required for fusion of two out of the seven rings that form the vulva, some hypodermal cells in the embryo and the lateral seam cells at the L4 stage. Following *aff-1*p::GFP transcriptional fusion revealed an intricate interplay between *aff-1* and *eff-1* expression which, sometimes occurs simultaneously in adjacent cells, possibly to prevent undesirable fusion (Sapir et al., 2007). Perhaps the most striking examples of this are the adjacent toroidal (ring-shaped) cells of the vulva. In addition, other rings in the pharynx and connecting to the intestine show a similar behavior. pm8 and vp1 are pharyngeal ring-shaped cells that express

AFF-1 and EFF-1, respectively (Rasmussen et al., 2008). Notch signaling in pm8 induces AFF-1 expression while repressing EFF-1. Thus, each cell expresses a different fusogen and fuses to itself to form a ring. Since AFF-1 and EFF-1 are homotypic fusogens, we propose that this is due to posttranslational regulation at the level of protein transport. AFF-1 and EFF-1 may contain transport/sorting signals that result in a different membrane localization (Avinoam et al., 2011; Podbilewicz et al., 2006; Sapir et al., 2007).

EFF-1 and AFF-1 revealed a novel form of cell fusion, namely, auto-fusion (e. g., when a single cell fuses with itself) (Oren-Suissa & Podbilewicz, 2010). Thus, membrane fusogens are not just utilized to sculpt whole tissues, by removing or merging cells, but intricate cell shapes such as toroids can be formed by auto-fusion (Nguyen et al., 1999; Oren-Suissa & Podbilewicz, 2010; Podbilewicz & White, 1994; Rasmussen et al., 2008; Sharma-Kishore et al., 1999). This may implicate membrane fusion in a variety of biological processes not yet completely understood, such as neuronal and immune function, angiogenesis of small capillaries, and long-range cell communication, possibly challenging the common view in these fields.

AFF-1 and EFF-1 display fundamental differences in their activity. For example, ectopic expression of AFF-1 seemed to result in a stronger fusion reaction, both *in vivo* and in heterologous cells (Sapir et al., 2007). This phenomenon was explored in tissue culture by correlating AFF-1 and EFF-1 levels of membrane expression, with their fusogenic activity. It was observed that low levels of AFF-1 on the cell surface are sufficient to mediate robust fusion compared to EFF-1. Consequently, artificially elevating the level of AFF-1 expression on the cell surface resulted in stronger fusion (Sapir et al., 2007).

However different, AFF-1 and EFF-1 also display similarities. AFF-1 expression is also required in both fusing partners (homotypic). This hypothesis is predominantly supported by the observation that *aff-1p::GFP* expression is always observed in at least two partners prior to fusion (e.g., *hyp5*, VulA, VulD, and AC-utse). In addition, we recently showed that AFF-1 can mediate homotypic cell fusion of baby hamster kidney cells (BHK; Avinoam et al., 2011). In summary, AFF-1 and EFF-1 emerge as a novel family of cellular fusogens operating via homotypic interactions between cells to mediate a fusion reaction through a canonical hemifusion intermediate (Podbilewicz et al., 2006; Sapir et al., 2007).

The lack of robust sequence homology between these paralogs and the conserved pattern of cysteins along the extracellular domain, suggest that their function is tightly associated with structure rather than sequence. A similar lack in sequence homology is common among viral fusogens, even those belonging to the same structural class (Roussel et al., 2006; Sapir et al., 2008; White et al., 2008). Therefore, identifying distant FF proteins in other organisms cannot rely on sequence homology alone.

D. Beyond the Nematode Phylum

The *C. elegans* functional paralogs EFF-1 and AFF-1 are the founding members of a cell–cell fusion protein family (Sapir et al., 2007). By 2011, 35 nematode species with putative FF orthologs were identified, suggesting that the FF family is conserved in the phylum nematoda. Remarkably, beyond this phylum putative orthologs could only be identified in two arthropods, a ctenophore, a chordate, and a protist (Fig. 1A) (Avinoam et al., 2011; Sapir et al., 2008). This unique dispersal can be explained by five alternative hypotheses. First, lack of sufficient sequence conservation, poor genome annotation, or coverage. Second, FFs may have been replaced by other fusogens through the course of evolution. Third, DNA or RNA from nematode parasites may have contaminated the genome/EST sequence libraries. Fourth, FF may have independent origins by convergent evolution to similar protein structures that can fuse cell membranes. Fifth, horizontal gene transfer (HGT) may have occurred. It is noteworthy that, the Syncytin family of fusion proteins (see Section IV), was apparently acquired in this manner from viruses and the possibility that FFs share structural similarities with viral fusogens cannot be ruled out (Avinoam et al., 2011; Kielian, 2006) and should be pursued experimentally.

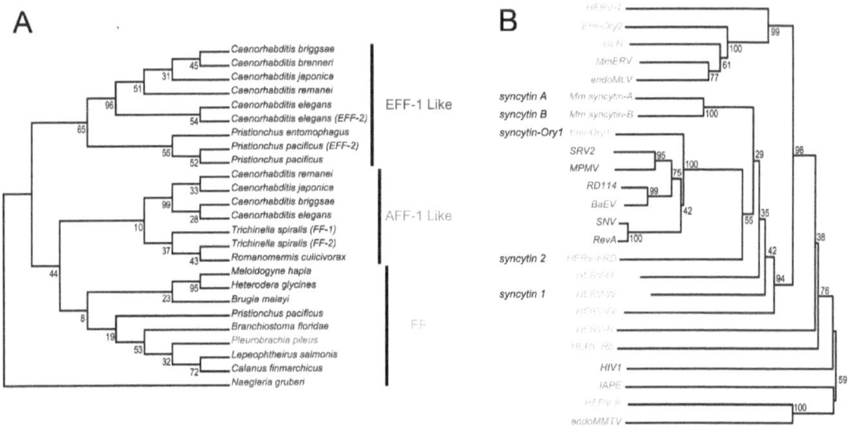

FIGURE 1 Phylogenetic relationships of FFs and Syncytins (A) Phylogenic relationships of FFs (Avinoam et al., 2011). From top to bottom - twenty nematodes, a chordate, a ctenophore, two arthropods, and a protist. [Black - Nematodes, Purple - Chordate, Green - ctenophore, Brown - arthropods, Blue - protest] (B) Phylogenic relationships of ERVs and viruses (Modified from (Heidmann et al., 2009)) are shown. BaEV - Baboon; HERV-T, GLN, HERV-FRD, HERV-H, HERV-W, HERV-Fc, HERV-R, HERV-Rb, HERV-K - Human; MmERV, endoMLV, Mm syncytin-A, Mm syncytin-B, IAPE, endoMMTV - Mouse; Env-Ory2, Env-Ory1 - Rabbit; SRV2, MPMV, RD114, SNV, RevA, HIV1 - Virus [Dark green - Baboon, Light green - Human, Orange - Mouse, Blue - Rabbit, Red - Virus]. (See Color Insert.)

FF proteins share several structural characteristics and a common organization of putative domains (Fig. 2). In general, FFs are type I transmembrane (TM) proteins with an extracellular portion approximately 520–540 aa long that contains four or five regions of elevated sequence homology (R1–R5; Fig. 2). R1 and R5 (CeAFF-1 Arg449–Leu538) are weakly conserved; however, they harbor an even number of conserved cysteines (Cys) that are probably important

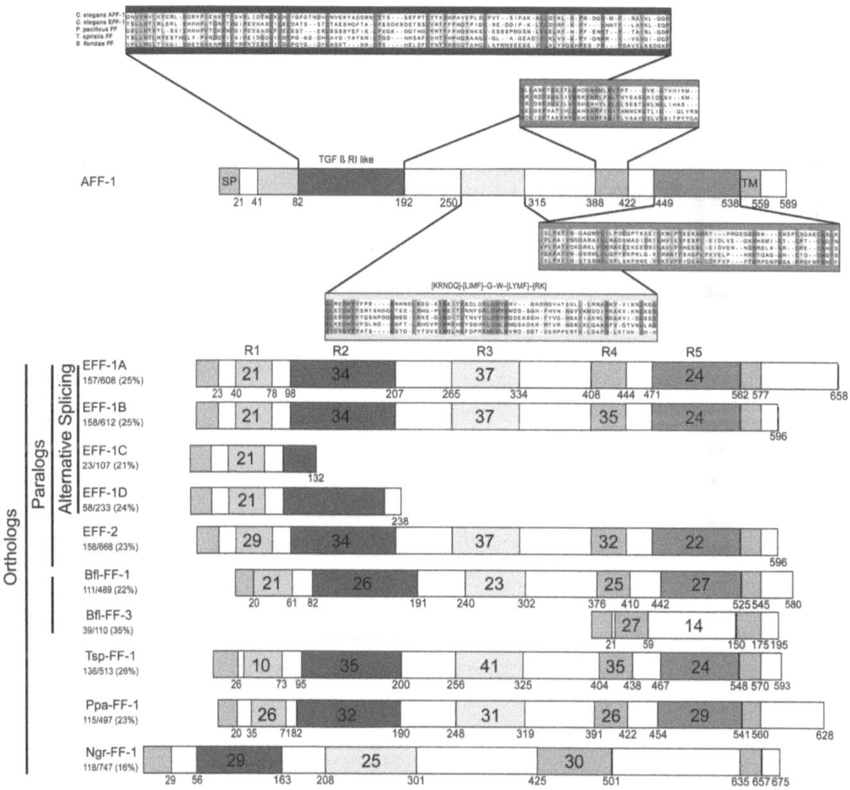

FIGURE 2 Distribution of conserved sequence motifs in FF paralogs and orthologs. Top - schematic representation of the multiple sequence alignment of FF proteins. Alignments of regions R2-R5 of five representative sequences are shown. SP - signal peptide, TM - transmembrane domain. Global sequence identity to AFF-1 is indicated under the name of the protein. Local sequence identity to AFF-1 is indicated within each domain (%). Sequence limits are indicated under the illustration unless identical to previously shown. EFF-2, Bfl- FF-1 and FF-3, Tsp-FF-1, Ppa-FF-1 and Ngr-FF-1 are according to (Avinoam et al., 2011). Multiple sequence alignment of conserved sequence motifs is shown. [Alignment color code was according to the Clustal X color scheme with 40% conservation color increment in Jalview software (Waterhouse et al., 2009). Green - Signal peptide (SP), Pink - R1, Brown - R2, Yellow- R3, Cyan - R4, Purple - R5, Ocher - Transmembrane domain (TM). (Modified from Avinoam et al., 2011).] (See Color Insert.)

for structure and function. R2 (CeAFF-1 Gln82–Pro192) is a Cys-rich domain, R3 (CeAFF-1 Gln250–Gln315) contains a conserved palindromic [KRDENQ]–[LMFI]–G–W–[LYMF]–[RK] pattern, and R4 (CeAFF-1 Ser388–Met422) is moderately hydrophobic (Fig. 2). In addition, *eff-1* encodes four alternative splicing variants (EFF-1 A to D) two of which lack the transmembrane domain (EFF-1 C and D). Both EFF-1C and EFF-1D lack R3–R5, conversely, while EFF-1D contains R1 and R2, EFF-1C only contains R1 and a part of R2. In addition, the *Branchiostoma floridae* genome encodes a truncated form containing a membrane anchored R4 (Fig. 2). It was shown that the two transmembrane isoforms of EFF-1 (EFF-1A and B) can fuse heterologous insect cells in culture, while the EFF-1C secreted isoform has no fusogenic activity (Podbilewicz et al., 2006). The role of these alternative isoforms in *C. elegans* is still unknown and the possibility remains that some are not involved in cell fusion.

To complete the picture, FFs from the human parasitic nematode *Trichinella spiralis* (*Tsp-ff-1*) and the chordate *B. floridae* (*Bfl-ff-1*) have been shown to induce fusion of BHK cells (Avinoam et al., 2011). TSP-FF-1 and BFL-FF-1 share only 26% and 22% sequence identity with AFF-1, respectively. *C. elegans* and *T. spiralis* are among some of the most distant species within nematoda (Mitreva & Jasmer, 2006). These results taken together suggest that FFs are functional orthologs that mediate cell fusion in different organisms. Furthermore, unidentified FFs may turn out to be the missing eukaryotic fusogens (Avinoam et al., 2011).

E. Are FFs Sufficient to Fuse Isolated Membranes?

By biochemical standards to define a protein as a fusogen it must be able to fuse isolated membranes *in vitro*. The experimental design should assure that the protein acts directly on the membranes themselves without additional cellular cofactors (Oren-Suissa & Podbilewicz, 2007). One approach is to reconstitute the candidate fusogen into liposomes and measure membrane fusion. An alternative approach, which has been extensively used in this context, is to express a candidate fusogen on the surface of a heterologous viral envelope. This process is generally referred to as pseudotyping (Fukushi et al., 2005; Matsuura et al., 2001; Schnell, Buonocore, Kretzschmar, Johnson, & Rose, 1996; Takada et al., 1997; Zavada, 1972). If a candidate fusogen is efficiently incorporated into the viral envelope and mediates fusion of the virus to target cells then the interpretation would be that it functions independently and directly on the membranes themselves. A cellular fusogen has never been shown to complement a virus probably because it is difficult to imagine that it would incorporate into a viral envelope. Furthermore, it is unknown whether viral–cell and cell–cell fusion share similar mechanisms.

Vesicular stomatitis virus (VSV) has been widely used to produce pseudotyped viruses presenting a variety of exogenous viral glycoproteins. This is likely because VSV can incorporate other membrane proteins into its envelope (Lodish & Porter, 1980; Schnell et al., 1996; Takada et al., 1997; Tamura et al., 2005). Recently, VSV pseudotyped viruses presenting AFF-1 and EFF-1 were generated. Both FF proteins were efficiently incorporated into the VSV envelope and mediated homotypic virus–cell fusion (Avinoam et al., 2011). AFF-1 pseudotyped viruses also infected EFF-1 expressing cells. This heterotypic interaction between AFF-1 and EFF-1 was independently demonstrated by showing that BHK cells expressing AFF-1 can fuse with cells expressing EFF-1 (Avinoam et al., 2011).

Virions also appear to be an attractive platform to study the ultrastructure of surface proteins. AFF-1 pseudoviruses were examined by negative stain transmission electron microscopy (N-TEM), cryo-TEM, and cryo electron tomography (Avinoam et al., 2011). These analyses revealed that AFF-1 forms distinct supercomplexes, seen as bulky spikes that resemble flowers or rosettes (Fig. 3A). These complexes are reminiscent of assemblies previously observed on viruses and are probably important for the fusogenic activity of these proteins (Gibbons et al., 2003).

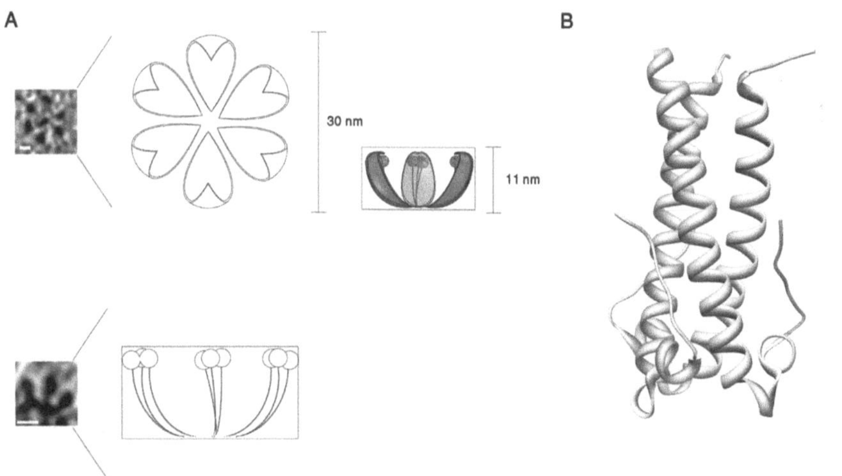

FIGURE 3 Structures of AFF-1 and Syncytin-2. (A) Model of AFF-1 based on the cryo electron tomography. Putative AFF-1 trimers form rosettes that resemble hexameric flowers. Top (left) and side view (right and bottom) (modified from Avinoam et al., 2011). (B) Ectodomain of the transmembrane subunit (residues 391–443) of Syncytin-2 based on chimera rendering of PDB 1Y4 M (Pettersen et al., 2004; Renard et al., 2005). Molecular graphics image was produced using UCSF Chimera package from the Resource for Biocomputing, Visualization, and Informatics at the University of California, San Francisco.

To summarize, AFF-1 and EFF-1 can fuse viral envelopes to cells. Fusion is dependent on the presence of either AFF-1 or EFF-1 in both fusing membranes suggesting that FF-mediated fusion depends on homotypic interactions between fusogens. These results further suggest that FF proteins share some similarities with viral fusogens, at least at the ultrastructural level. However, it remains to be seen whether these families are linked by convergent or divergent evolution.

F. *What Other Functions can Fusogens Perform? The Role of* eff-1 *in the Nervous System of* C. elegans

The possible implications of the surprising observation made in 2002 that *eff-1* RNA may be expressed in neurons, was not immediately realized. It is possible to speculate that this was due to the difficulty in conceiving the potential role that a cell–cell fusogen might be playing in the nervous system. The question of fusion of neuronal processes has received little attention so far and the potential functions of fusogens in sculpting membranes independently of fusion was not predicted. In addition, *aff-1* also showed neuronal expression with no noticeable overlap with *eff-1*. In 2010, a study describing the function of EFF-1 in four specialized mechanosensory neurons in *C. elegans* called the PVDs and FLPs was published (Oren-Suissa et al., 2010). The PVDs are an attractive model system to explore neuronal phenotypes. The PVDs are mechanoreceptors comprised of repetitive menorah-like structures containing stereotypic 30–80 nm (diameter) dendritic extensions or branches. The branches of the menorahs show right angle turns in wild-type animals. In *eff-1* mutant-animals these ordered structures are disrupted by excessive branching and variable turn angles. EFF-1 is expressed in the PVDs and its ectopic expression in *eff-1(hy21)* mutants partially rescues the disordered and excessive branching phenotypes. Strikingly, an increased dose leads to the reverse phenotype of hypobranching (reduction in the number of branches). Therefore, EFF-1 functions in a cell autonomous manner in the PVD neurons to restrict branching. Serial section TEM in wild-type *C. elegans* demonstrated that terminal dendrites at the ends of menorahs autofuse forming grids. The mechanosensory function of these "autofused dendrites" has not been determined. This is the first demonstration that neuronal branches fuse in wild-type animals (Oren-Suissa et al., 2010). By following the behavior of these extensions using three-dimensional live imaging, it was also observed that the dendritic branches undergo fission, retraction, and fusion. These processes depended on EFF-1 although its primary function in the context of the PVDs seemed to be retraction and not fusion (Oren-Suissa et al., 2010). Thus, FF proteins may yet reveal novel roles that fusogens fulfill; to sculpt cell membrane in parallel to fusion. These findings may also call for the reexamination of conventional doctrines concerning neuronal interconnectivity. The reticular theory conceived by Camillo Golgi over a hundred years ago, which suggested that neuronal processes fuse to form an elaborate syncytium may have been neglected prematurely. It is still quite possible that neurite–neurite fusion or its membrane intermediates play a role in neuronal function.

IV. SYNCYTINS: AN EXPANDING METAZOAN FUSION FAMILY

In mammals, several tissues and cell types are known to rely on cell–cell fusion for their function. Research has mostly focused on cell fusion that occurs between gametes during fertilization, myoblasts during muscle fiber formation, macrophages during bone maintenance, and trophoblasts during placenta development (Blobel, Myles, Primakoff, & White, 1990; Brodbeck & Anderson, 2009; Helming & Gordon, 2009; Huppertz, Bartz, & Kokozidou, 2006; Huppertz & Borges, 2008; Oren-Suissa & Podbilewicz, 2007; Pavlath, 2010; Primakoff & Myles, 2007; Rochlin et al., 2010; Schejter & Baylies, 2010; Shemer & Podbilewicz, 2000; Sutovsky, 2009; Vignery, 2000, 2005, 2008; Yanagimachi, 1988). The placenta is a vascular organ that develops in gestating females of most mammals, except marsupials and monotremes (Mess & Ferner, 2010). It connects the developing fetus to the uterine wall, thus, allowing nutrient uptake and waste elimination through the mother's circulation. Regardless of the compound difficulties that arise during the cytological analysis of an adult organ *in vivo*, it is the placenta (syncytiotrophoblast fusion) that emerges as an accessible organ for which the fusogens are well characterized (Huppertz et al., 2006; Huppertz & Borges, 2008; Knerr et al., 2004; Orendi, Gauster, Moser, Meiri, & Huppertz, 2010; Potgens, Drewlo, Kokozidou, & Kaufmann, 2004; Stoye, 2009).

The discovery of the syncytiotrophoblast is somewhat unique because it stands as one of the earliest accounts of multinucleated cells, dating back to the early decades of the nineteenth century. The placenta, as well as several other histological accounts, provoked the original question of whether these cells originate from successive mitoses without concomitant cytoplasmic divisions or from fusion of preexisting mononucleated cells. The emergence of electron microscopy techniques in the late 1930s and early 1940s, and their subsequent application in the study of biological material confirmed the syncytial nature of the syncytiotrophoblast. However, it was unclear whether fusion was involved until the 1970s (Boyd & Hughes, 1954; Carr, 1967; Cavicchia, 1971; Rhodin & Terzakis, 1962).

The molecular characterization of cell fusion did not advance significantly until the late 1990s when replication-competent *H*uman *E*ndogenous *R*etro-**V**iruses (HERV) were implicated in cell fusion. In 1998, Blond et al., found that a full length *env* gene, captured in the human genome, HERV-W, showed specific placental expression (Blond et al., 1999). This expression pattern was confirmed in parallel by an additional independent study (Mi et al., 2000). The possibility, that an ancient viral sequence, incorporated into the germ cell genomes of our ancestors, and adapted from its original function, was perceived immediately (Blond et al., 1999, 2000; Mi et al., 2000). Moreover, the fact that HERV-W encoded a putatively active envelope protein (ENV), that normally mediates viral fusion to target cells, suggested some role in the formation of the syncytiotrophoblast. Thus, HERV-W became the leading candidate fusogen of the syncytiotrophoblast and was hence called *Syncytin* (Mi et al., 2000). Further studies showed that *Syncytin* was sufficient to induce cell-to-cell fusion in several cell lines suggesting that it is capable of fusing plasma membranes of individual cells to form syncytia (Blond et al., 2000; Mi et al., 2000). *Syncytin* also mediates fusion of pseudotyped immunodeficiency viruses to target cells (Dewannieux, Blaise, & Heidmann, 2005). These results taken together suggest that *Syncytin* retained a structure and function that allows it to be efficiently incorporated into viral envelopes and mediate virus binding and fusion. In addition, antisense knockdown of *Syncytin* in primary cytotrophoblast cells reduces fusion (Frendo et al., 2003). This result demonstrates that *Syncytin* is necessary, at least in part, for fusion of cytotrophoblast cells.

Syncytins and class I viral fusogens share similar structures and undergo the same characteristic proteolytic processing; hence, it is considered as a *bona fide* class I viral fusogen directly descendant from viruses (Figs 1B and 3B) (Chang, Chen, Chang, Huang, & Chen, 2004; Cheynet et al., 2005; Gong et al., 2005; Renard et al., 2005).

In 2003, *Syncytin*-2 was discovered and *Syncytin*, now *Syncytin*-1, became the founding member of a family of virus-captured Hominoid fusogens (HERV-FRD; Blaise et al., 2003). *Syncytin*-1 and *Syncytin*-2 penetrated the primate genome recently, >25 and >40 million years (MY) ago, respectively; after the divergence of hominoids and old world monkeys (Goodman et al., 1998; Kim, Takenaka, & Crow, 1999; Voisset et al., 1999). Surprisingly, syncytial trophoblasts had also been observed in Muridae and Leporidae and *Syncytin* related genes such as *Syncytin* A, B (mouse), and *Syncytin*-OryI (Rabbit) have been identified (Fig. 1B) (Dupressoir et al., 2005; Heidmann et al., 2009). *Syncytin* A and B appear to be conserved in all Muridae and it is estimated that they entered the genome 20 MY ago. They are expressed at the time and place of cytotrophoblasts fusion and their expression is sufficient to fuse cells in culture (Dupressoir et al., 2005). *Syncytin* A antisense-knockdown and knockout mice display severe morphological defects in the maternal–fetal interface (Dupressoir et al., 2009; Gong et al., 2007). Taking into account its fusogenic function, these results strongly suggest that it is essential for syncytiotrophoblast morphogenesis in mouse. The discovery of two murine and one leporid ERVs, homologous but not orthologous, to the human *Syncytin* genes confirms that the acquisition of fusogens from viruses occurred repeatedly through the course of evolution. It is particularly noteworthy that despite the mechanistic differences between FFs

TABLE I
Comparison between Eukaryotic Cell–Cell Fusion Families

	Syncytins	FFs
Type I membrane glycoprotein	+	+
Secondary structure	Mainly alpha helices	Mainly beta sheets
Viral-class similarity	I	II, III, or novel
Mode of action	Unilateral (receptor mediated)	Homotypic
Hemifusion intermediate	+	+
Sequence similarity	HIV ENV	None
Loss of function phenotype (essential)	Maternal–fetal interface structural defects	Fusion failure in *C. elegans*
Gain of function phenotype (sufficient to fuse heterologous cells)	+	+
Can substitute a fusogen in a pseudotyped virus and mediates virus–cell fusion	+	+

and viral fusogens (homotypic/bilateral vs. unilateral mechanisms), it seems that the two major fusogenic families, FF proteins and *Syncytins* are viral-like or viral derived (Table I). This may open up new research avenues for future evolutionary and mechanistic studies. Do FFs also have viral origins or have they evolved separately and converged to similar structures? Are the missing fusogens in other systems, viral-like, or did novel mechanisms evolve? Furthermore, *Syncytins* represent 2 out of 16 coding *env* genes currently known in the human genome and *Syncytin*-mediated fusion has not been ruled out in other tissues (Soe et al., 2010). Although, we cannot exclude the possibility that the latter were inactivated or adapted to perform other functions, it seems likely that at least some of them are fusogens.

V. REGULATION OF CELL–CELL FUSION IN NEMATODES AND MAMMALS

Although the regulation of Syncytins and FF proteins is beyond the scope of this review, it is noteworthy that many major conserved molecular pathways have been implicated in the transcriptional regulation of FF proteins in *C. elegans*. Among these are the sex determination, Wnt signaling, Notch signaling, Ras/MAPK signaling, heterochronic pathway, chromatin remodeling pathways, and homeotic genes (Alper & Kenyon, 2002; Cassata et al., 2005;

Hedgecock & White, 1985; Koh et al., 2002; Koh & Rothman, 2001; Margalit et al., 2007; Podbilewicz et al., 2006; Rasmussen et al., 2008; Sapir et al., 2007; Shemer & Podbilewicz, 2002). Reviewed in Alper and Podbilewicz (2008), Friedlander-Shani and Podbilewicz (2011), and Shemer and Podbilewicz (2003). In addition, the c-FOS oncogenic paralog FOS-1 regulates the expression of AFF-1 in the invading AC (Sapir et al., 2007).

In parallel to transcriptional regulation several lines of evidence point to dynamic regulation at the protein level. *fus-1*, a gene encoding a vacuolar–ATPase complex subunit, represses EFF-1 (Kontani, Moskowitz, & Rothman, 2005; Kontani & Rothman, 2005). EFF-1 is localized predominantly in intracellular compartments associated with the apical membrane (Valansi, Smurova, Gonzalez, & Podbilewicz, unpublished results) suggesting tight regulation of EFF-1 dynamics or activity. In addition, del Campo et al. (2005) showed evidence that EFF-1 retention in the plasma membrane may depend on trans interactions with EFF-1 on opposing cell membranes. AFF-1 may also be subject to posttranslational regulation. *nsf-1*, the gene encoding a NEM-sensitive factor homolog in worms, is required for AFF-1-mediated fusion of the AC (Choi et al., 2006). However, the implications of these forms of regulation await further illumination. Nevertheless, it is tempting to speculate that cell–cell fusion in other organisms is regulated by orthologous pathways and that these findings will aid in the identification of the mammalian fusogens.

The regulation of Syncytins in the placenta may share some pathways with the transcriptional control of FF expression and activity. For example, the placental transcription factor GCM1 regulates Syncytin 2 and its receptor MFSD2A in human placenta (Liang et al., 2010). In addition, there is evidence for a role of chromatin remodeling and maternal imprinting in the regulation of syncytin expression (Chang, Chuang, Yu, Yao, & Chen, 2005; Smallwood et al., 2003).

VI. CONCLUDING REMARKS

Here we chronologically reviewed the major research strategies applied through the fruitful discovery of eukaryotic cell–cell fusogens in the past two decades. The major difference between these approaches is that while one relies on a reverse genetics approach to study the function of proviral genes in mammals (Syncytins) the other relies on a forward genetics approach to study fusion failure phenotypes in *C. elegans* (FFs). Nevertheless these fusogens are similar in being the minimal machineries necessary and sufficient for cell-to-cell membrane fusion. It is likely that other fusogens in eukaryotes will follow similar paradigms and that they only remain elusive because they are necessary for viability or are redundant. In addition, their function may rely on multi-protein mechanisms, or unknown cofactors, therefore, it is not possible to determine if they are sufficient for membrane fusion. However, we speculate that cell–cell fusion in other organisms and tissues will depend on similar protein architectures that arose through convergent evolution or linear decent from viruses.

Future work will determine how FFs and Syncytins act to fuse membranes in diverse epithelial cells (e.g., epidermis in nematodes and placenta in mammals) and how the expression and activities of these diverse fusogens are tightly regulated. Lessons learned from studies of syncytins and FFs may help to identify missing fusogens that mediate fusion of myoblasts, macrophages, stem cells, and gametes across the tree of life.

References

Albertson, D. G., & Thomson, J. N. (1976). The pharynx of *Caenorhabditis elegans*. *Philosophical Transactions of the Royal Society B: Biological Sciences, 275*, 299–325.

Alper, S., & Kenyon, C. (2002). The zinc finger protein REF-2 functions with the Hox genes to inhibit cell fusion in the ventral epidermis of *C. elegans*. *Development, 129*, 3335–3348.

Alper, S., & Podbilewicz, B. (2008). Cell fusion in *Caenorhabditis elegans*. *Methods in Molecular Biology, 475*, 53–74.

Ambrosi, D. J., & Rasmussen, T. P. (2005). Reprogramming mediated by stem cell fusion. *Journal of Cellular and Molecular Medicine, 9*, 320–330.

Avinoam, O., Fridman, K., Valansi, C., Abutbul, I., Zeev-Ben-Mordehai, T., Maurer, U. E., Sapir, A., Danino, D., Grunewald, K., & White, J. M., et al., (2011). Conserved eukaryotic fusogens can fuse viral envelopes to cells. *Science, 332*, 589–592.

Backovic, M., & Jardetzky, T. S. (2009). Class III viral membrane fusion proteins. *Current Opinion in Structural Biology, 19*, 189–196.

Bannert, N., & Kurth, R. (2004). Retroelements and the human genome: New perspectives on an old relation. *Proceedings of the National Academy of Sciences of the United States of America, 101* (Suppl 2), 14572–14579.

Barry, C., Key, T., Haddad, R., & Duncan, R. (2010). Features of a spatially constrained cystine loop in the p10 FAST protein ectodomain define a new class of viral fusion peptides. *The Journal of Biological Chemistry, 285*, 16424–16433.

Bate, M. (1990). The embryonic development of larval muscles in *Drosophila*. *Development, 110*, 781–804.

Baylies, M. K., Bate, M., & Ruiz Gomez, M. (1998). Myogenesis: A view from Drosophila. *Cell, 93*, 921–927.

Benit, L., Dessen, P., & Heidmann, T. (2001). Identification, phylogeny, and evolution of retroviral elements based on their envelope genes. *The Journal of Virology, 75*, 11709–11719.

Blaise, S., de Parseval, N., Benit, L., & Heidmann, T. (2003). Genomewide screening for fusogenic human endogenous retrovirus envelopes identifies syncytin 2, a gene conserved on primate evolution. *Proceedings of the National Academy of Sciences of the United States of America, 100*, 13013–13018.

Blaise, S., de Parseval, N., & Heidmann, T. (2005). Functional characterization of two newly identified human endogenous retrovirus coding envelope genes. *Retrovirology, 2*, 19.

Blobel, C. P., Myles, D. G., Primakoff, P., & White, J. M. (1990). Proteolytic processing of a protein involved in sperm-egg fusion correlates with acquisition of fertilization competence. *The Journal of Cell Biology, 111*, 69–78.

Blond, J. L., Beseme, F., Duret, L., Bouton, O., Bedin, F., & Perron, H., et al., (1999). Molecular characterization and placental expression of HERV-W, a new human endogenous retrovirus family. *The Journal of Virology, 73*, 1175–1185.

Blond, J. L., Lavillette, D., Cheynet, V., Bouton, O., Oriol, G., & Chapel-Fernandes, S., et al., (2000). An envelope glycoprotein of the human endogenous retrovirus HERV-W is expressed in the human placenta and fuses cells expressing the type D mammalian retrovirus receptor. *The Journal of Virology, 74*, 3321–3329.

Boyd, J. D., & Hughes, A. F. (1954). Observations on human chorionic villi using the electron microscope. *Journal of Anatomy, 88*, 356–362.

Brodbeck, W. G., & Anderson, J. M. (2009). Giant cell formation and function. *Current Opinion in Hematology, 16*, 53–57.

Carr, M. C. (1967). Biology of human trophoblast. *California Medicine, 107*, 338–343.

Cassata, G., Shemer, G., Morandi, P., Donhauser, R., Podbilewicz, B., & Baumeister, R. (2005). *ceh-16/engrailed* patterns the embryonic epidermis of *Caenorhabditis elegans*. *Development, 132*, 739–749.

Cavicchia, J. C. (1971). Junctional complexes in the trophoblast of the human full term placenta. *Journal of Anatomy, 108*, 339–346.

Chang, C., Chen, P. T., Chang, G. D., Huang, C. J., & Chen, H. (2004). Functional characterization of the placental fusogenic membrane protein syncytin. *Biology of Reproduction, 71*, 1956–1962.

Chang, C. W., Chuang, H. C., Yu, C., Yao, T. P., & Chen, H. (2005). Stimulation of GCMa transcriptional activity by cyclic AMP/protein kinase A signaling is attributed to CBP-mediated acetylation of GCMa. *Molecular and Cellular Biology, 25*, 8401–8414.

Chen, A., Leikina, E., Melikov, K., Podbilewicz, B., Kozlov, M. M., & Chernomordik, L. V. (2008). Fusion-pore expansion during syncytium formation is restricted by an actin network. *Journal of Cell Science, 121*, 3619–3628.

Chen, E. H., Grote, E., Mohler, W., & Vignery, A. (2007). Cell-cell fusion. *FEBS Letters, 581*, 2181–2193.

Chen, E. H., & Olson, E. N. (2005). Unveiling the mechanisms of cell-cell fusion. *Science, 308*, 369–373.

Chernomordik, L. V., & Kozlov, M. M. (2003). Protein-lipid interplay in fusion and fission of biological membranes. *Annual Review of Biochemistry, 72*, 175–207.

Chernomordik, L. V., & Kozlov, M. M. (2005). Membrane hemifusion: Crossing a chasm in two leaps. *Cell, 123*, 375–382.

Chernomordik, L. V., Zimmerberg, J., & Kozlov, M. M. (2006). Membranes of the world unite!. *The Journal of Cell Biology, 175*, 201–207.

Cheynet, V., Ruggieri, A., Oriol, G., Blond, J. L., Boson, B., & Vachot, L., et al., (2005). Synthesis, assembly, and processing of the Env ERVWE1/syncytin human endogenous retroviral envelope. *The Journal of Virology, 79*, 5585–5593.

Choi, J. (2006). Genetic study of cell fusion factors in *Caenorhabditis elegans*. In *Verna and Marrs McLean department of biochemistry and molecular biology*. Houston: Baylor College of Medicine.

Choi, J., Richards, K. L., Cinar, H. N., & Newman, A. P. (2006). *N*-ethylmaleimide sensitive factor is required for fusion of the *C. elegans* uterine anchor cell. *Developmental Biology, 297*, 87–102.

Clancy, E. K., Barry, C., Ciechonska, M., & Duncan, R. (2010). Different activities of the reovirus FAST proteins and influenza hemagglutinin in cell-cell fusion assays and in response to membrane curvature agents. *Virology, 397*, 119–129.

Coulson, A., Sulston, J., Brenner, S., & Karn, J. (1986). Toward a physical map of the genome of the nematode *Caenorhabditis elegans*. *Proceedings of the National Academy of Sciences of the United States of America, 83*, 7821–7825.

de Parseval, N., & Heidmann, T. (2005). Human endogenous retroviruses: From infectious elements to human genes. *Cytogenetic and Genome Research, 110*, 318–332.

de Parseval, N., Lazar, V., Casella, J. F., Benit, L., & Heidmann, T. (2003). Survey of human genes of retroviral origin: Identification and transcriptome of the genes with coding capacity for complete envelope proteins. *The Journal of Virology, 77*, 10414–10422.

del Campo, J. J., Opoku-Serebuoh, E., Isaacson, A. B., Scranton, V. L., Tucker, M., & Han, M., et al., (2005). Fusogenic activity of EFF-1 is regulated via dynamic localization in fusing somatic cells of *C. elegans*. *Current Biology, 15*, 413–423.

Dewannieux, M., Blaise, S., & Heidmann, T. (2005). Identification of a functional envelope protein from the HERV-K family of human endogenous retroviruses. *The Journal of Virology, 79*, 15573–15577.

Dupressoir, A., Marceau, G., Vernochet, C., Benit, L., Kanellopoulos, C., & Sapin, V., et al., (2005). Syncytin-A and syncytin-B, two fusogenic placenta-specific murine envelope genes of retroviral origin conserved in Muridae. *Proceedings of the National Academy of Sciences of the United States of America, 102*, 725–730.

Dupressoir, A., Vernochet, C., Bawa, O., Harper, F., Pierron, G., & Opolon, P., et al., (2009). Syncytin-A knockout mice demonstrate the critical role in placentation of a fusogenic, endogenous retrovirus-derived, envelope gene. *Proceedings of the National Academy of Sciences of the United States of America, 106*, 12127–12132.

Earp, L. J., Delos, S. E., Park, H. E., & White, J. M. (2005). The many mechanisms of viral membrane fusion proteins. *Current Topics in Microbiology and Immunology, 285*, 25–66.

Felix, M. A., De Ley, P., Sommer, R. J., Frisse, L., Nadler, S. A., & Thomas., et al., (2000). Evolution of vulva development in the Cephalobina (Nematoda). *Developmental Biology, 221*, 68–86.

Fire, A., Harrison, W. H., & Dixon, D. (1990). A modular set of *lacZ* fusion vectors for studying gene expression in Caenorhabditis elegans. *Gene, 93*, 189–198.

Frendo, J. L., Olivier, D., Cheynet, V., Blond, J. L., Bouton, O., & Vidaud, M., et al., (2003). Direct involvement of HERV-W Env glycoprotein in human trophoblast cell fusion and differentiation. *Molecular and Cellular Biology, 23*, 3566–3574.

Friedlander-Shani, L., & Podbilewicz, B. (2011). Heterochronic control of AFF-1-mediated cell-to-cell fusion in *C. elegans*. *In "Cell Fusion in Health and Disease*, vol. I: Cell Fusion in Health"(T. Dittmar and K. Zanker, eds.), pp. 200: Springer.

Fukushi, S., Mizutani, T., Saijo, M., Matsuyama, S., Miyajima, N., & Taguchi, F., et al., (2005). Vesicular stomatitis virus pseudotyped with severe acute respiratory syndrome coronavirus spike protein. *Journal of General Virology, 86*, 2269–2274.

Gattegno, T., Mittal, A., Valansi, C., Nguyen, K. C., Hall, D. H., & Chernomordik, L. V., et al., (2007). Genetic control of fusion pore expansion in the epidermis of *Caenorhabditis elegans*. *Molecular Biology of the Cell, 18*, 1153–1166.

Gibbons, D., Erk, I., Reilly, B., Navaza, J., Kielian, M., & Rey, F., et al., (2003). Visualization of the target-membrane-inserted fusion protein of Semliki Forest virus by combined electron microscopy and crystallography. *Cell, 114*, 573–583.

Gong, R., Huang, L., Shi, J., Luo, K., Qiu, G., & Feng, H., et al., (2007). Syncytin-A mediates the formation of syncytiotrophoblast involved in mouse placental development. *Cellular Physiology and Biochemistry, 20*, 517–526.

Gong, R., Peng, X., Kang, S., Feng, H., Huang, J., & Zhang, W., et al., (2005). Structural characterization of the fusion core in syncytin, envelope protein of human endogenous retrovirus family W. *Biochemical and Biophysical Research Communications, 331*, 1193–1200.

Goodman, M., Porter, C. A., Czelusniak, J., Page, S. L., Schneider, H., & Shoshani, J., et al., (1998). Toward a phylogenetic classification of Primates based on DNA evidence complemented by fossil evidence. *Molecular Phylogenetics and Evolution, 9*, 585–598.

Hall, D. H. (1996). MH27 binds to several types of epithelial junctions by EM-immunocytochemistry in the nematode *C. elegans*. *Microscopy and Microanalysis,* 26–27.

Harrison, S. C. (2005). Mechanism of membrane fusion by viral envelope proteins. *Advances in Virus Research, 64*, 231–261.

Harrison, S. C. (2008). Viral membrane fusion. *Nature Structural & Molecular Biology, 15*, 690–698.

Haywood, A. M. (2010). Membrane uncoating of intact enveloped viruses. *The Journal of Virology, 84*, 10946–10955.

Hedgecock, E. M., & White, J. G. (1985). Polyploid tissues in the nematode *Caenorhabditis elegans*. *Developmental Biology, 107*, 128–138.

Heidmann, O., Vernochet, C., Dupressoir, A., & Heidmann, T. (2009). Identification of an endogenous retroviral envelope gene with fusogenic activity and placenta-specific expression in the rabbit: A new "syncytin" in a third order of mammals. *Retrovirology, 6*, 107.

Helming, L., & Gordon, S. (2009). Molecular mediators of macrophage fusion. *Trends in Cell Biology, 19*, 514–522.

Hu, C., Ahmed, M., Melia, T. J., Sollner, T. H., Mayer, T., & Rothman, J. E. (2003). Fusion of cells by flipped SNAREs. *Science, 300*, 1745–1749.

Hughson, F. M., & Reinisch, K. M. (2010). Structure and mechanism in membrane trafficking. *Current Opinion in Cell Biology, 22*, 454–460.

Huppertz, B., Bartz, C., & Kokozidou, M. (2006). Trophoblast fusion: Fusogenic proteins, syncytins and ADAMs, and other prerequisites for syncytial fusion. *Micron, 37*, 509–517.

Huppertz, B., & Borges, M. (2008). Placenta trophoblast fusion. *Methods in Molecular Biology, 475*, 135–147.

Jena, B. P. (2009). Membrane fusion: Role of SNAREs and calcium. *Protein & Peptide Letters, 16*, 712–717.

Jones, D., Russnak, R. H., Kay, R. J., & Candido, E. P. (1986). Structure, expression, and evolution of a heat shock gene locus in *Caenorhabditis elegans* that is flanked by repetitive elements. *The Journal of Biological Chemistry, 261*, 12006–12015.

Kielian, M. (2006). Class II virus membrane fusion proteins. *Virology, 344*, 38–47.

Kim, H. S., Takenaka, O., & Crow, T. J. (1999). Isolation and phylogeny of endogenous retrovirus sequences belonging to the HERV-W family in primates. *Journal of General Virology, 80*(Pt 10), 2613–2619.

Kimble, J. (1981). Alterations in cell lineage following laser ablation of cells in the somatic gonad of *Caenorhabditis elegans*. *Developmental Biology, 87*, 286–300.

Kiontke, K., Barriere, A., Kolotuev, I., Podbilewicz, B., Sommer, R., & Fitch, D. H., et al., (2007). Trends, stasis, and drift in the evolution of nematode vulva development. *Current Biology, 17*, 1925–1937.

Knerr, I., Huppertz, B., Weigel, C., Dotsch, J., Wich, C., & Schild, R. L., et al., (2004). Endogenous retroviral syncytin: Compilation of experimental research on syncytin and its possible role in normal and disturbed human placentogenesis. *Molecular Human Reproduction, 10*, 581–588.

Koh, K., Peyrot, S. M., Wood, C. G., Wagmaister, J. A., Maduro, M. F., & Eisenmann., et al., (2002). Cell fates and fusion in the *C. elegans* vulval primordium are regulated by the EGL-18 and ELT-6 GATA factors—Apparent direct targets of the LIN-39 Hox protein. *Development, 129*, 5171–5180.

Koh, K., & Rothman, J. H. (2001). ELT-5 and ELT-6 are required continuously to regulate epidermal seam cell differentiation and cell fusion in *C. elegans*. *Development, 128*, 2867–2880.

Kolotuev, I., & Podbilewicz, B. (2004). *Pristionchus pacificus* vulva formation: Polarized division, cell migration, cell fusion and evolution of invagination. *Developmental Biology, 266*, 322–333.

Kolotuev, I., & Podbilewicz, B. (2008). Changing of the cell division axes drives vulva evolution in nematodes. *Developmental Biology, 313*, 142–154.

Kontani, K., Moskowitz, I. P., & Rothman, J. H. (2005). Repression of cell-cell fusion by components of the *C. elegans* vacuolar ATPase complex. *Developmental Cell, 8*, 787–794.

Kontani, K., & Rothman, J. H. (2005). Cell fusion: EFF is enough. *Current Biology, 15*, R252–R254.

Labouesse, M. (1997). Deficiency screen based on the monoclonal antibody MH27 to identify genetic loci required for morphogenesis of the *Caenorhabditis elegans* embryo. *Developmental Dynamics, 210*, 19–32.

Liang, C. Y., Wang, L. J., Chen, C. P., Chen, L. F., Chen, Y. H., & Chen, H. (2010). GCM1 regulation of the expression of syncytin 2 and its cognate receptor MFSD2A in human placenta. *Biology of Reproduction, 83*, 387–395.

Lodish, H. F., & Porter, M. (1980). Heterogeneity of vesicular stomatitis virus particles: Implications for virion assembly. *The Journal of Virology, 33*, 52–58.

Louvet-Vallee, S., Kolotuev, I., Podbilewicz, B., & Felix, M. -A. (2003). Control of vulval competence and centering in the nematode *Oscheius* sp.1 CEW1. *Genetics, 163*, 133–146.

Lu, X., & Kang, Y. (2009). Cell fusion as a hidden force in tumor progression. *Cancer Research, 69*, 8536–8539.

Margalit, A., Neufeld, E., Feinstein, N., Wilson, K. L., Podbilewicz, B., & Gruenbaum, Y. (2007). Barrier to autointegration factor blocks premature cell fusion and maintains adult muscle integrity in *C. elegans*. *The Journal of Cell Biology, 178*, 661–673.

Martens, S., & McMahon, H. T. (2008). Mechanisms of membrane fusion: Disparate players and common principles. *Nature Reviews Molecular Cell Biology, 9*, 543–556.

Mason, D. A., Rabinowitz, J. S., & Portman, D. S. (2008). dmd-3, a doublesex-related gene regulated by tra-1, governs sex-specific morphogenesis in *C. elegans*. *Development, 135*, 2373–2382.

Matsuura, Y., Tani, H., Suzuki, K., Kimura-Someya, T., Suzuki, R., & Aizaki, H., et al., (2001). Characterization of pseudotype VSV possessing HCV envelope proteins. *Virology, 286*, 263–275.

Mess, A. M., & Ferner, K. J. (2010). Evolution and development of gas exchange structures in Mammalia: The placenta and the lung. *Respiratory Physiology and Neurobiology, 173*(Suppl), S74–S82.

Mi, S., Lee, X., Li, X., Veldman, G. M., Finnerty, H., & Racie, L., et al., (2000). Syncytin is a captive retroviral envelope protein involved in human placental morphogenesis. *Nature, 403*, 785–789.

Mitreva, M., & Jasmer, D.P. (2006). Biology and genome of *Trichinella spiralis*. In The *C. elegans* research community (Ed.), *WormBook*. WormBook.

Mohler, W. A., Shemer, G., del Campo, J., Valansi, C., Opoku-Serebuoh, E., & Scranton, V., et al., (2002). The type I membrane protein EFF-1 is essential for developmental cell fusion in *C. elegans*. *Developmental Cell, 2*, 355–362.

Mohler, W. A., Simske, J. S., Williams-Masson, E. M., Hardin, J. D., & White, J. G. (1998). Dynamics and ultrastructure of developmental cell fusions in the *Caenorhabditis elegans* hypodermis. *Current Biology, 8*, 1087–1090.

Newman, A. P., & Sternberg, P. W. (1996). Coordinated morphogenesis of epithelia during development of the *Caenorhabditis elegans* uterine-vulval connection. *Proceedings of the National Academy of Sciences of the United States of America, 93*, 9329–9333.

Newman, A. P., White, J. G., & Sternberg, P. W. (1996). Morphogenesis of the *C. elegans* hermaphrodite uterus. *Development, 122*, 3617–3626.

Nguyen, C. Q., Hall, D. H., Yang, Y., & Fitch, D. H. A. (1999). Morphogenesis of the *Caenorhabditis elegans* male tail tip. *Developmental Biology, 207*, 86–106.

Onel, S. F., & Renkawitz-Pohl, R. (2009). FuRMAS: Triggering myoblast fusion in *Drosophila*. *Developmental Dynamics, 238*, 1513–1525.

Oren-Suissa, M., Hall, D. H., Treinin, M., Shemer, G., & Podbilewicz, B. (2010). The fusogen EFF-1 controls sculpting of mechanosensory dendrites. *Science, 328*, 1285–1288.
Oren-Suissa, M., & Podbilewicz, B. (2007). Cell fusion during development. *Trends in Cell Biology, 17*, 537–546.
Oren-Suissa, M., & Podbilewicz, B. (2010). Evolution of programmed cell fusion: Common mechanisms and distinct functions. *Developmental Dynamics, 239*, 1515–1528.
Orendi, K., Gauster, M., Moser, G., Meiri, H., & Huppertz, B. (2010). The choriocarcinoma cell line BeWo: Syncytial fusion and expression of syncytium-specific proteins. *Reproduction, 140*, 759–766.
Pavlath, G. K. (2010). Spatial and functional restriction of regulatory molecules during mammalian myoblast fusion. *Experimental Cell Research, 316*, 3067–3072.
Pettersen, E. F., Goddard, T. D., Huang, C. C., Couch, G. S., Greenblatt, D. M., & Meng, E. C., et al., (2004). UCSF Chimera—A visualization system for exploratory research and analysis. *Journal of Computational Chemistry, 25*, 1605–1612.
Podbilewicz, B. (2000). Membrane fusion as a morphogenetic force in nematode development. *Nematology, 2*, 99–111.
Podbilewicz, B. (2006). Cell fusion. In The *C. elegans* research community (Ed.), *WormBook*. WormBook.
Podbilewicz, B., Leikina, E., Sapir, A., Valansi, C., Suissa, M., & Shemer, G., et al., (2006). The *C. elegans* developmental fusogen EFF-1 mediates homotypic fusion in heterologous cells and in vivo. *Developmental Cell, 11*, 471–481.
Podbilewicz, B., & White, J. G. (1994). Cell fusions in the developing epithelia of *C. elegans*. *Developmental Biology, 161*, 408–424.
Potgens, A. J., Drewlo, S., Kokozidou, M., & Kaufmann, P. (2004). Syncytin: The major regulator of trophoblast fusion? Recent developments and hypotheses on its action. *Human Reproduction Update, 10*, 487–496.
Primakoff, P., & Myles, D. G. (2007). Cell-cell membrane fusion during mammalian fertilization. *FEBS Letters, 581*, 2174–2180.
Rasmussen, J. P., English, K., Tenlen, J. R., & Priess, J. R. (2008). Notch signaling and morphogenesis of single-cell tubes in the *C. elegans* digestive tract. *Developmental Cell, 14*, 559–569.
Renard, M., Varela, P. F., Letzelter, C., Duquerroy, S., Rey, F. A., & Heidmann, T. (2005). Crystal structure of a pivotal domain of human syncytin-2, a 40 million years old endogenous retrovirus fusogenic envelope gene captured by primates. *Journal of Molecular Biology, 352*, 1029–1034.
Rhodin, JA., & Terzakis, J. (1962). The ultrastructure of the human full-term placenta. *Journal of Ultrastructure Research, 6*, 88–106.
Rochlin, K., Yu, S., Roy, S., & Baylies, M. K. (2010). Myoblast fusion: When it takes more to make one. *Developmental Biology, 341*, 66–83.
Rodic, N., Rutenberg, M. S., & Terada, N. (2004). Cell fusion and reprogramming: Resolving our transdifferences. *Trends in Molecular Medicine, 10*, 93–96.
Roussel, A., Lescar, J., Vaney, M. C., Wengler, G., Wengler, G., & Rey, F. A. (2006). Structure and interactions at the viral surface of the envelope protein E1 of Semliki Forest virus. *Structure, 14*, 75–86.
Sapir, A., Avinoam, O., Podbilewicz, B., & Chernomordik, L. V. (2008). Viral and developmental cell fusion mechanisms: Conservation and divergence. *Developmental Cell, 14*, 11–21.
Sapir, A., Choi, J., Leikina, E., Avinoam, O., Valansi, C., & Chernomordik, L., et al., (2007). AFF-1, a FOS-1-regulated fusogen mediates fusion of the anchor-cell in *C. elegans*. *Developmental Cell, 12*, 683–698.
Schejter, E. D., & Baylies, M. K. (2010). Born to run: Creating the muscle fiber. *Current Opinion in Cell Biology, 22*, 566–574.

Schnell, M. J., Buonocore, L., Kretzschmar, E., Johnson, E., & Rose, J. K. (1996). Foreign glycoproteins expressed from recombinant vesicular stomatitis viruses are incorporated efficiently into virus particles. *Proceedings of the National Academy of Sciences of the United States of America, 93*, 11359–11365.

Scott, I., & Youle, R. J. (2010). Mitochondrial fission and fusion. *Essays in Biochemistry, 47*, 85–98.

Sharma-Kishore, R., White, J. G., Southgate, E., & Podbilewicz, B. (1999). Formation of the vulva in *C. elegans*: A paradigm for organogenesis. *Development, 126*, 691–699.

Shemer, G., Kishore, R., & Podbilewicz, B. (2000). Ring formation drives invagination of the vulva in *Caenorhabditis elegans*: Ras, cell fusion, and cell migration determine structural fates. *Developmental Biology, 221*, 233–248.

Shemer, G., & Podbilewicz, B. (2000). Fusomorphogenesis: Cell fusion in organ formation. *Developmental Dynamics, 218*, 30–51.

Shemer, G., & Podbilewicz, B. (2002). LIN-39/Hox triggers cell division and represses EFF-1/fusogen-dependent vulval cell fusion. *Genes & Development, 16*, 3136–3141.

Shemer, G., & Podbilewicz, B. (2003). The story of cell fusion: Big lessons from little worms. *BioEssays, 25*, 672–682.

Shemer, G., Suissa, M., Kolotuev, I., Nguyen, K. C., Hall, D. H., & Podbilewicz, B. (2004). EFF-1 is sufficient to initiate and execute tissue-specific cell fusion in *C. elegans*. *Current Biology, 14*, 1587–1591.

Smallwood, A., Papageorghiou, A., Nicolaides, K., Alley, M. K., Jim, A., & Nargund, G., et al., (2003). Temporal regulation of the expression of syncytin (HERV-W), maternally imprinted PEG10, and SGCE in human placenta. *Biology of Reproduction, 69*, 286–293.

Soe, K., Andersen, T. L., Hobolt-Pedersen, A. S., Bjerregaard, B., Larsson, L. I., & Delaisse, J. M. (2010). Involvement of human endogenous retroviral syncytin-1 in human osteoclast fusion. *Bone, 48*, 837–846.

Sternberg, P. W. (2005). Vulval development. *WormBook,* 1–28.

Stoye, J. P. (2009). Proviral protein provides placental function. *Proceedings of the National Academy of Sciences of the United States of America, 106*, 11827–11828.

Sulston, J. E., & Horvitz, H. R. (1977). Postembryonic cell lineages of the nematode *Caenorhabditis elegans*. *Developmental Biology, 56*, 110–156.

Sulston, J. E., Schierenberg, E., White, J. G., & Thomson, J. N. (1983). The embryonic cell lineage of the nematode *Caenorhabditis elegans*. *Develoomental Biology, 100*, 64–119.

Sulston, J. E., & White, J. G. (1980). Regulation and cell autonomy during postembryonic development of *Caenorhabditis elegans*. *Developmental Biology, 78*, 577–597.

Sutovsky, P. (2009). Sperm-egg adhesion and fusion in mammals. *Expert Reviews in Molecular Medicine, 11*, e11.

Takada, A., Robison, C., Goto, H., Sanchez, A., Murti, K. G., & Whitt, M. A., et al., (1997). A system for functional analysis of Ebola virus glycoprotein. *Proceedings of the National Academy of Sciences of the United States of America, 94*, 14764–14769.

Tamura, K., Oue, A., Tanaka, A., Shimizu, N., Takagi, H., & Kato, N., et al., (2005). Efficient formation of vesicular stomatitis virus pseudotypes bearing the native forms of hepatitis C virus envelope proteins detected after sonication. *Microbes and Infection, 7*, 29–40.

The-C.-elegans-Sequencing-Consortium(1998). Genome sequence of the nematode *C. elegans*: A platform for investigating biology. *Science, 282*, 2012–2018.

Vida, T. A., & Emr, S. D. (1995). A new vital stain for visualizing vacuolar membrane dynamics and endocytosis in yeast. *The Journal of Cell Biology, 128*, 779–792.

Vignery, A. (2000). Osteoclasts and giant cells: Macrophage-macrophage fusion mechanism. *International Journal of Experimental Pathology, 81*, 291–304.

Vignery, A. (2005). Macrophage fusion: The making of osteoclasts and giant cells. *The Journal of Experimental Medicine, 202*, 337–340.

Vignery, A. (2008). Macrophage fusion: Molecular mechanisms. *Methods in Molecular Biology, 475*, 149–161.

Villesen, P., Aagaard, L., Wiuf, C., & Pedersen, F. S. (2004). Identification of endogenous retroviral reading frames in the human genome. *Retrovirology, 1*, 32.

Voisset, C., Blancher, A., Perron, H., Mandrand, B., Mallet, F., & Paranhos-Baccala, G. (1999). Phylogeny of a novel family of human endogenous retrovirus sequences. *HERV-W, in humans and other primates. AIDS Research and Human Retroviruses, 15*, 1529–1533.

Waterhouse, A. M., Procter, J. B., Martin, D. M., Clamp, M., & Barton, G. J. (2009). Jalview Version 2—A multiple sequence alignment editor and analysis workbench. *Bioinformatics, 25*, 1189–1191.

Weissenhorn, W., Hinz, A., & Gaudin, Y. (2007). Virus membrane fusion. *FEBS Letters, 581*, 2150–2155.

Westermann, B. (2010). Mitochondrial fusion and fission in cell life and death. *Nature Reviews Molecular Cell Biology, 11*, 872–884.

White, J. G., Southgate, E., Thomson, J. N., & Brenner, S. (1986). The structure of the nervous system of *Caenorhabditis elegans. Philosophical Transactions of the Royal Society B: Biological Sciences, 314*, 1–340.

White, J. M., Delos, S. E., Brecher, M., & Schornberg, K. (2008). Structures and mechanisms of viral membrane fusion proteins: Multiple variations on a common theme. *Critical Reviews in Biochemistry and Molecular Biology, 43*, 189–219.

Wickner, W., & Schekman, R. (2008). Membrane fusion. *Nature Structural & Molecular Biology, 15*, 658–664.

Yanagimachi, R. (1988). Sperm-egg fusion. N. Duzgunes, and F. Bonner, eds.), *32*, pp. 3–43. Academic Press, Inc, Orlando, FL.

Zavada, J. (1972). Pseudotypes of vesicular stomatitis virus with the coat of murine leukaemia and of avian myeloblastosis viruses. *Journal of General Virology, 15*, 183–191.

CHAPTER 10

Invasive Podosomes and Myoblast Fusion

Elizabeth H. Chen

Department of Molecular Biology and Genetics, Johns Hopkins University School of Medicine, Baltimore, Maryland, USA

I. OVERVIEW

Myoblast fusion is critical for the development, maintenance, and regeneration of skeletal muscles. Despite the identification of many fusion-related molecules in the past decades, the cellular mechanics of myoblast membrane fusion have just begun to be understood. Recent studies using the fruit fly *Drosophila* as a model system have revealed an asymmetric fusogenic synapse composed of an actin-enriched, invasive podosome-like structure (PLS) in one fusion partner and a thin sheath of actin underlying the apposing membrane in the other. The PLS consists of ring of cell adhesion molecules encircling the F-actin-enriched focus

Current Topics in Membranes, Volume 68

0065-230X/10 $35.00
DOI: 10.1016/B978-0-12-385891-7.00001-5

which propels finger-like protrusions from one cell to the other, and PLS invasion is critical for fusion pore formation. The discovery of the PLSs by both light and electron microscopy (EM) allows for the first time the identification of the sites of fusion at the ultrastructural level. These new findings pose challenges to a classic model on myoblast fusion, and provide the foundation for a new conceptual framework for understanding the cellular mechanisms of myoblast fusion *in vivo*.

II. INTRODUCTION

A fascinating feature of skeletal muscle development is the fusion of mononucleated muscle cells (myoblasts) to form multinucleated myotubes (Hauschka, 1994). Myoblast fusion is not only critical for myogenesis during embryonic development but also indispensible for skeletal muscle maintenance and repair during adult life (Bischoff, 1994). In addition, the fusogenic property of myoblasts has been exploited in cell-based gene therapy for the treatment of muscle degenerative disorders (Wagers & Conboy, 2005). Thus, understanding the mechanisms underlying myoblast fusion will provide significant insights into skeletal muscle biology and muscle regeneration.

Unlike epithelial cells that are tightly attached via adherence junctions, myoblasts are migratory mesodermal cells that do not have permanent adhesion partners. Accordingly, while fusion between epithelial cells "simply" involves targeting fusogenic proteins to the adherence junctions, fusion between muscle cells requires more sophisticated interactions between fusion partners in order to create a tight adhesive junction, followed by targeted transport of fusogenic materials to the sites of fusion, and ultimately cell membrane merger. Much of what we have learned about myoblast fusion to date concerns how muscle cells recognize their fusion partners and how they build a tight adhesive junction to promote membrane fusion.

III. MYOBLAST FUSION IN VERTEBRATES

Ever since the seminal finding that multinucleated skeletal muscle fibers are formed by fusion of individual myoblasts during myogenesis (Mintz & Baker, 1967), numerous studies have aimed to identify mechanisms underlying myoblast fusion. Early studies were mostly performed with cultured vertebrate myoblasts, including both primary cultures and transformed myoblast cell lines. Studies using inhibitory chemicals, antibodies and antisense oligos have provided valuable information on the types of macromolecules involved in myoblast fusion. These include cell adhesion molecules, phospholipases, protein kinases, metalloproteases and calcium binding proteins (for review see

Knudsen, 1992; Wakelam, 1985). Recent studies with cultured mammalian myoblasts continued to reveal additional components involved in myoblast fusion (for review see Pavlath, 2010; Rochlin, Yu, Roy, & Baylies, 2010). Many of these molecules have been localized along the broad contact zone of muscle cells, consistent with their playing a role in myoblast fusion. However, it remains unclear how the fusion machinery is assembled by these molecules at the actual sites of membrane merger. In addition, the *in vivo* functions of many of these implicated molecules remain to be elucidated.

Besides studies identifying molecular components involved in myoblast fusion, electron microscopy (EM) has been employed to detect cellular changes during myoblast fusion at the ultrastructural level. Classic EM studies revealed several intriguing structures along the muscle cell contact zone, including vesicles with or without electron-dense cores (Engel, Egar, & Przybylski, 1986; Kalderon & Gilula, 1979), gap junction-like structures (Rash & Fambrough, 1973), electron-dense plaques (Engel et al., 1986), and single or multiple openings along apposing membranes (Kalderon & Gilula, 1979; Lipton & Konigsberg, 1972; Shimada, 1971). The relevance of these intriguing structures to the actual fusion process remains uncertain, since the precise sites of fusion were not pinpointed by molecular or cellular markers in these studies. Thus, a comprehensive understanding of the fusion mechanism requires the identification of the sites of fusion at the ultrastructural level and the localization of implicated molecules relative to these sites.

IV. *DROSOPHILA* AS A MODEL SYSTEM TO STUDY MYOBLAST FUSION

Since the discovery of the first myoblast fusion mutant in *Drosophila* (Rushton, Drysdale, Abmayr, Michelson, & Bate, 1995), the fruit fly has been used as a model system to study this fusion process *in vivo*. In *Drosophila*, the larval body wall muscles start to develop during embryogenesis (Bate, 1990). Through the function of a set of transcription factors and a lateral inhibition process, two types of muscle cells are specified in the embryonic mesoderm, muscle founder cells and fusion competent myoblasts (FCMs) (for review see Baylies, Bate, & Ruiz Gomez, 1998). While muscle founder cells are divided into subgroups, each of which expresses a unique combination of transcription factors (muscle identity genes), all FCMs are specified by a single transcription factor, Lame duck (Lmd) (Duan, Skeath, & Nguyen, 2001). In the embryo, muscle founder cells reside in the outer mesodermal layer and act as sources of attraction for FCMs, which reside in deeper mesodermal layers. FCMs are attracted by founder cells, migrate toward, adhere and fuse with them to form multinucleated myotubes, each of which has its specific position, orientation, and size (Bate, 1990; Beckett & Baylies, 2007). As a consequence, the nucleus of a fused FCM adopts the transcription profile of the founder cell with which it has fused. The entire process of embryonic muscle development takes 5–6 h and each much fiber contains 3–25 nuclei (Bate, 1990), making this process a relatively simple system to dissect myoblast fusion *in vivo*.

At the early stage of *Drosophila* studies of myoblast fusion, it was not immediately apparent if mechanisms identified in flies would be readily applicable to vertebrates. This is because the first set of fusion-related genes identified in *Drosophila* did not correspond to the specific macromolecules revealed by myoblast culture studies. However, it rapidly became clear that most of the *Drosophila* fusion-related genes have vertebrate homologues, many of which have since then been demonstrated to play conserved roles in myoblast culture and/or zebrafish/mouse models (Bach et al., 2010; Kim et al., 2007; Laurin et al., 2008; Moore, Parkin, Bidet, & Ingham, 2007; Nowak, Nahirney, Hadjantonakis, & Baylies, 2009; Pajcini, Pomerantz, Alkan, Doyonnas, & Blau, 2008; Srinivas, Woo, Leong, & Roy, 2007; Vasyutina, Martarelli, Brakebusch, Wende, & Birchmeier, 2009). Thus, myoblast fusion in *Drosophila* shares highly conserved molecular components with vertebrates and the basic mechanisms revealed by fly studies appear to be readily applicable to higher organisms.

V. MYOBLAST FUSION IN *DROSOPHILA*

For any migratory cell to fuse with a partner, a key event is to establish a tight adhesive junction where the two cell membranes can be brought to intimate proximity. Studies of myoblast fusion in *Drosophila* to date suggest that formation of close membrane juxtaposition between muscle founder cells and migratory FCMs is a two-step process that involves the initial muscle cell attachment via cell adhesion molecules and a subsequent reinforcement of membrane apposition via the action of the actin cytoskeleton.

A. Initial Recognition/Attachment Between Muscle Cells

The initial recognition between founder cells and FCMs is mediated by cell type-specific cell adhesion molecules containing Ig domains (Fig. 1). In founder cells, a pair of paralogues, Dumbfounder (Duf)/Kin of IrreC (Kirre) and Rougest (Rst)/IrreC, have redundant functions during myoblast fusion (Ruiz-Gomez, Coutts, Price, Taylor, & Bate, 2000; Strunkelnberg et al., 2001). Thus, loss of either *duf* or *rst* does not lead to a fusion defect, and expression of either *duf* or *rst* can rescue the fusion defect in the *duf rst* double mutant embryos. Rst is expressed in both types of muscle cells, but is only demonstrated to function in

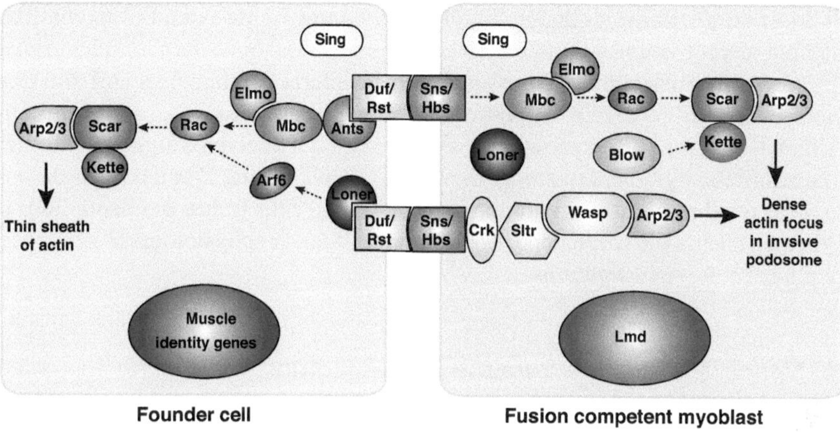

FIGURE 1 Signaling pathways controlling myoblast fusion in *Drosophila*. A simplified diagram describing the signaling pathways in founder cell and fusion competent myoblasts. See text for details.

founder cells (Shelton, Kocherlakota, Zhuang, & Abmayr, 2009; Strunkelnberg et al., 2001). FCMs specifically express a second pair of paralogues, Sticks and stones (Sns) (Bour, Chakravarti, West, & Abmayr, 2000) and Hibris (Hbs) (Artero, Castanon, & Baylies, 2001; Dworak, Charles, Pellerano, & Sink, 2001), which play partially redundant functions during myoblast fusion (Shelton et al., 2009). While the *hbs* mutant appears largely wild type (Artero et al., 2001; Dworak et al., 2001), the *sns* mutant exhibits a significant fusion defect (Bour et al., 2000), which can be partially rescued by Hbs overexpression (Shelton et al., 2009).

Among these cell adhesion molecules, the localization and function of Duf and Sns have been extensively characterized and both proteins have been localized to muscle cell contact sites. By confocal microscopy, Duf and Sns appear to colocalize at these contact sites despite the fact that they are expressed in different cell types (Galletta, Chakravarti, Banerjee, & Abmayr, 2004; Sens et al., 2010). This is likely due to the close membrane apposition at muscle cell contact sites and the limited resolution of the light microscopy (the thickness of cell membranes is ~10 nm and the resolution of light microscopy is 200 nm). Duf physically interacts with Sns, and this interaction has been proposed to mediate the adhesion between founder cells and FCMs (Galletta et al., 2004). Interestingly, ectopic expression of Duf (or Rst) in epithelial cells can attract FCMs to migrate toward these cells (Ruiz-Gomez et al., 2000; Strunkelnberg et al., 2001), demonstrating that Duf (or Rst) expression is sufficient to attract the Sns-expressing FCMs and that founder cells are the organizing centers for myoblast fusion. It remains unclear, however, how Duf-expressing cells attract

FCMs. One possibility is that both founder cells and FCMs extend filopodia-like membrane protrusions to probe their environment, and once the filopodia from a founder cell contacts those from an FCM, the interaction between Duf and Sns would lead to polarized migration of the FCM toward the founder cell. Alternatively, the Duf-expressing cells may send out a secreted signal that forms a concentration gradient to attract the FCMs. In this regard, Duf has been shown to be cleaved to release its ectodomain in cultured cells (Chen & Olson, 2001). Whether such a cleavage occurs *in vivo* and, if so, its physiological relevance requires future investigations.

B. Signal Transduction from the Cell Membrane to the Actin Cytoskeleton

Upon the engagement of transmembrane adhesion molecules, such as Duf and Sns, the cytoplasmic domains of these molecules are involved in transducing the signal for cell fusion to the actin cytoskeleton (Bulchand, Menon, George, & Chia, 2010; Galletta et al., 2004; Kocherlakota, Wu, McDermott, & Abmayr, 2008) (Fig. 1). In founder cells, Duf recruits Antisocial (Ants)/Rolling pebbles (Rols), an adaptor protein containing ankyrin repeat, TPR repeat, and coiled-coil domains (Chen & Olson, 2001; Menon & Chia, 2001; Rau et al., 2001), to sites of fusion via protein–protein interactions (Chen & Olson, 2001). In turn, Ants/Rols has been shown to be involved in recycling Duf to sites of fusion (Menon, Osman, Chenchill, & Chia, 2005). A potential link between Ants/Rols and the actin cytoskeleton was suggested by the biochemical interaction between Ants and Myoblast city (Mbc) (Chen & Olson, 2001), an actin cytoskeleton-associated protein required for fusion (Erickson et al., 1997; Rushton et al., 1995). Mbc is the *Drosophila* homologue of DOCK180, which, in complex with Elmo, functions as a bipartite guanine nucleotide exchange factor (GEF) for the small GTPase, Rac (Brugnera et al., 2002). Both Elmo and Rac have been shown to be required for myoblast fusion (Geisbrecht et al., 2008; Hakeda-Suzuki et al., 2002). Rac regulates a variety of cellular processes by promoting site-specific actin polymerization (for review see Etienne-Manneville & Hall, 2002). For example, Rac is thought to activate and/or localize Scar (also known as WAVE), a nucleation promoting factor (NPF) for the Arp2/3 complex, which is a seven-protein complex with actin nucleation activity (for review see Kurisu & Takenawa, 2009; Stradal & Scita, 2006; Takenawa & Suetsugu, 2007). Scar resides in a heteropentameric Scar complex with Kette (Nap1/Hem), Sra1 (PIR121/CYFIP), Abi, and HSPC300 (for review see Ibarra, Pollitt, & Insall, 2005). Both the Scar and Arp2/3 complexes are essential for myoblast fusion (Berger et al., 2008; Richardson, Beckett, Nowak, & Baylies, 2007; Schroter et al., 2004). In particular, the Scar complex has been shown to localize in a broad domain at muscle cell contact sites and is required in both founder cells and FCMs (Richardson et al., 2007; Sens et al., 2010). In addition to Mbc, a second GEF, Loner, has been shown to partially colocalize with Ants and is required for myoblast fusion (Bulchand et al., 2010; Chen, Pryce, Tzeng, Gonzalez, & Olson, 2003). Loner belongs to the Arf family of GEFs and has specific activity toward Arf6 *in vitro* (Chen et al., 2003). Although an *Arf6* mutant did not exhibit a fusion defect (Dyer et al., 2007), overexpression of a dominant negative form of Arf6 partially blocked fusion (Chen et al., 2003), suggesting that there may be functional redundancy among Arf family members and that Loner may act through other Arfs in the absence of Arf6. A link between the Loner/Arf module and the actin cytoskeleton is shown by the ability of Loner to regulate the proper localization of Rac (Chen et al., 2003). Interestingly, Loner is also expressed in FCMs (Richardson et al., 2007), although its precise function in this cell population requires further investigations.

In FCMs, two Arp2/3 NPFs, Scar and WASP, function together to activate actin polymerization (Fig. 1). The recruitment of Scar to the site of fusion in FCMs may be mediated by Rac, since Scar's enrichment at the leading edge of FCM disappears in *rac* mutant embryos (Gildor, Massarwa, Shilo, & Schejter, 2009). Notably, a member of the Scar complex, Kette, has been shown to genetically interact with a PH domain-containing protein, Blown fuse (Blow) (Doberstein, Fetter, Mehta, & Goodman, 1997), although the molecular function of Blow remains unknown (Schroter et al., 2004). While Scar is functionally required in both cell populations, WASP appears to function only in FCMs, since overexpression of a dominant-negative form of WASP ($WASP^{DN}$) generated a fusion defect when $WASP^{DN}$ was expressed in all muscle cells, but not in founder cells alone (Schafer et al., 2007). Moreover, the WASP-interacting protein, Solitary (Sltr)/dWIP/Vrp1, is only expressed in FCMs (Kim et al., 2007), and is recruited to sites of fusion by the FCM-specific adhesion molecule Sns (Kim et al., 2007; Massarwa, Carmon, Shilo, & Schejter, 2007). The cytoplasmic domain of Sns binds Crk (Kim et al., 2007), a small SH2 and SH3 domain-containing adaptor protein (Galletta, Niu, Erickson, & Abmayr, 1999). Crk, in turn, interacts with Sltr, which forms a tight complex with WASP (Kim et al., 2007) and recruits WASP to sites of fusion (Massarwa et al., 2007; Schafer et al., 2007). The FCM-specific function of the WASP–Sltr complex raises an interesting possibility that FCMs may have a different pattern of actin polymerization than founder cells.

C. *An Invasive Podosome-Like Structure at the Site of Fusion*

1. F-Actin Foci are Cell-Type Specific

The recruitment of actin cytoskeleton-associated proteins to sites of fusion suggests that there may be active actin polymerization at these sites. Indeed, a dense, F-actin-enriched structure (referred to as an F-actin focus or plug) is observed between pairs of founder cells/myotubes and FCMs and is associated with the cell adhesion molecules Duf and Sns (Kesper et al., 2007; Kim et al., 2007; Richardson et al., 2007). Live imaging analyses revealed that these F-actin foci are transient structures that appear and dissolve at the site of fusion prior to each fusion event, with an average lifespan of ~11.9 min (ranging from 5.7 to 29.5 min) (Richardson et al., 2007). Initially, the F-actin foci were thought to be symmetrically localized across the adherent muscle cell membranes (Richardson et al., 2007). However, the extensive colocalization of the F-actin foci with the FCM-specific protein Sltr (Kim et al., 2007) suggests that the majority of the actin polymerization many occur in FCMs. Indeed, by expressing GFP-actin in specific muscle cell types and comparing the accumulation and localization of GFP-actin versus total F-actin (stained with phalloidin), it became clear that these actin-enriched structures were exclusively generated in FCMs (Sens et al., 2010). This conclusion was further supported by live imaging experiments in which dynamic GFP-actin foci were only observed when GFP-actin was specifically expressed in FCMs, but not in founder cells (Sens et al., 2010).

2. FCM-Specific Actin Foci are Invasive

A striking feature of the FCM-specific actin focus is its invasion into the territory of the apposing founder cell, which creates a V- or U-shaped dent on the founder cell membrane (Sens et al., 2010). Three-dimensional reconstruction of the F-actin foci showed that they reside in cup-shaped dents, with the wall of the cup lined with the cell adhesion molecules Duf and Sns (Fig. 2) (Sens et al., 2010). Thus, when viewed along the vertical/longitudinal axis of the cup, an actin focus appears to be encircled by the cell adhesion molecules, Duf and Sns inner (Fig. 2) (Kesper et al., 2007; Sens et al., 2010). Live imaging of these F-actin foci revealed extensive size and shape changes during their lifespan (Sens et al., 2010), suggesting a dynamic actin polymerization process. Taken together, these light microscopic studies suggest that the dynamic, FCM-specific actin foci can invade into the founder cell and deform the founder cell membrane.

Ultrastructural studies by EM also revealed an F-actin-enriched structure between founder cells/myotubes and FCMs (Sens et al., 2010). This structure is localized exclusively at the tip of FCMs, not in founder cells. The area of F-actin enrichment is similar to that of an actin focus under light microscopy. Most strikingly, the tip of the FCM-specific actin focus is composed of multiple finger-like protrusions invading into the apposing founder cell (Fig. 2) (Sens et al., 2010). The remarkable similarities in the sidedness, size, and invasiveness between the F-actin-enriched structures revealed by light microscopy and EM demonstrate that they correspond to the same cellular structure. Since the F-actin foci observed by light microcopy effectively mark the sites of

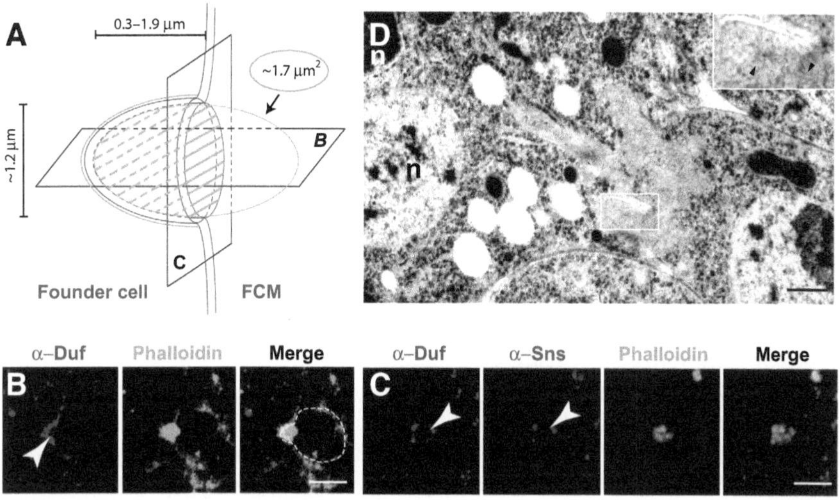

FIGURE 2 The FCM-specific F-actin focus invades the apposing founder cell. (A) Schematic drawing of the asymmetric muscle cell adhesion junction. An F-actin focus (green oval) forms at the tip of the FCM (right) and invades the apposing founder cell (left) to create a cup-shaped dimple. The inner wall of the cup is lined with Sns (blue), and the outer wall with Duf (red). Numbers show average actin foci size (1.7 μm^2), diameter of the adhesive rings (1.2 μm), and depth of invasion (0.3–1.9 μm). (B and C) Confocal images of wild-type embryos showing horizontal (B) and perpendicular (C) pairs of founder cell/myotube and FCM. Embryos labeled with α-Duf (red), phalloidin (green) and α-Sns (blue). (B) An F-actin focus in an FCM (outlined) invading a founder cell. Arrowhead indicates the inward curvature on the founder cell membrane. (C) An F-actin focus encircled by cell adhesion molecules (arrowheads). (D) Ultrastructural details of an invasive F-actin focus. An FCM (pseudo colored pink) projects multiple F-actin-enriched invasive fingers into a binucleated myotube in a wild-type (wt) embryo fixed by HPF/FS. The F-actin-enriched area within the FCMs (boundary marked by dashed green lines) is identified by the light gray coloration and lack of ribosomes and intracellular organelles. Magnified inset shows faint actin filaments (arrowheads) within an invasive finger. n: nuclei in myotube. Scale bars: (B and C) 5 μm; (D) 500 nm. (See Color Insert.)

fusion, the invasive F-actin-enriched structures identified by EM, therefore, provide the first unambiguous cellular marker for myoblast fusion sites at the ultrastructural level.

3. Invasive F-Actin Focus Resembles a Podosome

The actin-enriched foci that form at the junction between fusing muscle cells resemble podosomes, which have been identified in a variety of cultured cells, including monocytic, endothelial, smooth muscle, and skeletal muscle cells (for review see Gimona, Buccione, Courtneidge, & Linder, 2008; Linder, 2009). Podosomes, also called invadopodia in cancer cells, are bipartate structures consisting of an F-actin-enriched core encircled by cell adhesion molecules. They are associated with extracellular matrix (ECM) protease secretion and play important roles in cell adhesion, migration, and invasion (for review see Linder, 2009; Weaver, 2006). Tens or even hundreds of podosomes are often present in a belt-like pattern at the periphery of a single cultured cell. The FCM-specific actin focus encircled by Sns and Duf resembles a single podosome, judged by its size, dynamics, morphology, and protrusive behavior, and is therefore named a podosome-like structure (PLS) (Sens et al., 2010). The FCM-specific PLS represents the first PLS identified in an intact developing tissue. It will be interesting to determine whether this PLS is also associated with ECM protease secretion.

4. A Thin Sheath of Actin in Founder Cell

Founder cells do not form invasive PLSs during myoblast fusion. In fact, wild-type embryos expressing GFP-actin in founder cells do not show visible GFP-actin enrichment at sites of fusion (Sens et al., 2010). It is possible that actin polymerization at the sites of fusion in founder cells is transient and/or masked by the overexpressed GFP-actin. Consistent with this, in certain mutants where fusion is blocked, actin enrichment has been observed as a thin sheath that underlies the founder cell membrane apposing the FCM-specific actin focus (Sens et al., 2010). Thus, founder cells respond to the PLS invasion by making a transient thin sheath of actin at the sites of fusion (Fig. 2).

5. Molecular Components of the PLS

The formation of the F-actin foci with in the FCM-specific PLSs requires both the Scar and WASP complexes, based on the fact that the number and size of actin foci are greatly diminished in *scar sltr*, as well as in *sltr kette*, double mutant embryos (Sens et al., 2010; Sens & Chen, unpublished data). These results are in contrary to a previous report showing enlarged F-actin foci when both the Scar and WASP complexes were disrupted (Gildor et al., 2009). This discrepancy might be due to the differences in the way by which the Scar and WASP function were disrupted (mutant alleles versus overexpression of dominant negative forms of proteins) (Gildor et al., 2009; Sens et al., 2010). In support of the conclusion that both Scar and WASP play a role in F-actin foci formation, Sltr precisely colocalizes with the enlarged actin foci in *scar* or *kette* single mutant embryos (Sens et al., 2010), suggesting that the WASP–Sltr complex promotes actin polymerization in the absence of Scar or Kette. Conversely, recruitment of Scar to cell contact sites is greatly elevated in *sltr* mutant embryos (Sens et al., 2010), and is likely responsible for promoting actin foci formation in the absence of WASP–Sltr. Taken together, Scar and WASP are the major Arp2/3 NPFs activating actin polymerization at sites of fusion.

Although Scar and WASP play redundant roles in actin foci formation, overexpressing one of them in the absence of the other does not rescue the fusion defect in embryos (Sens et al., 2010), suggesting that Scar and WASP have distinct functions during the fusion process. Indeed, the WASP-Sltr complex, but not the Scar complex, promotes PLS invasion. Thus, while F-actin foci in *scar* and *kette* mutants are as invasive as wild type, those in *wasp* and *sltr* mutants are defective in their invasion (Sens et al., 2010). The latter has been confirmed by ultrastructural analysis, which revealed actin-enriched fingers from FCMs folding upon each other without protruding into the apposing founder cell (Sens et al., 2010). On the other hand, the Scar complex plays an independent role in founder cells to promote the formation of the transient thin sheath of actin along the F-actin focus (Sens et al., 2010).

D. Invasive PLSs Promote Fusion Pore Formation

Genetic and cell biological studies suggest that the invasive PLSs promote fusion pore formation. In *sltr* mutant embryos where PLS invasion is defective, cytoplasmic GFP failed to diffuse between myotubes and attached FCMs (Kim et al., 2007; Sens et al., 2010). Furthermore, EM analysis of *sltr* mutant embryos revealed the absence of pores on the cell membranes abutting the F-actin-enriched foci, which correspond to muscle cell contact sites, further supporting this conclusion (Sens et al., 2010). Two conflicting reports showing cytoplasmic GFP transfer from founder cells to FCMs in *dWIP* and *wasp* mutants led to the interpretation that the WASP complex was not required for fusion pore formation, but was required for fusion pore expansion (Gildor et al., 2009; Massarwa et al., 2007). However, these results can be explained by leaky GFP expression in FCMs by the founder cell driver (*rP298-GAL4*) (Sens et al., 2010). In addition to the invasive PLS, regulated actin polymerization by the Scar complex in founder cells also promotes fusion pore formation, since founder cell-expressed GFP does not diffuse into attached FCMs in *kette* mutant embryos (Gildor et al., 2009; Sens et al., 2010).

How does an invasive PLS promote fusion pore formation? Several potential mechanisms are considered here. First, the invasive PLS in the FCM and the thin sheath of actin in the apposing founder cell could mechanically push the two cell membranes against each other and bring them to very close proximity, thus priming them for fusion. Interestingly, podosome-like protrusions generated by leukocytes are thought to bring the inner leaflets of the apical and basal membranes of endothelial cells together for the formation of transcellular pores (Carman et al., 2007). Since PLSs with folded, noninvasive fingers in *sltr* mutant embryos fail to promote fusion pore formation, it is likely that the invasive tips of the protruding fingers in wild-type embryos are fusogenic. Second, the FCM-specific podosome could be associated with protease secretion, give that all podosomes studied so far are associated with protease activity (Linder, 2009). Secreted proteases may degrade ECM proteins and/or the ectodomains of the Ig domain-containing cell adhesion molecules, to remove any physical barriers between the two cell membranes, in order to achieve tight membrane juxtaposition. And third, rapid polymerization at the barbed ends of several actin filaments could generate a local membrane curvature at the tip of an invasive finger. In light of the proposed role of membrane curvature in fusion pore formation (Kozlov, McMahon, & Chernomordik, 2010), a sharply bent local curvature created by the corporative extension of several actin filaments may facilitate membrane fusion. These potential mechanisms are likely to be used in combination to regulate fusion pore formation.

E. A Single-Channel MACRO Fusion Pore Mediates Myoblast Fusion

Fusion pore formation is the hallmark of any membrane fusion event. A classic EM study of *Drosophila* myoblast fusion has led to the conclusion that fusion pores are multiple membrane discontinuities (MMDs) formed along the muscle cell contact zone (Doberstein et al., 1997). Each pore is 50–100 nm in width, and there are no cytoplasmic materials in the lumen (Doberstein et al., 1997). Several groups have since reproduced such fusion pore morphology (Berger et al., 2008; Kim et al., 2007; Massarwa et al., 2007), thus making it a widely accepted view. However, these EM studies were performed using conventional chemical fixation method at room temperature, which is known to generate artifacts including membrane discontinuities (McDonald & Auer, 2006; Zhang & Chen, 2008). Indeed, the same fixation method also resulted in the presence of MMDs between cells that do not normally fuse, as well as between muscle cells in fusion mutants (Sens et al., 2010). In contrast, wild-type and mutant embryos fixed by the high-pressure freezing and freeze substitution (HPF/FS) method, which provides better preservation of the cell membrane (McDonald & Auer, 2006; Zhang & Chen, 2008), do not show any MMDs (Sens et al., 2010). Additional EM studies using the HPF/FS method revealed single-channel macro fusion pores with a diameter between 300 nm and 1.5 μm (Sens et al., 2010). The lumens of these single-channel fusion pores are filled with evenly distributed cytoplasmic materials, indicating active cytoplasmic exchange between the two fusing cells. Interestingly, pores smaller than 300 nm have not been detected thus far (Sens et al., 2010), likely due to the rapid expansion of nascent fusion pores with a diameter smaller than 200 nm (Kaplan, Zimmerberg, Puri, Sarkar, & Blumenthal, 1991; Plonsky, Cho, Oomens, Blissard, & Zimmerberg, 1999; Plonsky & Zimmerberg, 1996).

It is unclear at this point how many nascent fusion pores form during each fusion event. Since each F-actin focus is composed of multiple (~4.3 on average) finger-like protrusions (Sens et al., 2010), multiple nascent fusion pores could potentially form during each fusion event. However, all protrusive fingers extend to different directions and exhibit different depth of invasion, making it unlikely that they would all promote fusion pore formation simultaneously. Alternatively, once a single nascent fusion pore has formed at one of the protrusive finger tips, downstream events, such as actin depolymerization and fusion pore expansion, may proceed and prevent the initiation of additional pores. If the latter is the case, why does each PLS generate multiple invasive fingers? It is conceivable that muscle cell membranes contain specific fusogenic domains of particular lipids and/or proteins required for fusion. Close contact between these fusogenic domains across the apposing cell membranes is likely to be a stochastic event mediated by the protrusive fingers. This possibility is consistent with the wide range of time (between 5.7 and 29.5 min) required for each PLS to promote fusion pore formation (Richardson et al., 2007). Thus, the presence of multiple exploratory fingers in each PLS ensure the successful engagement of fusogenic materials across the two cell membranes, and ultimately fusion pore initiation. Future studies are required to test this hypothesis.

What mediates the expansion of nascent fusion pores? Based on the observation of MMDs in wild-type and fusion mutant embryos (*dWIP* and *wasp*) prepared by chemical fixation, it was proposed that fusion pore expansion is mediated by membrane vesiculation and/or actin polymerization (Doberstein et al., 1997; Gildor et al., 2009; Massarwa et al., 2007). However, these proposals need to be reevaluated in light of the recent finding of invasive PLSs and single-channel macro fusion pores revealed by HPF/FS EM. The previously reported MMDs in mutant embryos of Arp2/3 NPFs (Gildor et al., 2009; Massarwa et al., 2007) were not associated with F-actin enrichment, which marks the muscle cell contact sites, consistent with the idea that these MMDs may have been caused by fixation artifacts on random segments of the cell membrane. Furthermore, single-channel macro fusion pores in wild-type embryos are not associated with vesicles/membrane sacs or F-actin (Sens et al., 2010), suggesting that neither membrane vesiculation nor actin polymerization plays a role in fusion pore expansion, at least for pores with a diameter of over 300 nm. These results are consistent with the findings by Chen et al. (2008) on virus (gp64)-induced fusion of Sf9 cells, which suggest that fusion pore expansion is not mediated by membrane vesiculation or actin polymerization.

F. The Elusive Fusogen

Although actin polymerization is essential for myoblast fusion and may play a role in bringing the adherent membranes into close proximity, it remains likely that additional proteins with transmembrane domains are involved in initiating the actual membrane merger. To date, none of the transmembrane proteins implicated in *Drosophila* myoblast fusion fulfill the definition of a "fusogen," which should be both necessary and sufficient to induce membrane fusion.

There are several plausible explanations for why a fusogen for myoblast fusion, if there is one, has not yet been identified by genetic approaches. First, genetic screens for fusion mutants have not been saturated. Second, the putative fusogenic proteins may be maternally contributed, which would be make it difficult to be identified by a zygotic screen in embryos. Third, there may be more than one fusogenic proteins that have redundant functions, thus a simple loss-function screen would not be sufficient to identity them. Finally, the putative fusogen may play an essential role during early embryonic development, such that mutants in the gene may not survive till the mid-embryogenesis stage to allow functional analyse in myoblast fusion.

If there is indeed a fusogen required for fusing muscle cell membranes, how is it delivered to the sites of fusion? Since all known fusogens are transmembrane proteins, the myoblast fusogen(s) is likely transported to the sites of fusion by vesicles trafficking. Early EM studies revealed clusters of paired vesicles with electron-dense rims aligned along the muscle cell contact sites (Doberstein et al., 1997). It was proposed that these sites correspond to sites of fusion and that exocytosis of these vesicles may release fusogenic materials to trigger fusion (Doberstein et al., 1997). However, two observations raised questions about the relevance of these vesicles to the sites of fusion in wild-type embryos. First, the frequency of the vesicle clusters shown in Doberstein et al. far exceeds that of the F-actin-enriched foci at any developmental stage in wild-type embryos (Kim et al., 2007; Zhang & Chen, 2008). Second, Doberstein et al. (1997) described that these vesicles are present during stage 13 and disappear by stage 14 in wild-type embryos, yet most of the fusion events occurs during stage 14 (Beckett & Baylies, 2007). The recent discovery of invasive PLSs that mark the sites of fusion at the ultrastructural level provides an opportunity to address these questions. Strikingly, the previously reported clusters of vesicles (Doberstein et al., 1997) are not associated with the F-actin-enriched invasive PLSs, and are therefore unlikely to be localized to the sites of fusion. Instead, these vesicles may have been mistargeted to other membrane adhesion sites, for example, between FCMs, as in *sltr* mutant embryos (Kim et al., 2007). Indeed, clusters of paired vesicles are not frequently observed in wild-type embryos (Estrada et al., 2007; Kim et al., 2007; Zhang & Chen, 2008), and the few vesicles seen in wild-type embryos are mostly localized in the cytoplasm but rarely at the PLS (Sens et al., 2010). Taken together, clusters of paired vesicles do not appear to mark the sites of fusion in wild-type embryos as previously suggested, and vesicle accumulation at ectopic sites only occurs in certain fusion mutants.

Although vesicles do not accumulate to the sites of fusion in wild-type embryos, several pieces of evidence suggest that vesicle trafficking plays a role in myoblast fusion. First, there is a gradual increase in the amount of the Ig domain-containing cell adhesion molecules at each site of fusion (Menon et al., 2005), indicating a continuous transport of these molecules to these sites.

Second, a MARVEL domain protein, Singles Bar (Sing) is required for myoblast fusion (Estrada et al., 2007). Interestingly, MARVEL domain proteins in mammals have been implicated in vesicular trafficking and several members of the MARVEL protein family are components of transport vesicles (Sanchez-Pulido, Martin-Belmonte, Valencia, & Alonso, 2002). Although Sing does not appear to be required for transporting cell adhesion molecules to sites of fusion (Estrada et al., 2007), it could be a functional component of vesicles that trafficking other fusion-related components. And third, vesicles with electron-dense rims have been observed at the vicinity of the Golgi and associated with microtubules that point toward the muscle cell periphery (Kim et al., 2007), suggesting that the Golgi-derived vesicles are likely transported by microtubules to the cell membrane in adherent muscle cells. If these vesicles are involved in myoblast fusion, why are they scarcely observed in muscle cells of wild-type embryos? The simplest explanation would be that they undergo rapid exocytosis once they reach the site of fusion, without accumulating in the cytoplasm or at the cell membrane. Additional studies will be required to clearly define the components and trafficking of these intriguing vesicles.

G. Challenge to an Old Model

For the past decade, a prevailing model has been proposed to describe the ultrastructural events associated with myoblast fusion in *Drosophila* (Doberstein et al., 1997). Upon the adhesion of a founder cell and an FCM, pairs of prefusion vesicles from the two muscle cells align across the apposed cell membranes. This prefusion complex resolves into dense membrane plaques between apposed cells. The adherent cells then establish cytoplasmic continuity by fusing the cell membranes at multiple positions along the cell contact zone, followed by membrane vesiculation and fusion pore expansion. However, in light of the unambiguous marking of the sites of fusion by the invasive F-actin foci at the ultrastructural level, much of this classic model needs to be revised.

First, despite the presence of prefusion vesicles in muscle cells, they do not appear to accumulate, singularly or as pairs, at the site of fusion in wild-type embryos. Extensive accumulation of these vesicles only occurs in certain mutant embryos and at membrane adhesion sites without enriched F-actin, presumably due to defects in vesicle targeting and/or exocytosis in these particular mutants. Second, the dense membrane plaques, if they exist in wild-type embryos, may result from the exocytosis of continuously trafficked prefusion vesicles, instead of a synchronized release from accumulated paired vesicles, given the scarcity of the latter. Third, fusion pores have a single-channel morphology, instead of containing multiple neighboring membrane openings. And finally, fusion pore expansion is not likely to be mediated by membrane vesiculation. Given the multiple inaccuracies of the classic model, a new conceptual framework is required to describe the fusogenic structure in *Drosophila* (see below).

H. The Asymmetric Fusogenic Synapse

Based on the recent light and EM studies, the fusogenic structure mediating myoblast fusion can be described as follows (Fig. 3A). It is an asymmetric adhesive junction composed of an F-actin-enriched, invasive PLS from the FCM and a thin sheath of actin underlying the apposing founder cell membrane. Within the boundary of the adhesive rings (or "cups" viewed in 3D space), the PLS extends multiple protrusive fingers to palpitate the founder cell membrane. Meanwhile, prefusion vesicles that are trafficked from the Golgi to the cell membrane via microtubules continuously deposit cell adhesion molecules and perhaps fusogenic proteins and/or lipids to the fusion site. At the tip of one of the invasive fingers, the cell membranes are brought into intimate proximity, such that the putative fusogen can be engaged across the adherent cell membranes, leading to fusion pore formation. Once a nascent fusion pore is formed, rapid

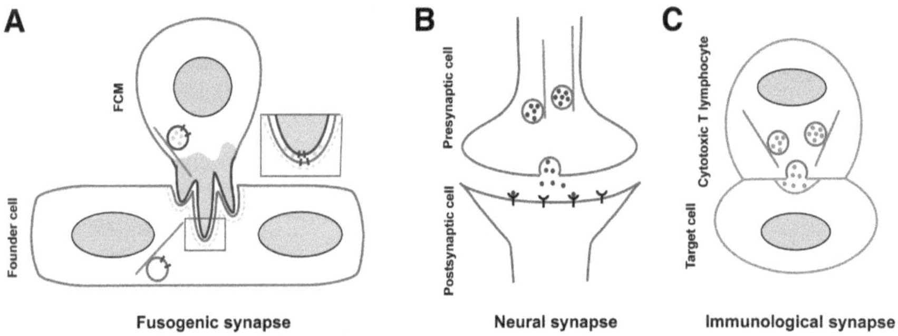

FIGURE 3 The fusogenic synapse shares common features with neural synapse and immunological synapse. Fusogenic synapse (A), neural synapse (B), and immunological synapse (C) are all relatively stable adhesive junctions that are targets for vesicle trafficking. In (A), Ig domain-containing cell adhesion molecules, Duf in the founder cell and Sns in the FCM, mediate the adhesion between the two muscle cells. Engagement of Duf and Sns leads to the formation of asymmetrically localized actin-based structures, an invasive PLS in the FCM (shaded area) and a thin sheath of actin (dashed line) in the founder cell. Adhesion molecules for neural and immunological synapses are not shown in (B and C). In (A–C), vesicles are trafficked along microtubules (light gray lines) to the adhesive junctions. (A) Vesicles in muscle cells may contain fusogenic protein(s) (short dark lines) and putative ECM proteases (gray dots) associated with the PLS. Boxed area is enlarged to show the putative fusogen on cell membranes and the ECM proteases in the intermembrane space. (B) Neural transmitters (gray dots) are transported to the synaptic cleft between the pre- and postsynaptic cells. (C) Pore-forming proteins and esterases (gray dots) are transported to the cleft between the cytotoxic T lymphocyte and the target cell.

actin depolymerization occurs to allow fusion pore expansion. Ultimately, the FCM integrates into the founder cell/myotube and completes a round of fusion. The asymmetric fusogenic structure mediating myoblast fusion is analogous to two known synapses, neural synapse, and immunological synapse (Fig. 3) (Billadeau, Nolz, & Gomez, 2007; Dillon & Goda, 2005; Dustin, 2005; Salinas & Price, 2005; Stinchcombe & Griffiths, 2007). They are all relatively stable adhesive structures that are targets for polarized vesicle trafficking, despite the differences in cell type, adhesion molecule, and physiological function. Since the adhesive junction in myoblast fusion eventually leads to cell membrane fusion, it is accordingly referred to as a "fusogenic synapse" (Fig. 3).

Why is the fusogenic synapse asymmetric? In principle, following the adhesion between a founder cell and an FCM, the cell type-specific adhesion molecules could have organized symmetrical PLSs to push the membranes to closer proximity. However, a potential disadvantage of this configuration is that the two actively protruding structures may not be able to coordinate their activities to effectively respond to the invasion from the apposing cell. On the other hand, an invasive structure from only one of the fusing partners can create a "key and lock" configuration, if its fusion partner only needs to respond to its invasion. Thus, the tight membrane juxtaposition generated by an asymmetric invasion may be a more efficient way to initiate fusion pore formation.

I. Same Cellular Mechanism Underlying Each Fusion Event

An important yet unresolved question in myoblast fusion is whether each fusion event is mediated by the same molecular and cellular mechanisms. Previous studies have shown that myoblast fusion in *Drosophila* embryos proceeds in two temporal phases. In the first phase, bi- or tri-nucleated muscle precursor cells form from fusion between FCMs and mononucleated founder cells or bi-nucleated myotubes (Bate, 1990). These muscle precursor cells appear to pause for a period of time before starting the second phase of fusion, in which they fuse with additional FCMs to form multinucleated myotubes (Beckett & Baylies, 2007). Two different models have been proposed to describe the mechanisms underlying the two temporal phases of fusion based on genetic observations. One model states that the two phases of fusion utilize distinct molecular mechanisms, since some fusion mutants seem to block fusion at the stage of bi- or tri-nucleated myotubes (Rau et al., 2001). However, detailed analyses of these fusion mutants revealed that not all founder cells form bi- or tri-nucleated myotubes (some remain mononucleated) and the limited rounds of fusion are delayed in fusion mutants (Beckett, Rochlin, Duan, Nguyen, & Baylies, 2008). These observations led to a second model in which all fusion-related genes are equally required in the two temporal phases of fusion and that the formation of occasional bi- or tri-nucleated myotubes in fusion mutants is due to maternal contributions in the zygotic mutants (Beckett & Baylies, 2007).

The discovery of the asymmetric fusogenic synapse provides an opportunity to address this question at a cellular level and the new evidence so far lends support to the second model described above. EM studies with serial sections have shown that invasive PLSs are present not only between FCMs and multinucleated myotubes but also between FCMs and mononucleated founder cells (Luo and Chen, unpublished data; Sens et al., 2010). This is consistent with the live-imaging observation that each fusion event is mediated by an F-actin focus (Richardson et al., 2007). In addition, the formation of bi- or tri-nucleated myotubes in strictly zygotic fusion mutants is likely due to inefficient functional compensations by other components of the fusion machinery. For example, in *sltr* mutant, elevated level of the Scar complex promotes the formation of F-actin foci that only exhibit occasional invasive behavior toward founder cells with aberrant fingers (fewer, shorter, and wider) (Sens et al., 2010). These abnormal invasive fingers likely account for the delayed, inefficient, and somewhat random pattern of fusion (resulting in the formation of miniature myotubes with one to four nuclei) (Kim et al., 2007). Taken together, these findings support and extend the model that all fusion events in wild-type embryos (including those in both first and second temporal phases) use the same molecular and cellular machineries to induce myoblast fusion. Thus, FCMs do not need to reinvent themselves after a temporal delay (between the two phases of fusion) to create a molecularly and/or morphologically distinct fusogenic synapse to promote additional rounds of fusion.

VI. RELEVANCE TO VERTEBRATE MYOBLAST FUSION

The identification of PLSs at the sites of fusion in *Drosophila* myoblast fusion raises an interesting question of whether a similar structure is involved in promoting fusion pore formation during vertebrate skeletal muscle differentiation. Loss-function studies with RNAi in cultured mouse C2C12 myoblasts have demonstrated an essential role of the WASP and Scar complexes in myoblast fusion (Kim et al., 2007; Nowak et al., 2009). Given the highly conserved molecular function of the two complexes, it is conceivable that the fly and vertebrate myoblast fusion share a similar cellular mechanism underlying fusion pore initiation.

Studies using mammalian myoblast cultures have yet to pinpoint specific sites of fusion with molecular or cellular markers. Although many proteins, especially cell adhesion molecules, have been localized to the broad muscle cell contact zone along the long axes of the adherent, spindle-shaped muscle cells, none of them appears to be enriched at specific focal points within the contact zone (for review see Pavlath, 2010). A recent live imaging study with mouse C2C12 cells showed that fusion occurred in a region enriched with phosphotidylinositol 4,5-bisphosphate (PI(4,5)P2) (Nowak et al., 2009), although the region is still relatively large and no specific sites of fusion could be distinguished. With regard to the actin cytoskeleton, no significant changes were observed in cultured "wild-type" muscle cells, despite the presence of discrete F-actin-enriched foci in Kette/Nap1 RNAi cells that failed to fuse (Nowak et al., 2009). Taken together, these studies suggest that although the Arp2/3 NPFs are essential for myoblast fusion in cultured cells, actin cytoskeletal rearrangements may occur at a subtler scale in these cells compared with the formation of dense F-actin foci during myoblast fusion in intact embryos. Such a difference in actin cytoskeletal rearrangement could be due to the different cellular environment in cells residing *in vitro* versus *in vivo*. It is well known that cultured myoblasts adopt a spindle-like, elongated shape when they adhere to a 2D culture dish, whereas myoblasts *in vivo* are tear drop-shaped cells, as seen in the 3D environment of *Drosophila* embryos. As a consequence, cultured muscle cells adhere along a broad contact zone, whereas myoblasts *in vivo* attach to myotubes at a focal point. Different modes of cell adhesion, in turn, may lead to differential distributions of the actin cytoskeletal machinery and distinct cytoskeletal responses to fusion signals. Alternatively and/or in addition, the difference in actin cytoskeletal rearrangement could be due to differences in gene expression when muscle cells were transformed to become immortal. In this regard, it is known that transformed cultured myoblasts, which are widely used for myoblast fusion studies, are less efficient in fusion than the primary myoblasts (Wakelam, 1985), indicating a compromise in the fusogenic potential in the immortal cell lines. Thus, to gain a clear understanding of the cellular mechanism underlying myoblast fusion in a physiologically relevant condition, it would be important to examine the actin cytoskeletal changes during mouse embryogenesis and/or muscle repair in 3D tissues *in vivo*.

VII. CONCLUDING REMARKS

Our understanding of myoblast fusion has advanced significantly in recent years with the application of a versatile toolbox including genetics, immunohistochemistry, live imaging, EM, and biochemistry. The discovery of an invasive PLS allows, for the first time, a clear identification of the sites of fusion at the ultrastructural level and provides an unprecedented view of the asymmetric fusogenic synapse. This exciting new development also leads to many important questions for future investigations. For example, what is required for the formation of the invasive fingers and what controls the dynamics of the PLS invasion? Is there an ECM protease activity associated with myoblast fusion?

How is vesicle trafficking regulated and what are the biochemical components of the vesicles? What controls fusion pore expansion? Are similar PLSs used in vertebrate myoblasts fusion *in vivo*? Answers to these questions will undoubtedly lead to significant new insights into the molecular and cellular mechanisms underlying myoblast fusion in flies and vertebrates.

Acknowledgments

I apologize to colleagues whose work could not be cited because of space limitations. I thank Drs. Eric Grote and Duojia Pan and members of the Chen lab for comments on the manuscript. Research in the Chen lab has been supported in part by grants from National Institute of Health, American Heart Association, Muscular Dystrophy Association, and the David and Lucile Packard Foundation.

References

Artero, R. D., Castanon, I., & Baylies, M. K. (2001). The immunoglobulin-like protein Hibris functions as a dose-dependent regulator of myoblast fusion and is differentially controlled by Ras and Notch signaling. *Development, 128*(21), 4251–4264.

Bach, A. S., Enjalbert, S., Comunale, F., Bodin, S., Vitale, N., Charrasse, S., et al. (2010). ADP-ribosylation factor 6 regulates mammalian myoblast fusion through phospholipase D1 and phosphatidylinositol 4,5-bisphosphate signaling pathways. *Molecular Biology of the Cell, 21* (14), 2412–2424.

Bate, M. (1990). The embryonic development of larval muscles in Drosophila. *Development, 110* (3), 791–804.

Baylies, M. K., Bate, M., & Ruiz Gomez, M. (1998). Myogenesis: A view from Drosophila. *Cell, 93* (6), 921–927.

Beckett, K., & Baylies, M. K. (2007). 3D analysis of founder cell and fusion competent myoblast arrangements outlines a new model of myoblast fusion. *Developmental Biology, 309*(1), 113–125.

Beckett, K., Rochlin, K. M., Duan, H., Nguyen, H. T., & Baylies, M. K. (2008). Expression and functional analysis of a novel Fusion Competent Myoblast specific GAL4 driver. *Gene Expression Patterns, 8*(2), 87–91.

Berger, S., Schafer, G., Kesper, D. A., Holz, A., Eriksson, T., Palmer, R. H., et al. (2008). WASP and SCAR have distinct roles in activating the Arp2/3 complex during myoblast fusion. *Journal of Cell Science, 121*(Pt 8), 1303–1313.

Billadeau, D. D., Nolz, J. C., & Gomez, T. S. (2007). Regulation of T-cell activation by the cytoskeleton. *Nature Reviews Immunology, 7*(2), 131–143.

Bischoff, R. (1994). The satellite cell and muscle regeneration. *In* "Myogenesis" (A. G. Engel, and C. Franszini-Armstrong, (Eds.), *Myogenesis* pp. 97–118. McGraw-Hill, New York:.

Bour, B. A., Chakravarti, M., West, J. M., & Abmayr, S. M. (2000). Drosophila SNS, a member of the immunoglobulin superfamily that is essential for myoblast fusion. *Genes & Development, 14* (12), 1498–1511.

Brugnera, E., Haney, L., Grimsley, C., Lu, M., Walk, S. F., Tosello-Trampont, A. C., et al. (2002). Unconventional Rac-GEF activity is mediated through the Dock180-ELMO complex. *Nature Cell Biology, 4*(8), 574–582.

Bulchand, S., Menon, S. D., George, S. E., & Chia, W. (2010). The intracellular domain of Dumbfounded affects myoblast fusion efficiency and interacts with Rolling pebbles and Loner. *PLoS One, 5*(2), e9374.

Carman, C. V., Sage, P. T., Sciuto, T. E., de la Fuente, M. A., Geha, R. S., Ochs, H. D., et al. (2007). Transcellular diapedesis is initiated by invasive podosomes. *Immunity, 26*(6), 784–797.

Chen, A., Leikina, E., Melikov, K., Podbilewicz, B., Kozlov, M. M., & Chernomordik, L. V. (2008). Fusion-pore expansion during syncytium formation is restricted by an actin network. *Journal of Cell Science, 121*(Pt 21), 3619–3628.

Chen, E. H., & Olson, E. N. (2001). Antisocial, an intracellular adaptor protein, is required for myoblast fusion in Drosophila. *Developmental Cell, 1*(5), 705–715.

Chen, E. H., Pryce, B. A., Tzeng, J. A., Gonzalez, G. A., & Olson, E. N. (2003). Control of myoblast fusion by a guanine nucleotide exchange factor, loner, and its effector ARF6. *Cell, 114*(6), 751–762.

Dillon, C., & Goda, Y. (2005). The actin cytoskeleton: Integrating form and function at the synapse. *Annual Review of Neuroscience, 28*, 25–55.

Doberstein, S. K., Fetter, R. D., Mehta, A. Y., & Goodman, C. S. (1997). Genetic analysis of myoblast fusion: Blown fuse is required for progression beyond the prefusion complex. *Journal of Cell Science, 136*(6), 1249–1261.

Duan, H., Skeath, J. B., & Nguyen, H. T. (2001). Drosophila Lame duck, a novel member of the Gli superfamily, acts as a key regulator of myogenesis by controlling fusion-competent myoblast development. *Development, 128*(22), 4489–4500.

Dustin, M. L. (2005). A dynamic view of the immunological synapse. *Seminars in Immunology, 17* (6), 400–410.

Dworak, H. A., Charles, M. A., Pellerano, L. B., & Sink, H. (2001). Characterization of Drosophila hibris, a gene related to human nephrin. *Development, 128*(21), 4265–4276.

Dyer, N., Rebollo, E., Dominguez, P., Elkhatib, N., Chavrier, P., Daviet, L., et al. (2007). Spermatocyte cytokinesis requires rapid membrane addition mediated by ARF6 on central spindle recycling endosomes. *Development, 134*(24), 4437–4447.

Engel, L. C., Egar, M. W., & Przybylski, R. J. (1986). Morphological characterization of actively fusing L6 myoblasts. *European Journal of Cell Biology, 39*(2), 360–365.

Erickson, M. R., Galletta, B. J., & Abmayr, S. M. (1997). Drosophila myoblast city encodes a conserved protein that is essential for myoblast fusion, dorsal closure, and cytoskeletal organization. *Journal of Cell Biology, 138*(3), 589–603.

Estrada, B., Maeland, A. D., Gisselbrecht, S. S., Bloor, J. W., Brown, N. H., & Michelson, A. M. (2007). The MARVEL domain protein, Singles Bar, is required for progression past the prefusion complex stage of myoblast fusion. *Developmental Biology, 307*(2), 328–339.

Etienne-Manneville, S., & Hall, A. (2002). Rho GTPases in cell biology. *Nature, 420*(6916), 629–635.

Galletta, B. J., Chakravarti, M., Banerjee, R., & Abmayr, S. M. (2004). SNS: Adhesive properties, localization requirements and ectodomain dependence in S2 cells and embryonic myoblasts. *Mechanisms of Development, 121*(12), 1455–1468.

Galletta, B. J., Niu, X. P., Erickson, M. R., & Abmayr, S. M. (1999). Identification of a Drosophila homologue to vertebrate Crk by interaction with MBC. *Gene, 228*(1–2), 243–252.

Geisbrecht, E. R., Haralalka, S., Swanson, S. K., Florens, L., Washburn, M. P., & Abmayr, S. M. (2008). Drosophila ELMO/CED-12 interacts with Myoblast city to direct myoblast fusion and ommatidial organization. *Developmental Biology, 314*(1), 137–149.

Gildor, B., Massarwa, R., Shilo, B. Z., & Schejter, E. D. (2009). The SCAR and WASp nucleation-promoting factors act sequentially to mediate Drosophila myoblast fusion. *EMBO Reports, 10* (9), 1043–1050.

Gimona, M., Buccione, R., Courtneidge, S. A., & Linder, S. (2008). Assembly and biological role of podosomes and invadopodia. *Current Opinion in Cell Biology, 20*(2), 235–241.

Hakeda-Suzuki, S., Ng, J., Tzu, J., Dietzl, G., Sun, Y., Harms, M., et al. (2002). Rac function and regulation during Drosophila development. *Nature, 416*(6879), 438–442.

Hauschka, S. D. (1994). The embryonic origin of muscle. *In* "Myology" (A. G. Engel, and C. Franzini-Armstrong, (Eds.), *Myology* pp. 3–73. McGraw Hill Inc., New York.

Ibarra, N., Pollitt, A., & Insall, R. H. (2005). Regulation of actin assembly by SCAR/WAVE proteins. *Biochemical Society Transactions, 33*(Pt 6), 1243–1246.

Kalderon, N., & Gilula, N. B. (1979). Membrane events involved in myoblast fusion. *Journal of Cell Biology, 81*(2), 411–425.

Kaplan, D., Zimmerberg, J., Puri, A., Sarkar, D. P., & Blumenthal, R. (1991). Single cell fusion events induced by influenza hemagglutin: Studies with rapid-flow, quantitative fluorescence microscopy. *Experimental Cell Research, 195*(1), 137–144.

Kesper, D. A., Stute, C., Buttgereit, D., Kreiskother, N., Vishnu, S., Fischbach, K. F., et al. (2007). Myoblast fusion in *Drosophila melanogaster* is mediated through a fusion-restricted myogenic-adhesive structure (FuRMAS). *Developmental Dynamics, 236*(2), 404–415.

Kim, S., Shilagardi, K., Zhang, S., Hong, S. N., Sens, K. L., Bo, J., et al. (2007). A critical function for the actin cytoskeleton in targeted exocytosis of prefusion vesicles during myoblast fusion. *Developmental Cell, 12*(4), 571–586.

Knudsen, K. A. (1992). Fusion of myoblasts. *In* "Membrane fusion" (J. Wilschut, and D. Hoekstra, (Eds.), *Membrane fusion* Marcel Decker, Inc, New York, NY.

Kocherlakota, K. S., Wu, J. M., McDermott, J., & Abmayr, S. M. (2008). Analysis of the cell adhesion molecule sticks-and-stones reveals multiple redundant functional domains, protein-interaction motifs and phosphorylated tyrosines that direct myoblast fusion in *Drosophila melanogaster. Genetics, 178*(3), 1371–1383.

Kozlov, M. M., McMahon, H. T., & Chernomordik, L. V. (2010). Protein-driven membrane stresses in fusion and fission. *Trends in Biochemical Sciences, 35*(12), 699–706.

Kurisu, S., & Takenawa, T. (2009). The WASP and WAVE family proteins. *Genome Biology, 10*(6), 226.

Laurin, M., Fradet, N., Blangy, A., Hall, A., Vuori, K., & Cote, J. F. (2008). The atypical Rac activator Dock180 (Dock1) regulates myoblast fusion in vivo. *Proceedings of the National Academy of Sciences of the United States of America, 105*(40), 15446–15451.

Linder, S. (2009). Invadosomes at a glance. *Journal of Cell Science, 122*(Pt 17), 3009–3013.

Lipton, B. H., & Konigsberg, I. R. (1972). A fine-structural analysis of the fusion of myogenic cells. *Journal of Cell Science, 53*(2), 348–364.

Massarwa, R., Carmon, S., Shilo, B. Z., & Schejter, E. D. (2007). WIP/WASp-based actin-polymerization machinery is essential for myoblast fusion in Drosophila. *Developmental Cell, 12*(4), 557–569.

McDonald, K. L., & Auer, M. (2006). High-pressure freezing, cellular tomography, and structural cell biology. *Biotechniques, 41*(2) 137, 139, 141 passim.

Menon, S. D., & Chia, W. (2001). Drosophila rolling pebbles: A multidomain protein required for myoblast fusion that recruits D-Titin in response to the myoblast attractant Dumbfounded. *Developmental Cell, 1*(5), 691–703.

Menon, S. D., Osman, Z., Chenchill, K., & Chia, W. (2005). A positive feedback loop between Dumbfounded and Rolling pebbles leads to myotube enlargement in Drosophila. *Journal of Cell Biology, 169*(6), 909–920.

Mintz, B., & Baker, W. W. (1967). Normal mammalian muscle differentiation and gene control of isocitrate dehydrogenase synthesis. *Proceedings of the National Academy of Sciences of the United States of America, 58*(2), 592–598.

Moore, C. A., Parkin, C. A., Bidet, Y., & Ingham, P. W. (2007). A role for the Myoblast city homologues Dock1 and Dock5 and the adaptor proteins Crk and Crk-like in zebrafish myoblast fusion. *Development, 134*(17), 3145–3153.

Nowak, S. J., Nahirney, P. C., Hadjantonakis, A. K., & Baylies, M. K. (2009). Nap1-mediated actin remodeling is essential for mammalian myoblast fusion. *Journal of Cell Science, 122*(Pt 18), 3282–3293.

Pajcini, K. V., Pomerantz, J. H., Alkan, O., Doyonnas, R., & Blau, H. M. (2008). Myoblasts and macrophages share molecular components that contribute to cell-cell fusion. *Journal of Cell Biology, 180*(5), 1005–1019.

Pavlath, G. K. (2010). Spatial and functional restriction of regulatory molecules during mammalian myoblast fusion. *Experimental Cell Research, 316*(18), 3067–3072.

Plonsky, I., Cho, M. S., Oomens, A. G., Blissard, G., & Zimmerberg, J. (1999). An analysis of the role of the target membrane on the Gp64-induced fusion pore. *Virology, 253*(1), 65–76.

Plonsky, I., & Zimmerberg, J. (1996). The initial fusion pore induced by baculovirus GP64 is large and forms quickly. *Journal of Cell Biology, 135*(6 Pt 2), 1831–1839.

Rash, J. E., & Fambrough, D. (1973). Ultrastructural and electrophysiological correlates of cell coupling and cytoplasmic fusion during myogenesis in vitro. *Developmental Biology, 30*(1), 166–186.

Rau, A., Buttgereit, D., Holz, A., Fetter, R., Doberstein, S. K., Paululat, A., et al. (2001). rolling pebbles (rols) is required in Drosophila muscle precursors for recruitment of myoblasts for fusion. *Development, 128*(24), 5061–5073.

Richardson, B. E., Beckett, K., Nowak, S. J., & Baylies, M. K. (2007). SCAR/WAVE and Arp2/3 are crucial for cytoskeletal remodeling at the site of myoblast fusion. *Development, 134*(24), 4357–4367.

Rochlin, K., Yu, S., Roy, S., & Baylies, M. K. (2010). Myoblast fusion: When it takes more to make one. *Developmental Biology, 341*(1), 66–83.

Ruiz-Gomez, M., Coutts, N., Price, A., Taylor, M. V., & Bate, M. (2000). Drosophila dumbfounded: A myoblast attractant essential for fusion. *Cell, 102*(2), 189–198.

Rushton, E., Drysdale, R., Abmayr, S. M., Michelson, A. M., & Bate, M. (1995). Mutations in a novel gene, myoblast city, provide evidence in support of the founder cell hypothesis for Drosophila muscle development. *Development, 121*(7), 1979–1988.

Salinas, P. C., & Price, S. R. (2005). Cadherins and catenins in synapse development. *Current Opinion in Neurobiology, 15*(1), 73–80.

Sanchez-Pulido, L., Martin-Belmonte, F., Valencia, A., & Alonso, M. A. (2002). MARVEL: A conserved domain involved in membrane apposition events. *Trends in Biochemical Sciences, 27* (12), 599–601.

Schafer, G., Weber, S., Holz, A., Bogdan, S., Schumacher, S., Muller, A., et al. (2007). The Wiskott-Aldrich syndrome protein (WASP) is essential for myoblast fusion in Drosophila. *Developmental Biology, 304*(2), 664–674.

Schroter, R. H., Lier, S., Holz, A., Bogdan, S., Klambt, C., Beck, L., et al. (2004). kette and blown fuse interact genetically during the second fusion step of myogenesis in Drosophila. *Development, 131*(18), 4501–4509.

Sens, K. L., Zhang, S., Jin, P., Duan, R., Zhang, G., Luo, F., et al. (2010). An invasive podosome-like structure promotes fusion pore formation during myoblast fusion. *Journal of Cell Biology, 191* (5), 1013–1027.

Shelton, C., Kocherlakota, K. S., Zhuang, S., & Abmayr, S. M. (2009). The immunoglobulin superfamily member Hbs functions redundantly with Sns in interactions between founder and fusion-competent myoblasts. *Development, 136*(7), 1159–1168.

Shimada, Y. (1971). Electron microscope observations on the fusion of chick myoblasts in vitro. *Journal of Cell Biology, 48*(1), 128–142.

Srinivas, B. P., Woo, J., Leong, W. Y., & Roy, S. (2007). A conserved molecular pathway mediates myoblast fusion in insects and vertebrates. *Nature Genetics, 39*(6), 781–786.

Stinchcombe, J. C., & Griffiths, G. M. (2007). Secretory mechanisms in cell-mediated cytotoxicity. *Annual Review of Cell and Developmental Biology, 23*, 495–517.

Stradal, T. E., & Scita, G. (2006). Protein complexes regulating Arp2/3-mediated actin assembly. *Current Opinion in Cell Biology, 18*(1), 4–10.

Strunkelnberg, M., Bonengel, B., Moda, L. M., Hertenstein, A., de Couet, H. G., Ramos, R. G., et al. (2001). rst and its paralogue kirre act redundantly during embryonic muscle development in Drosophila. *Development, 128*(21), 4229–4239.

Takenawa, T., & Suetsugu, S. (2007). The WASP-WAVE protein network: Connecting the membrane to the cytoskeleton. *Nature Reviews Molecular Cell Biology, 8*(1), 37–48.

Vasyutina, E., Martarelli, B., Brakebusch, C., Wende, H., & Birchmeier, C. (2009). The small G-proteins Rac1 and Cdc42 are essential for myoblast fusion in the mouse. *Proceedings of the National Academy of Sciences of the United States of America, 106*(22), 8935–8940.

Wagers, A. J., & Conboy, I. M. (2005). Cellular and molecular signatures of muscle regeneration: Current concepts and controversies in adult myogenesis. *Cell, 122*(5), 659–667.

Wakelam, M. J. (1985). The fusion of myoblasts. *Biochemical Journal, 228*(1), 1–12.

Weaver, A. M. (2006). Invadopodia: Specialized cell structures for cancer invasion. *Clinical & Experimental Metastasis, 23*(2), 97–105.

Zhang, S., & Chen, E. H. (2008). Ultrastructural analysis of myoblast fusion in Drosophila. *In* "Cell fusion: Overviews and methods" (E. H. Chen (Ed.), *Cell fusion: Overviews and methods* pp. 275–297. Humana Press, New Jersey.

PART 5

THEORETICAL MODELING

CHAPTER 11

Lipid Acrobatics in the Membrane Fusion Arena

Albert J. Markvoort[1] and Siewert J. Marrink[2]

[1]Institute for Complex Molecular Systems & Biomodeling and Bioinformatics Group, Eindhoven University of Technology, Eindhoven, The Netherlands

[2]Groningen Biomolecular Sciences and Biotechnology Institute & Zernike Institute for Advanced Materials, University of Groningen, Groningen, The Netherlands

I. OVERVIEW

In this review, we describe the recent contribution of computer simulation approaches to unravel the molecular details of membrane fusion. Over the past decade, fusion between apposed membranes and vesicles has been studied using a large variety of simulation methods and systems. Despite the variety in techniques, some generic fusion pathways emerge that predict a more complex

Current Topics in Membranes, Volume 68 0065-230X/10 $35.00

DOI: 10.1016/B978-0-12-385891-7.00011-8

picture beyond the traditional stalk–pore pathway. Indeed the traditional pathway is confirmed in particle-based simulations, but in addition alternative pathways are observed in which stalks expand linearly rather than radially, leading to inverted-micellar or asymmetric hemifusion intermediates. Simulations also suggest that the first barrier to fusion is not the formation of the stalk, but rather, the formation of a lipid bridge consisting of one or two lipids only. Fusion occurring during the fission process involves other intermediates, however, and is not just fusion reversed. Finally, recent progress in simulations of peptide and protein-mediated fusion shows how fusion proceeds in a more biologically relevant scenario.

II. INTRODUCTION

The fusion and fission of membranes is an essential process in cell biophysics, occurring during exo- and endocytosis, intracellular trafficking and enveloped virus infection. Membrane fusion is also important in a range of biomedical applications such as in gene or drug delivery. A wide range of regulatory protein complexes exists *in vivo*. Due to the wide variety and complexity of fusion protein arrays, the molecular picture of protein-mediated fusion and fission is largely unclear. The basic mechanism, however, is believed to be primarily determined by the physics of lipid–lipid interactions (Chernomordik & Kozlov, 2008; Jahn & Grubmuller, 2002; Lentz, Malinin, Haque, & Evans, 2000).

This basic mechanism seems simple enough: two vesicles touch and merge to form a single vesicle. This process is understood to follow various discrete stages. It all starts with two separate vesicles that somehow come into close contact. An initial contact is formed involving lipid mixing of the contacting monolayers, leaving the remainder of the vesicle intact, a stage called hemifusion. Once the vesicle contents mix, full fusion is reached. These stages are macroscopic phenomena, relatively easy to discern. However, the structures of these stages and their transitions on a microscopic scale are hard to verify experimentally and have therefore been subject of theoretical and simulation studies. Theoretical studies have traditionally been applied using continuum models, providing insight into the energetics and nature of possible fusion intermediates (for reviews see Kozlovsky, Chernomordik, & Kozlov, 2002; Muller, Katsov, & Schick, 2006). Whereas continuum models depend on the assumption of fusion pathways rather than to be able to predict them, the simulation approach is essentially free of any assumptions about the character and sequence of fusion intermediates. Moreover, once properly parameterized, computer simulations can facilitate the direct visualization of the fusion process and may be used to estimate energy barriers. In the past decade, simulation studies have contributed valuable insights to the fusion field; these include the prediction of alternative fusion pathways, determination of the rate-limiting steps, and, most recently, exploration of novel roles for fusion proteins.

This review aims to give an overview of the current computational perspective on membrane fusion. We consider particle-based simulations of apposed bilayers, bilayers and vesicles, or between vesicles. We do not try to be comprehensive, but rather give a representative overview of the field. Additional information can be found in a number of other reviews (Marrink, De Vries, & Tieleman, 2009; Muller et al., 2006; Shillcock & Lipowsky, 2006). Our review is organized as follows. We set out with a short overview of the historical background of fusion-related simulation studies, introducing the main particle-based simulation techniques used. We proceed with a detailed description of the various fusion pathways that are observed in these simulations, and the energetics involved. Special attention is devoted to discuss the accumulating evidence that the first barrier to fusion is in fact the splaying of a single lipid tail. Subsequently, we describe the process of vesicle fission, showing that it is not just fusion reversed. The final section deals with the growing body of simulations probing the effect of fusion peptides and proteins on the fusion process. A short outlook section concludes this review.

III. HISTORICAL BACKGROUND

The simplest geometry for simulating membrane fusion seems to be provided by the fusion of two planar membranes. As simulations of flat membranes only require a relatively small simulation box using periodic boundary conditions, they are computationally least demanding and have thus been used in the earliest studies on fusion relevant events, such as lamellar to nonlamellar phase transitions (Knecht, Mark, & Marrink, 2006; Marrink & Mark, 2004), pore formation (Leontiadou, Mark, & Marrink, 2004; Tieleman, Leontiadou, Mark, & Marrink, 2003; Tolpekina, den Otter, & Briels, 2004a), disorientation of lipid tails in closely apposed bilayers (Ohta-Iino et al., 2001), and the evolution of hydrophobic defects (Tieleman & Bentz, 2002). An early simulation of full fusion of two flat membranes was by means of lattice Monte Carlo (MC) (Muller, Katsov, & Schick, 2002, 2003), where lipids were described as amphiphilic diblock copolymers and the surrounding solvent as a melt of hydrophilic homopolymers, each representing a group of solvent molecules.

Simulations of fusion of two planar membranes, however, have an implicit problem in that the boundary conditions may influence the simulations; the volume of solvent between the two fusing membranes remains constant until a fusion pore opens to allow the solvent to flow into the spaces behind the fusing membranes. This may influence the dynamics and require the two membranes to be initially already so close together that they are effectively dehydrated along their whole surface. Fusing two vesicles or a vesicle with a planar membrane avoids this difficulty as the solvent can flow around the vesicle in response to any local fluctuation of its shape. However, because of the larger system sizes required and the concomitant increase in number of molecules involved, the simulation of complete vesicles, and their fusion, comes at the expense of much higher computational cost and is thus only feasible using less detailed models and/or very small vesicles. The first simulation study of the fusion of two vesicles was performed by Noguchi and Takasu (2001). They studied the fusion pathway of two lipid vesicles using Brownian dynamics (BD) simulations. In these simulations no solvent molecules are present and the amphiphilic molecules were modeled as small rigid rods, built up of three particles, one hydrophilic and two hydrophobic.

Although solvent-free models have a computational advantage through not having to simulate the bulk solvent, the absence of solvent is a disadvantage as the solvent will influence possible membrane conformations as well as the dynamics of the membrane. These disadvantages were overcome by using another simulation technique, dissipative particle dynamics (DPD), which includes explicit solvent particles. This technique has been shown to reproduce the correct hydrodynamic forces appropriate to a fluid, and is capable of exploring the phase behavior of lipid molecules and simulating vesicle formation (Li & Liu, 2005; Shillcock & Lipowsky, 2005). Li and Liu (2005) used DPD to study the fusion of two flat membranes consisting of linear lipids and Shillcock and Lipowsky (2005) used this technique to study the fusion of a flat membrane with a small (28 nm diameter) vesicle. Tensionless vesicles and membranes did not fuse but adhered to each other, whereas fusion could be induced by stretching the membranes. Although their lipid model was already more detailed in having two hydrophobic chains, they did not attempt to represent specific lipid molecules. Instead, they aimed to capture the essential features of tension-induced fusion, arguing that membrane fusion is a ubiquitous process observed in diverse amphiphilic systems, having quite distinct microscopic interactions.

The final technique used, which at the expense of higher computational cost can provide a more accurate chemical representation, is molecular dynamics (MD). Because of the computational cost coarse-grained (CG) models were introduced into the fusion field by different groups, for example, Marrink (Marrink & Mark, 2003), Stevens (Stevens, Hoh, & Woolf, 2003), and Smeijers (Smeijers, Markvoort, Pieterse, & Hilbers, 2006). The use of MD allowed, for instance, to discriminate between different types of lipids. Interesting differences were found in the fusion pathways between, for example, PC and PE lipids, or lysolipids. More recently also simulations at atomistic detail have been reported of full vesicle fusion. The first such study was performed by Knecht and Marrink (2007) on a highly fusogenic mixture of

DPPC and palmitic acid. Because of the high computational cost of such atomistic MD simulations, they considered the self-fusion of a very small ~15-nm vesicle with its periodic image. Noteworthy is also the introduction of distributed computing to the fusion arena by Kasson et al. (2006). The availability of thousands of processor nodes allows for systematic studies on fusion pathway statistics.

IV. FUSION PATHWAYS AT THE MOLECULAR LEVEL

The protein-free fusion process has been studied using a variety of simulation techniques, that not only vary in the coarseness of the representations of both the lipids and the solvent but also in the geometry of the fusing membranes and in the way membranes are brought into sufficiently close juxtaposition to initiate fusion. Despite all differences in the simulation techniques, similar processes and fusion intermediates appear, suggesting that they are robust against details of the simulation methods, thus providing strong evidence of their universality. Based on these simulations we identify three fusion pathways: one traditional pathway that we denote the *symmetric stalk expansion pathway*, and two novel, alternative pathways, the *inverted micelle pathway* and the *stalk–pore complex pathway*. Details of these pathways are presented below, together with an evaluation of their composition dependency. Representative snapshots from simulations, illustrating these pathways, are shown in Fig. 1.

A. Symmetric Stalk Expansion Pathway

A widely observed pathway is the *symmetric stalk expansion pathway* that has been suggested from continuum-elasticity models (Kozlovsky et al., 2002) and was first described in Kozlov and Markin (1983). In this pathway, first an hourglass-like connection is made between the outer (*cis*) monolayers of both membranes. This early hemifusion connection is referred to as the fusion stalk. At this stage the outer monolayers can start mixing whereas the two inner (*trans*) monolayers are still completely separated. This stalk then expands radially (axially symmetric), either resulting in the direct formation of a fusion pore or in the formation of a –symmetric– hemifusion diaphragm (HD). The HD is denoted symmetric as it is formed by the two *trans* monolayers of the fusing membranes. The *trans* monolayers still do not mix until a pore is formed in this HD. Upon formation of this pore full fusion is obtained and the inner monolayers as well as the vesicle contents can start mixing.

In this pathway there is thus, apart from usual flip–flop (which, however, is too slow to be observed in realistic simulation models), no exchange of lipids

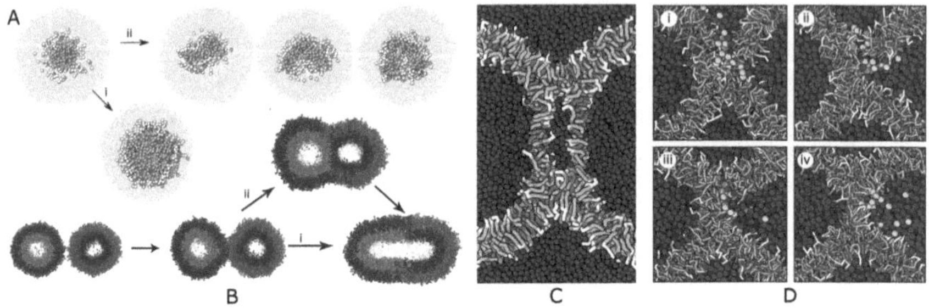

FIGURE 1 Traditional and alternative fusion pathways. (A) All fusion pathways start with the formation of a stalk. This stalk can then expand either radially (i) or anisotropically (ii) (Smeijers, Markvoort, et al., 2006). Based on the fusion intermediates we discern three fusion pathways. (B) In the first pathway, the traditional symmetric stalk expansion pathway, radial stalk formation is followed by direct pore formation (i) or proceeds via a hemifusion diaphragm intermediate (ii) (reproduced with permission from Kasson & Pande, 2007). (C) In the second pathway, the inverse micelle pathway, anisotropic stalk expansion results in formation of an inverse micellar intermediate (adapted from Marrink et al., 2009). (D) In the third pathway, the stalk–pore complex pathway, the stalk nucleates a pore in its vicinity which the stalk subsequently encircles by expanding anisotropically, resulting in the formation of an asymmetric hemifusion diaphragm (reprinted with permission from Smeijers, Markvoort, et al., 2006). Lipids in panels A, C, D are represented with white head groups and gray tails; in panel B lipids are depicted with different shades of gray between *cis* and *trans* monolayers, and between vesicles. Water is depicted either as spheres or omitted for clarity. All snapshots show cross sections through two fusing vesicles, either perpendicular (A) or parallel (B–D) to the fusion coordinate. (See Color Insert.)

between the *cis* and *trans* leaflets of the vesicles and there is no mixing between the vesicle interior and its exterior. The direct pore formation was first observed in BD simulations (Noguchi & Takasu, 2001), which used rigid lipids and implicit solvent. Later this direct pore pathway was also observed with more detailed CG models using DPD (Gao, Lipowsky, & Shillcock, 2008; D. W. Li & Liu, 2005) and MD (Kasson et al., 2006) as well as in atomistic (Kasson, Lindahl, & Pande, 2010) simulations (see Fig. 1A and B). The stalk–HD–pore transition was first observed by Marrink and coworkers, both in CG (Marrink & Mark, 2003) and later in atomistic (Knecht & Marrink, 2007) MD simulations. Using CG MD simulation of over 10,000 fusion events (Kasson et al., 2006) showed that both pathways can happen, randomly, in the same system.

B. Alternative Pathways

Many simulations also demonstrate other fusion pathways, which never have been suggested and analyzed by continuum approaches. A similarity between these fusion pathways is that they are much less symmetric than the priorly described traditional mechanism. Although the first stage, the stalk formation, is the same once the stalk is formed instead of growing radially, it undergoes anisotropic growth forming an elongated (linear) connection (see Fig. 1A).

In the *inverted micelle pathway* this expansion proceeds in a bent manner, forming into a banana-shaped stalk. Upon closing on further bending, a circular stalk is formed, giving rise to a fusion intermediate in which an inverted micelle is trapped in between two vesicles. Such an inverted micellar intermediate (IMI) has for example been reported in the fusion of two membranes by Li and Liu (2005) in DPD simulations in case of lipids with sufficiently negative spontaneous curvature, and also by Marrink, Fuhrmans, Risselada, and Periole (2008) in CG MD simulations of mixed PC/PE vesicles (Fig. 1C). Formation of a pore through rupture of one of the membranes separating the vesicle interiors from the water bubble, then leads to the hemifused state. Note that, in contrast to the traditional pathway, in this case the HD is asymmetric, consisting of the *cis* and *trans* monolayer of only one of the fusing membranes (Muller, Katsov, & Schick, 2003). A similar linear stalk expansion is also seen in simulations where two vesicles, or a vesicle and a membrane, do not fuse at the central location, that is, the initial point of closest approach, but where first a (large) flattened contact is formed. This is seen both in some CG MD (Stevens et al., 2003) as well as DPD (Grafmuller, Shillcock, & Lipowsky, 2007, 2009) simulations. The stalk is formed and expands along the strained membrane at the contact edge; this is attributed to the high curvature at the edge with lipids already in a tilted or splayed conformation. Via a bent stalk an asymmetric HD is also formed here.

The third pathway, that is, *the stalk–pore complex pathway*, also starts with the stalk expanding linearly. However, in this mechanism a key feature is the destabilization of the contacting membranes by the elongated stalk, as a result of which pores appear in the fusing bilayers in its vicinity. The stalk then starts encircling these pores. When pores nucleate in both membranes the bent stalk aligns them, forming an incomplete fusion pore, and the fusion process is completed by propagation of the stalk along the edges of the aligned pores to produce a fusion pore. This pathway was first observed in the BD simulations at high temperatures (Noguchi & Takasu, 2001) and lattice MC simulations (Muller et al., 2002, 2003), and later also seen in CG MD simulations (Smeijers, Markvoort, et al., 2006). If, on the other hand, the stalk encircles a pore in only one of the membranes, an asymmetric HD is formed again (Fig. 1D). Full fusion is then completed by the formation of a fusion pore upon rupturing of this HD as in the traditional pathway. This pathway was also observed in the before-mentioned MC simulations (Muller et al., 2003) as well as in CG MD simulations (Marrink & Mark, 2003; Smeijers, Markvoort, et al.,

2006) and later confirmed by atomistic simulations (Knecht & Marrink, 2007). Somewhat analogous is the pathway observed by Gao et al. (2008) using DPD, where high tension caused rupture of one of the membranes in the contact zone, also resulting in an HD consisting of the two monolayers of the still intact bilayer.

Thus, simulations of a large number of fusion events show some common fusion pathways, which involve much more disordered and less symmetric intermediate states than is typically assumed (continuum models routinely assume that the fusion intermediates have axially symmetric shapes). Moreover, simulations reveal that the conformations of individual lipids play a crucial role in membrane processes, which although at a very small scale cannot be neglected. Surprisingly, a number of simulations showed that a given system can follow several pathways. A first example is the before-mentioned study by Kasson et al. (2006) where systems of exactly the same size and composition follow the stalk–pore pathway at random with or without the HD intermediate. Even stronger, a given system can fuse following both radial and anisotropic stalk expansion, porous and nonporously (Marrink & Mark, 2003; Smeijers, Markvoort, et al., 2006), where the pathway again seems to be chosen rather stochastically.

How to distinguish, experimentally, between these different pathways? As stated before, the HDs in these linear stalk expansion pathways are not classical ones in the sense that they are comprised of both leaflets of one intact bilayer instead of by the two *trans* monolayers. This implies mixing between *trans* lipids from one membrane and *cis* lipids from the other. Incorporation of lipids with sufficiently slow flip–flop rates allows one, in principle, to probe this mechanism. Linear stalk expansion also results in a partially confined solvent cavity between the two fusing membranes. Upon full encircling this originally external solvent is thus internalized. As suggested in Risselada, Knutzer, and Grubmuller (2011), reasonably large fluorescent molecules, that are too large to diffuse through the leakage pores, may get encapsulated this way offering a route to experimental testing. Moreover, when (transient) holes emerge in the vesicles before the stalk completely encircles the contact site, the contents of the vesicles may mix with the exterior phase. The pathways with linear stalk expansion are thus a leaky way to fuse vesicles, offering a molecular explanation for those effects seen in experimental fusion assays (Frolov, Dunina-Barkovskaya, Samsonov, & Zimmerberg, 2003; Muller et al., 2003) and providing the possibility of lysis, which the standard pathways do not. At the same time, leaky fusion appears not advantageous for living cells as content separation constitutes the primary purpose of vesicle transport. This suggests another potential role for fusion proteins, that is, to avoid the stalk from expanding linearly.

C. Composition Dependence

Although simulations show that the choice of fusion pathway is stochastic, that is, the same system can follow multiple pathways, membrane composition does play a role in the prevalence of one pathway over the other. As evidenced from experimental data, the formation of different fusion intermediates can be correlated to the lipids, spontaneous curvature (reviewed in Chernomordik & Kozlov, 2003). This dependence is also apparent from various simulation studies. For instance, DPD simulations (Li & Liu, 2005) showed a stalk–pore complex or IMI for lipids possessing a small head group versus traditional stalks for lipids with larger head groups. Using MD simulations the compositional dependence was studied in more detail. An increased speed of stalk formation on increased PE over PC content is observed by Marrink and Mark (2004) in a study of the lamellar to inverted hexagonal phase transition, a transition that proceeds through stalk formation and subsequent stalk elongation. This study also shows that for a narrow composition/hydration range the stalks do not elongate but remain stable instead. Such a phase, denoted rhombohedral or stalk phase, has also been observed experimentally (Yang & Huang, 2002). Concerning the fusion between small vesicles, Marrink and Mark (2003) found that for a pure POPE system the hemifused state is stable, whereas for mixed PC:PE systems a fusion pore appears quickly and for pure DPPC formation of a HD was never observed. By systematically changing the PC:PE ratio, Kasson and Pande (2007) showed the dependence on the lipid composition of the relative prevalence of a stalk–HD–fusion pore pathway over a less frequent direct transition from stalk to fusion pore. Increased PC content causes increased activation energies for formation of the initial stalk-like intermediate for fusion and of hemifusion intermediates, resulting in many unfused vesicles and an increased utilization of the direct fusion pathway. Increased PE content, on the other hand, not only increases the speed of stalk formation but also stabilizes hemifused states, such that a 1:1 PC:PE ratio offered the fastest rate to full fusion. Stabilization of the HD by PE lipids is confirmed by recent CG MD simulations of Nishizawa and Nishizawa (2010b). These authors showed that DOPE accumulates at the high curvature regions of a preformed HD.

For vesicles with asymmetric lipid composition the rate of fusion not only depends on the overall lipid composition but also on the distribution of lipid types over the monolayers. Marrink and Mark (2003) for instance showed that PE, having a negative spontaneous curvature, increases the fusogenicity when present in the outer monolayer, and stabilizes hemifused states when present in the inner monolayer. Oppositely, lysoPC, having a high positive spontaneous curvature, decreases the likeliness to fuse when present in the outer monolayer, whereas only present in the inner monolayer, it is found to accelerate fusion by strongly destabilizing the HD. However, more simulation data are needed to get a better understanding of the relationship between lipid composition and fusion intermediates. Clear is that the propensity to fuse is strongly dependent on lipid composition, consistent with evidence from a large number of experimental studies (e.g., reviewed in Chernomordik & Kozlov, 2003). This fusion dependence is usually explained based on the effective molecular shapes of lipids (spontaneous curvature), reflecting the propensity to bend into fusion intermediates. But, as we will discuss in more detail in Section V, simulation studies show that there are more ways in which lipids can increase the fusion rate, not so easily understood in terms of the curvature concept. We hope to provide some insight by taking a closer look at the energy landscape of the fusion arena.

V. ENERGY LANDSCAPE ALONG THE FUSION PATHWAY

In many simulation studies, it is apparent that both the stalk and the HD can be metastable states. In general, high energies of these intermediates are avoided by tilting of the tails, avoiding empty voids, although regions of lower density can be observed. The major energy barriers in these pathways that thus can be identified are (i) in the transition from unfused membranes to a stalk, (ii) in the transition from the stalk to the HD, and (iii) in the initiation of a pore in this HD. Initially, energy estimates for these various intermediate fusion stages and the associated energy barriers were obtained from continuum or field theoretical models (Katsov, Muller, & Schick, 2004; Katsov, Muller, & Schick, 2006; Kozlovsky et al., 2002; Kuzmin, Zimmerberg, Chizmadzhev, & Cohen, 2001; Markin & Albanesi, 2002). Nowadays, exploration of the energy landscape is amenable to more detailed calculations using molecular simulation techniques. In this section, we first discuss recent estimates pertaining the stalk and hemifused states, and then focus on the initial fusion steps involving some true lipid acrobatics. In the last part of this section, we present a schematic overview of the energy landscape emerging from *in silico* studies.

A. *Stalk and Hemifusion Diaphragm Intermediates*

Because of the height of the energy barriers most membranes do not fuse spontaneously in the (submillisecond) time scales reachable with molecular simulations. Experimentally it has been uncovered that the fusogenicity of liposomes depends on their size and the membrane tension (e.g., Finkelstein, Zimmerberg, & Cohen, 1986; Ohki, 1984). Most simulations are, because of the limited system sizes, on very small vesicles with a size, 15–30 nm in diameter, which is at the lower limit of experimentally producible vesicles. Although such highly curved membranes already fuse more easily than flat bilayers, in order to obtain fusion within the reachable time frame, fusion often still needs to be induced by initial placement of the membranes/vesicles in close vicinity, which might include partial dehydration of the lipid head groups, the addition of a (polyethylene glycol) cross-linker to keep the membranes together, pushing the membranes toward each other, elevating the temperature, stretching a membrane, or pulling a selected lipid across the inter-bilayer space. This procedure eventually results in stalk formation and possible formation of the HD. Noteworthy, simulations show that the high energy barriers are not formed by the stalk and HD intermediates themselves, but by transitions between such different intermediates.

Using their CG MD simulations and their large-scale distributed computing approach, Kasson et al. (2006) for instance looked at over 10,000 separate fusion events, which allowed them to construct a Markovian state model by which free energies of fusion intermediates as well as rates could be determined. In this way, they obtained decreasing free energies for unfused, stalk, hemifused, and fully fused states in small (14 nm in diameter) POPE vesicles, with a difference in free energy between the stalk and the unfused state of 6 k_BT. The free energy of stalks has been calculated from some other simulations as well. Norizoe, Daoulas, and Muller (2010) for instance derived, using a thermodynamic integration method, an excess free energy for the stalk of 16 k_BT for a solvent-free, CG model of two apposed planar bilayers. A similar energy difference of 10 k_BT was derived from self-consistent field calculations of a polymeric diblock system by Katsov et al. (2004). Using CG MD simulations, Smirnova, Marrink, Lipowsky, and Knecht (2010) derive a free energy of 3 k_BT for the stalk between two planar DOPC bilayers at low hydration, with an energy barrier from the unconnected state of 20 k_BT as will be discussed in Section V.B. Precise values of the stalk free energy thus depend on the lipid composition and the level of dehydration of the unfused reference state; the stalk free energy can be negative for lipids with negative spontaneous curvature (e.g., PEs), and positive for lamellar forming lipids (e.g., PCs). However, simulations indicate that the stalk is not a transition state, but metastable or on the downhill slope of the fusion process.

In the pathways with an HD, full fusion is obtained by the formation of a pore in the HD. Pores can be formed in membranes by several agents, but can also form spontaneously or triggered by electric fields or membrane tension. Such pore formation has been studied extensively in simulations of periodic membranes and has been reviewed recently for instance in Marrink et al. (2009) and Gurtovenko, Anwar, and Vattulainen (2010). It is generally found that the shapes of the pores are toroidal, with the lipid head groups lining the pore wall, quite independent of the details of the models used. Using MD simulations,

Tolpekinaet al. (2004a) report extensive free energy calculations on pores, showing an energy barrier of 15–20 k_BT for pore opening, while no barrier to pore closure was found although a small barrier has been predicted from atomistic MD simulations (Marrink, Lindahl, Edholm, & Mark, 2001). The same authors (Tolpekina, den Otter, & Briels, 2004b) also stretched bilayers to study the stability of pores establishing a phase diagram of pores, identifying regions where pores are stable, metastable, or unstable. The HD being metastable, with the fully fused state having the lowest energy, the formation of a local defect is the rate-limiting step in the decay of the HD. The energy barrier may, however, be lower than in a periodic bilayer as in the HD these defects appear most often close to the rim (Gao et al., 2008; Smeijers, Markvoort, et al., 2006) as the membrane is already distorted there (cf. Fig. 3, snapshot of the pore formation barrier).

B. Lipid Splaying as First Step

There is a growing body of evidence from simulations that the first, and often rate-limiting, step in the fusion process is the splaying of a single lipid tail rather than the formation of the stalk. Already in some of the first MD simulations of vesicle fusion (Marrink & Mark, 2003; Stevens et al., 2003) it was observed that stalk formation was preceded by the movement of the tails of just one or two lipids. The process was described in most detail in Stevens et al. (2003). The authors found that, upon pushing two small vesicles toward each other, the bilayers dilate at the contact edge producing a tilting of the individual molecules. Fusion is initiated when some of these tilted lipids splay their aliphatic tails, such that the molecules are shared between the opposing leaflets. Multiple splayed lipids subsequently associate with their aliphatic tails in contact, which produces a new hydrophobic core leading to the stalk state as described in Section IV. Later the initiation of fusion by the splaying of a single lipid tail has also been observed in unconstrained vesicles that fused spontaneously without having tilted lipids at the edge of a flattened contact zone (Smeijers, Markvoort, et al., 2006).

Quite recently, a number of more quantitative studies have been performed in order to elucidate the energetic barriers of the initial fusion event(s), confirming the importance of the splayed-tail intermediate. In a study by Smirnova et al. (2010), free energy calculations were performed for apposed POPC membranes, using the tail of a single lipid as reaction coordinate. Pulling the tail out of the one leaflet toward the other eventually leads to spontaneous stalk formation, as shown in Fig. 2. Stalk formation is initiated by the establishment of a localized hydrophobic contact between the bilayers. This contact is either formed by two partially splayed lipids or a single fully splayed one leading to the formation of a (metastable) splayed lipid bond intermediate. Whereas the stalk structure has

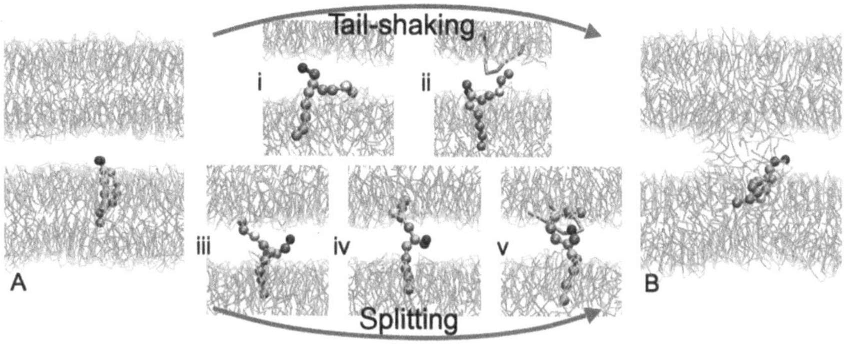

FIGURE 2 Lipid acrobatics during the initial fusion event. Snapshots are taken from a CG simulation of stalk formation between POPC bilayers (Smirnova et al., 2010). (A) Initial state with two closely apposed bilayers; (B) final state in which a stalk has formed. (i–ii) Tail-shaking pathway in which lipid splaying (i) is followed by the establishment of a hydrophobic contact with a lipid tail from the other membrane (ii). (iii–v) Splitting pathway, again initiated with splaying of a single lipid (iii) that subsequently inserts its tail into the other membrane adopting a metastable, split, conformation (iv), followed by the attraction of another lipid tail (v). Lipid tails and glycerol groups are shown in stick representation. Lipid heads and water are not shown for clarity. The lipid triggering the initial fusion event is displayed in sphere representation. (See Color Insert.)

only a small free energy difference, of 3 k_BT, with respect to the initial state of two unconnected bilayers, the intermediate states were found to have a significant free energy of 20 k_BT. These findings indicate that, at least under conditions of low hydration, early membrane fusion kinetics is not determined by the stalk energy but by the energy of prestalk transition states involving solvent-exposed lipid tails. A similar conclusion was drawn from Mirjanian, Dickey, Hoh, Woolf, and Stevens (2010) based on CG simulations using a similar force field. In this study, fusion between pairs of liposomes was simulated for four systems: DPPC, DOPC, a 3:1 mixture of DPPC/DPPE, and an asymmetric lipid tail system in which one tail of DPPC was reduced to half the length. The most prominent molecular detail of barrier crossing in all cases examined was, again, the splaying of lipid tails. It was further concluded that the tail splay appears to be closely connected to the energetics of the process. For example, the high barrier for the asymmetric lipid is the result of a smaller distance between terminal methyl groups in the splayed molecule. The shortening of this distance requires the liposomes to be closer together, which significantly increases the cost of water removal and bilayer deformation. Before tail splay can initiate fusion, contact must occur between a tail end and the external water. In the work of Grafmuller, Shillcock, and Lipowsky (2009), the fusion of vesicles with planar lipid bilayers was studied with DPD simulations. Again, it was found that the fusion process starts with individual lipids assuming a splayed tail configuration with one tail inserted in each membrane. To determine the corresponding energy barrier, the average work to displace one lipid tail from one bilayer to the other was measured. This energy barrier is estimated to lie in the range of 8–15 k_BT, depending on the strength of the interaction between the lipid tail and lipid head. Furthermore, it was found that the energy barrier for flipping of the lipid tails decreases with increasing tension of the membrane.

The studies cited above are all based on CG representations of lipids. The obvious question is whether the neglect of the atomistic degrees of freedom results in an overestimation of the importance of splayed-lipid intermediates. First evidence that this is not the case comes from fully atomistic studies of Ohta-Iino et al. (2001) already a while ago. These authors simulated two DMPC bilayers separated by only a thin water layer, in the range of 0–5 waters/lipid. Under these conditions, lipid tails were observed to penetrate into the interbilayer region on a subnanosecond time scale. A more elaborate study on fusion was performed by Kasson et al. (2010). Here, million-atom MD simulations of vesicle fusion were performed using distributed computing. Committor analysis was used to identify a transition state for fusion stalk formation. This transition state was found to occur when the bulk properties of each lipid bilayer remain in a lamellar state but a few hydrophobic tails bulge into the hydrophilic interface layer and make contact to nucleate a stalk.

Thus, both CG and atomistically detailed simulations point at the key roles of lipid tail dynamics at the onset of membrane fusion. The transition state toward stalk formation appears to be either a single lipid splayed between the two fusing membranes (for which we propose the name "*splitting*"), or the merging of the tails of two lipids, one from each membrane (coined as "*tail-shaking*"). See Fig. 2 for illustrative snapshots of these states. Note that the importance of the splitting conformation in triggering membrane fusion was already recognized in the work of Kinnunen and Holopainen (2000), based on indirect experimental evidence. The *in silico* confirmation of the importance of lipid tail splaying has a number of important implications for the role of lipid composition during the initial stages of fusion, namely (i) only the *cis*-monolayer lipid composition plays a role, and (ii) lipid tails, rather than lipid headgroups, are important. In this light, one would predict that especially polyunsaturated lipids accelerate the initial fusion step toward stalk formation, provided they are present in the contacting monolayers. Polyunsaturated tails are both more soluble and are easier to splay compared to saturated ones, favorable conditions for the splayed intermediate. There is experimental evidence that polyunsaturated lipids or fatty acids are indeed important for fusion (e.g., Glaser & Gross, 1994; Stillwell & Wassall, 2003). Likewise, oxidized lipids are expected to trigger fusion, as the oxidized tails have a high propensity to bend toward the water phase where the oxygen atoms form hydrogen bonds with water and the lipid headgroup (Wong-Ekkabut et al., 2007). Lipids with three tails, for example, triglycerides, could also favor tail splaying due to the large tail volume compared to their head group size. Lysolipids, on the other hand, are unable to adopt a splitting conformation but may still form the tail-shaking intermediate. According to the simulations of Smirnova et al. (2010), both intermediates have a similar free energy. However, the tail-shaking conformation might result in the actual desorption of the single tail lipid, and therefore the presence of lysolipids in the contacting monolayer is predicted to impede the very first step of fusion. Lipids that are unable to form any of the splayed intermediates are cyclic lipids (also denoted tetraether, or bola-lipids); indeed there is experimental evidence that vesicles composed of such lipids do not easily fuse (Relini et al., 1994).

The importance of splayed lipid intermediates may also suggest possible roles for other agents which are known to promote fusion. One example is Ca^{2+} which promotes fusion both *in vivo* and *in vitro* (Papahadjopoulos, Nir, & Duzgunes, 1990). Simulations (Issa, Manke, Jena, & Potoff, 2010) of apposed, dehydrated, membranes in the presence of Ca^{2+} provide some relevant insight. Whereas Ca^{2+} induces an overall ordering of the lipid tails due to the in-plane binding of the PC headgroups, local disordering is observed in regions where the ion bridges lipids from the opposing membranes. In other words, in addition to its role in keeping membranes bound together, Ca^{2+} might be able to destabilize the very lipids that are being bridged, facilitating the formation of splayed lipid intermediates. As we will see in Section VII, fusion peptides might operate in a similar way.

Note that the importance of splayed lipid intermediates at the onset of membrane fusion will be strongly dependent on the level of hydration. Simulations of stalk formation in bilayer stacks at different hydration levels, for instance, show that the rate of stalk formation decreases with increasing width of the hydration layer (Marrink & Mark, 2004). Most of the studies cited above probe conditions of low hydration ($\sim$5 waters/lipid) for which the splayed lipid constitutes the main barrier. The relative free energy of the stalk state with respect to the splayed lipid state will depend on the separation distance; it is not clear that membrane bridging by a splayed lipid molecule would lead to stalk formation in case there is no strong dehydration over an extended region or strong membrane stress through, for example, bending, both favoring the stalk formation. Whether or not these strongly dehydrated conditions resemble the fusion process *in vivo* is not known.

C. Many Barriers to Cross

Based on the combined data of particle-based simulation studies on membrane fusion, we composed a schematic energy landscape, Fig. 3, presenting the main intermediates and barriers involved. The main intermediates separating the unfused and fused states are (i) the adsorbed state, (ii) the stalk state, and

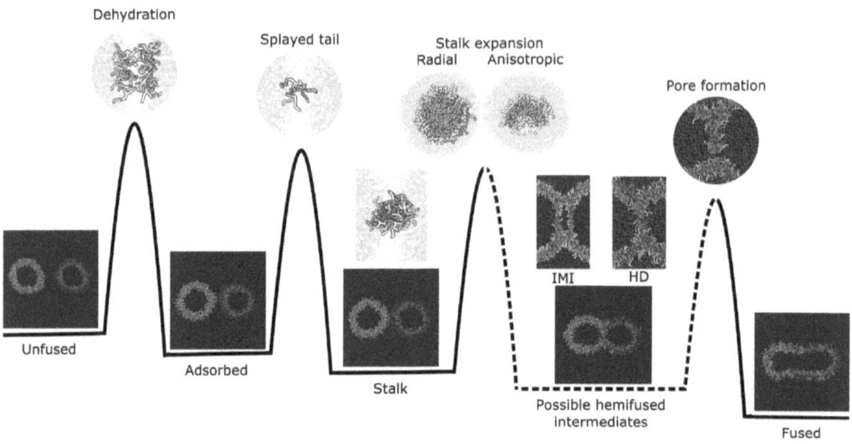

FIGURE 3 Schematic representation of the fusion energy landscape. Stable or metastable intermediate states leading from the "unfused" to the "fused" membranes are indicated, namely the "adsorbed," "stalk," and a number of "hemifused" states including the inverted micellar intermediate (IMI) and hemifusion diaphragm (HD). These states are separated by energy barriers involving subsequently "dehydration," lipid tail "splaying," either radial or anisotropic "stalk expansion," and "pore formation." Snapshots, obtained from CG MD simulations (Smeijers, Markvoort, et al., 2006) show cross-sections through two fusing vesicles, either perpendicular or parallel to the fusion coordinate. Color scheme as in Fig. 1. Note that the relative heights of the barriers and intermediate states have no meaning in this figure, as they will depend strongly on the fusion conditions. (See Color Insert.)

(iii) the IMI and HD states. These last two states are not always observed, that is, fusion may proceed directly from the stalk to the fused state without forming a stable intermediate, or only form the HD state but not the IMI (see Section IV.A). Whether the first, adsorbed, state is (meta)stable seems to depend on the contact area of the fusing membranes; for small vesicles it is not observed but for larger patches the dehydrated, bound state appears as a real intermediate. The barriers in between of the intermediate states differ substantially in their nature, reflecting the complexity of the fusion process. The first barrier is encountered upon bringing the membranes in close proximity, and involves overcoming the undulatory and hydration forces arising from the collective interactions of the lipids in the contact patch (see recent simulation studies on hydration forces, Eun and Berkowitz (2009); Gentilcore, Michaud-Agrawal, Crozier, Stevens, and Woolf (2010)). On the contrary, the second barrier involves splaying of just one or a few lipid tails, as discussed in Section V.B. The third barrier, stalk elongation, involves multiple lipids, and is dominated by the ability of lipids to pack in nonlamellar geometries. The final barrier is the nucleation of the fusion pore, again a localized process involving few lipids only.

Naturally, the relative stabilities of the states depend entirely on the state conditions, that is, composition, temperature, ionic strength, hydration, etc. The energy levels drawn in Fig. 3 do not have any quantitative meaning; for instance, the overall fusion process might be downhill, as observed for some systems *in vitro* which spontaneously fuse, but require substantial energy input (e.g., by means of a protein machinery) under *in vivo* conditions. The *in silico* studies point to some generic mechanisms by which the energy landscape can be modulated. This is summarized in Table I. Here we consider the effect of changing lipid composition, either modifying the head group (PC vs. PE), or the tails (saturated vs. unsaturated lipids and lysolipids), or the effect of adding fusion peptides or fusion proteins. The anticipated effects are only indicated in a qualitative sense, that is, either as stabilizing or destabilizing a particular (transition) state. Even then, Table I should be considered with care as the combined lipid/peptide fusion process is much too complicated to be captured in such a simplistic way. Nevertheless, we hope that Table I may provide some guideline for the understanding of fusion experiments, be it *in silico*, *in vitro*, or *in vivo*. In the following paragraph, we briefly discuss the data in Table I.

The *dehydration* of the interface is expected to be facilitated by lipids that do not swell much, such as PEs. Adsorption of peptides at the interface also lowers the hydration repulsion. An increase in temperature increases the undulation repulsion, and therefore makes it harder for membranes to come into close contact. An increase in membrane tension, on the other hand, suppresses undulations and results in the opposite effect. The stability of the *adsorbed state* is likely governed

TABLE I

Effect of State Conditions on Energy of Fusion Barriers and Intermediates, as Predicted by Simulation Studies[a]

	PC	PE	Sat	Unsat	Lyso	Pep	Prot	Temp	Tens
Dehydration	+	–	o	o	o	–	–	+	–
Bound	+	–	o	o	o	–	–	+	–
Splaying	o	o	o	–	+	–		–	–
Stalk	+	–	o	–	+			–	+
Elongation						–		–	
Hemifused	+	–	o	–	+			–	+
Poration	–	+	o	+	–	–	–	–	–

[a]An increase in the energy is denoted by "+," a decrease by "–." A "o" indicates that no effect is expected, and a blank that the effect is unknown. The following conditions are considered: increased ratio of PC and PE lipids, lipids with saturated (sat) or unsaturated (unsat) tails, and lysolipids (lyso), or presence of fusion peptides (pep) or fusion proteins (prot). For the proteins, the effect beyond that of the fusion peptide is considered. The effect of increased temperature (temp) or membrane tension (tens) is also listed (see text for details).

by the same factors. Especially proteins are required at this step to bring and keep the membranes close together. *Lipid splaying* is controlled to a large extent by the lipid tails; notably the presence of unsaturated tails will lower this barrier. In addition, fusion peptides are able to facilitate lipid splaying as will be discussed in Section VII.A. Increased temperature and tension result in more tail disorder and are therefore also predicted to favor tail splaying. The *stalk state* is stabilized by lipids with negative spontaneous curvature, such as PEs and unsaturated lipids. Lipids with positive curvature, such as single tail lipids (lysolipids) or lipids with large head groups (PC), destabilize the stalk. The effect of fusion peptides, or proteins, is actually less clear. An increase in temperature induces more negative curvature (through disordering of the tails), and therefore results in lowering of the stalk energy. Membranes under tension, however, are less likely to form (meta) stable stalks as the lipids will be sucked back into the membrane. *Stalk expansion*, subsequently, can proceed via different pathways as discussed in Section IV.A; it is not yet clear from simulation studies which conditions favor this step. There are some indications that fusion peptides may have a role here promoting bending of the stalk or formation of stalk–pore complexes (see Section VII.A). Both *hemi-fused states*, the IMI and HD, still have an overall negative curvature. Indeed simulations point to a stabilization of these states by the same type of lipids that stabilize stalks. Finally, *pore formation* is triggered by lipids with positive curvature, that is, PCs rather than PEs, and saturated rather than unsaturated tails. Short, single tail lysolipids are especially efficient at this stage. Fusion peptides may also play a role in formation of the fusion pore, inducing positive curvature stress due to their interfacial adsorption (see Section VII.A). Both increased temperature and tension also favor pore formation, destabilizing the lamellar state. Possibly also proteins are involved in this step putting the HD under additional tension.

VI. FISSION PATHWAYS IN MOLECULAR DETAIL

The opposite process of fusion, where two membranes fuse to a single membrane, is the division of a single membrane into two. This process, which is called membrane fission, is a crucial stage during exo- and endocytosis as well as in the formation of all kinds of intracellular carriers, for instance by the Golgi apparatus. Like fusion, the fission process can also be simulated using particle-based methods. The fission process is here described in two stages. First, the membrane has to be deformed, forming a narrow neck to allow it to self-fuse. This process of curvature generation, that is often denoted as budding, can be triggered by a number of mechanisms as described below. Second, the actual fission takes place as a self-fusion of the membrane in the neck according to a pathway that could, in principle, be the same as observed in regular fusion. We will see that this is, however, not the case.

A. Budding/Neck Formation

Whereas for fusion two separate membranes have to come into close juxtaposition, an important prerequisite for fission is the deformation of a membrane or vesicle into a shape that can easily be divided into two parts, like a budded shape where a spherical vesicle is connected to a flat membrane or another vesicle by a narrow neck. As the undeformed state has the lowest free energy, a driving force for such membrane budding is needed. This driving force can arise from a number of different (biological) mechanisms, which may involve some kind of protein machinery, but which may also be completely lipid based.

A first example of a protein-based mechanism is the deformation mediated by proteins binding to the membrane and functioning as a template. Atomistic MD simulations for instance showed how so-called BAR domains can induce a strong local curvature upon binding to a negatively charged membrane (Blood & Voth, 2006), whereas multiscale methods (Arkhipov, Yin, & Schulten, 2008; Ayton, Blood, & Voth, 2007) can show the cooperative nature of such BAR domains in inducing global membrane bending. Another nice example of how curved proteins can imply their curvature to an initially flat membrane, finally resulting in a vesicle budding off, is shown by Reynwar et al. (2007). A second mechanism is membrane deformation by a cytoskeletal element capable of either pushing or pulling on the membrane. An example of a simulation study mimicking an external force on a membrane is a BD study (Noguchi & Takasu, 2002b) where a force is applied on the membrane by pulling on nanoparticles inside the vesicle, like in an optical tweezer experiment, which resulted in the deformation and finally fission of the vesicle. A third mechanism comprises the enwrapping of a (extra)cytoplasmic particle, driven by the adhesion of the membrane to an already curved particle. An example of such fission by adhesion to nanoparticles is shown by Noguchi and Takasu (2002a) as well. A final mechanism is asymmetric redistribution of lipids by flipases. Ramachandran, Kumar, and Laradji (2008) and Ramachandran, Laradji, and Kumar (2009) for instance showed bud formation in a flat membrane upon asymmetric flip–flop, where the asymmetric flip–flop was driven by such flipase proteins.

Fission may, however, also occur without such external interaction or presence of proteins. A first example of a lipid-based deformation mechanism is given by membranes formed from multiple lipid components, which undergo lateral phase separation into coexisting liquid phases, or domains, with distinct compositions (Fig. 4A). As already demonstrated before experimentally as well as using continuum models (Baumgart, Hess, & Webb, 2003; Kumar, Gompper, & Lipowsky, 2001; Lipowsky, 1992), minimization of line tension between these domains may then result in membrane deformations and, depending on membrane rigidity and volume constraints, even in parts of the membrane

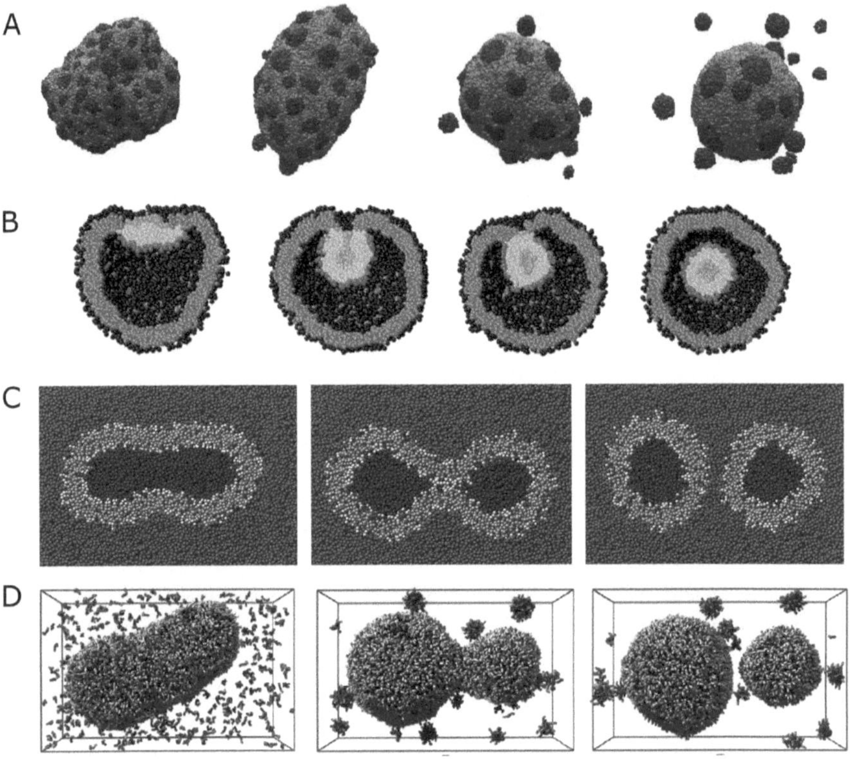

FIGURE 4 Four lipid-based fission driving mechanisms. (A) Phase separation driven budding (reprinted with permission from Laradji & Kumar, 2004). The two lipid types are shown in different shades of gray. (B) Inward budding by phase separation in combination with an asymmetric distribution of lipids over the leaflets (reprinted with permission from Yang et al., 2009). Cross-section, lipids in the bud are shown with light tails, other lipids with darker tails. (C) Fission induced by changing hydration of lipids in outer leaflet (Markvoort, Smeijers, Pieterse, van Santen, & Hilbers, 2007). Cross-section, lipids shown in stick representation with white heads, gray tails, solvent as spheres. (D) Fission induced by asymmetric membrane growth upon uptake of new membrane material in outer leaflet (reprinted with permission from Markvoort et al., 2010). (See Color Insert.)

budding off. Cooke, Kremer, and Deserno (2005) used an implicit solvent model to investigate the kinetics of domain formation and the budding process in membranes made of two lipid types. In their model, the two types of lipids have exactly the same shape, only experiencing a different mutual interaction that drives the phase separation. Laradji and Kumar (2004, 2005, 2006) used DPD with explicit solvent and found that area-to-volume ratio determined final shapes, something that cannot be studied with the implicit solvent models as in such models the vesicle interior volume can freely adapt. Such budding was also shown with other DPD studies by a number of groups (Fuchslin, Maeke, & McCaskill, 2009; Hong, Qiu, Zhang, & Yang, 2007; Illya, Lipowsky, & Shillcock, 2006; Yamamoto & Hyodo, 2003; Yang, Shao, & Ma, 2009; Zheng, Liu, Li, & Zhang, 2010), where phase separation is obtained by changing interaction parameters between two types of lipids and/or using lipids with different tail lengths (Stevens, 2005). Drawback of these DPD models is that the phase behavior exhibited cannot be mapped directly to that of real lipids. Studies using more detailed models capable of domain formation (e.g., Risselada & Marrink, 2008) are warranted.

Apart from phase separation, a second lipid-based membrane deformation mechanism comprises of an asymmetric lipid distribution over the two leaflets, something that is well known to be the case in biological membranes. Yamamoto and Hyodo (2003) for instance found that if line tension alone is not sufficient for a domain to bud off, an asymmetric transversal distribution of the lipid types can favor budding and fission. The basis behind this mechanism is the notion of a membrane as two monolayers that can slide over each other. In order to align the edges of the two unequally sized domains in the two leaflets, thus minimizing the interaction energy between unlike lipids, the membrane may bend to comply with this area difference (as shown earlier theoretically (Miao, Seifert, Wortis, & Döbereiner, 1994; Seifert, Berndl, & Lipowsky, 1991)). In such a way, the domain may bud off even in case of relatively low edge tension. The same effect was shown by Laradji and Kumar (2006), who also found that such an asymmetry tremendously slows down growth of the curved domains, and Yang et al. (2009), who showed that with a combination of phase separation and an asymmetric distribution of lipids also budding to the inside (endocytic fission) can occur (Fig. 4B).

Also in absence of phase separation, asymmetry between the two leaflets may still drive a vesicle to fission. Markvoort, van Santen, and Hilbers (2006) for instance showed, using their CG MD simulations, that slightly changing the hydration of the lipid head groups in one of the leaflets can result in dramatic shape deformations from prolate to oblate ellipsoid vesicles as well as to cup-shaped, pear-shaped and budded vesicles, where these budded vesicles could eventually split into two, more or less, equal daughter cells (Markvoort, Smeijers, Pieterse, van Santen, & Hilbers, 2007). The influence of the change in hydration can either be explained as a change in area difference between the two leaflets, which is relaxed by (slightly) bending the whole membrane, that is, deforming the vesicle, or as a change in the membranes spontaneous curvature. As the change in head group hydration for instance can be the result of a changing pH or ion concentration outside the vesicle, this mechanism can even drive single component vesicles to fission (Fig. 4C). Li, Liu, Wang, Deng, and Liang (2009) showed fission in a similar fashion in DPD simulations of triblock copolymers.

A final way to introduce an asymmetry between the leaflets (in a single component vesicle) is by the addition of new material to one of the leaflets. The area difference between the two monolayers can, for example, grow by addition of new membrane constituents from the outside (or inside) only (Markvoort et al., 2010; Yang & Ma, 2009), when the rate of addition is fast compared to relaxation by means of flip–flop. The resulting area difference can then again be released by vesicle deformation to budded states, continuing to full fission into two, more or less, equal daughter cells. This mechanism thus provided a good explanation for the so-called matrix effect (Blochliger, Blocher, Walde, & Luisi, 1998) in the reproduction of fatty acid vesicles (Fig. 4D).

B. Fission not Just Fusion Reversed

After the membrane is deformed, by means of one of the above driving mechanisms, in such a way that a narrow neck is formed, the fission process continues with further narrowing of this neck. Abandoning of internal solvent from this neck provides a division of the internal solvent into two parts. As already predicted using a continuum model (Kozlovsky & Kozlov, 2003), at this point the inner monolayer self-fuses. As the lipid head groups in the neck follow the two separating solvent compartments, this self-fusion results in a hemifused state with a continuous outer leaflet and two separate inner leaflets. The outer membrane then forms a circular stalk and breakage of this stalk completes fission. Although fusion and fission are each others inverse processes, the observed pathways are thus not always simply each others reverse. In the first place, a driving mechanism is needed to deform the membrane such that a sufficiently narrow neck is formed. In the second place, whereas the fusion process can follow a variety of both leaky and nonleaky pathways, the fission process follows a nonleaky mechanism that is most like the reverse of the direct stalk–pore fusion mechanism of the symmetric stalk expansion pathway (see Section IV.A). Cross-sections of fission intermediates are for instance shown in Markvoort et al. (2007, 2010). Common fusion intermediates as extended HDs and anisotropically extended stalks are, at least to date, not observed in the simulated fission processes.

Another difference is in the possible shapes of the stalk. In case of phase-separation driven budding the fission process may be triggered by cleavage along the domain boundary, hardly resulting in a stalk. In absence of phase separation, on the other hand, the stalk that is formed can be longer than in case of fusion. Whereas in case of fusion the stalk is limited in length as it is only formed when two membranes are already close, the fission stalk can grow much longer. This can arise especially in case of a relatively long neck, when the inner leaflets follow the solvent parts that separate when becoming more spherical in reducing their surface tension, or in case the two daughter vesicles are pulled apart (Noguchi & Takasu, 2002b). In this respect also simulations of micelle fission are of interest, mimicking the final stage of the fission of a membrane where a stalk between the two daughter membranes still needs to be broken. Micelle fission has for instance been studied using CG MC simulations (Pool & Bolhuis, 2006, 2007), but, contrary to complete membrane fission, also using atomic scale MD (Sammalkorpi, Karttunen, & Haataja, 2008). The main observation in these atomistic resolution simulations is that micelle fission progresses through a dumbbell-like morphology involving the formation of a long and narrow stalk in which the surfactants are highly interdigitated. The rod-like interdigitating structure allows the fission stalk to elongate over a much longer distance while providing partial shielding of the hydrocarbon tails from the solvent and leading to a stable intermediate.

VII. PEPTIDE MODULATED FUSION

Fusion may be intrinsic to lipids, but peptides and proteins are required to control and or facilitate the process. They might act passively to bring (and hold) the bilayers together, or have a more active role in stabilizing any of the intermediate steps. This is clear from *in vivo* as well as *in vitro* studies (e.g., reviewed in Jahn & Grubmuller, 2002; Sapir, Avinoam, Podbilewicz, & Chernomordik, 2008), but now also from *in silico* studies.

A. The Role of Fusion Peptides

Fusion peptides are moderately hydrophobic segments of viral and nonviral membrane fusion proteins that enable these proteins to disrupt and connect two closely apposed biological membranes (Epand, 2003; Tamm & Han, 2000). Most fusion peptides are located at the extreme N-termini of the transmembrane subunits of the fusion proteins. Deletion of the fusion peptide and, in many cases, even relatively conservative single amino acid changes in the fusion peptide completely abolish the ability of fusion proteins to fuse membranes. *In vitro* assays furthermore show that even isolated fusion peptides alone can support membrane fusion in model systems. The combined experimental evidence (e.g., Tamm & Han, 2000) points to three effects of the fusion peptides on membrane structure, namely changes in hydration, curvature, and lipid chain order. Furthermore, fusion peptides are observed to induce inverted lipid phases, suggesting stabilization of the stalk intermediate which has a dominant negative curvature. The relation between peptide-induced membrane changes and peptide-induced membrane fusion at the molecular level remains controversial, however.

Most simulation studies to date have concentrated on the hemagglutinin (HA) fusion peptide and its mutants. Full fusion assays are computationally still very challenging, therefore, most of these studies address the binding of a single fusion peptide or analogue, either to a lipid bilayer or micelle. Atomistic simulation studies along these lines have been reported by many groups (Huang, Chen, & Herrmann, 2004; Lague, Roux, & Pastor, 2005; Li, Das, & Zhou, 2010; Nishizawa & Nishizawa, 2010a; Vaccaro et al., 2005; Volynsky, Polyansky, Simakov, Arseniev, & Efremov, 2005). Despite the simplicity of the set-up, a number of key points are revealed: (i) the HA fusion peptide adopts a kinked conformation. The kinking is in agreement with the NMR structure of the peptide in detergent micelles (Han, Bushweller, Cafiso, & Tamm, 2001), revealing a boomerang shape. (ii) The peptide inserts obliquely into the lipid membrane, in line with experimental data (Tamm & Han, 2000). Insertion into the interface is mainly with the N-terminal helix. (iii) The peptide locally induces disorder, increasing the chain gauche population and area of adjacent lipids in the same binding leaflet. Thus, it induces local thinning of the bilayer and disordering of acyl chains of lipids in close proximity to the binding site. A more quantitative analysis of the disordering effect of the HA fusion peptide was performed in the study of Kasson et al. (2010). Lipid tail protrusions were monitored for a POPC bilayer patch at a peptide:lipid ratio of 3:500. Lipids within 5 Å of the peptides exhibited significantly increased protrusion frequencies (approximately fourfold) compared to lipids more than 20 Å away.

The membrane binding of nonfusogenic mutants of the HA fusion peptide was also investigated (Li et al., 2010; Nishizawa & Nishizawa, 2010a; Vaccaro et al., 2005). Compared to the wild type, Vaccaro et al. (2005) observed that most of them equilibrate parallel to the interface plane and do not adopt a tilted conformation. In Li et al. (2010), a total of six single point mutants, all with no fusion or hemifusion activity, were examined systematically. All mutants show a strong tendency toward a linear alpha-helix conformation, with the initial kink structure in the wild-type broken. One of the key hydrophobic residues around the initial kink region, Phe-9, is found to flip away from the membrane surface in most of these mutants. Also in the simulation study of Nishizawa and Nishizawa (2010a), comparison among the mutants supports a view that the oblique orientation of the wild type is required to facilitate perturbation of the lipid–water interface. Moreover, these authors show that in closely apposed bilayers, the wild-type fusion peptide is able to trigger stalk formation under conditions where the pure bilayer is stable. However, when the peptide is forced to adopt a conformation characteristic of some of the mutants, this ability disappears.

Other fusion peptides have received less attention. The structures of the 16-residue fusion peptide of the HIV gp41 fusion protein, two of its mutants, and a shortened peptide were studied by MD simulation in an explicit POPE bilayer (Kamath & Wong, 2002). The simulations reveal that the active wild-type peptide inserts into the bilayer at an oblique angle, whereas the inactive mutants and the shortened fragment lie on the bilayer surface. Furthermore, the simulations show the structure of the wild type to be remarkably similar to that of the HA fusion peptide. Likewise, the gp41 fusion peptide also has a disruptive effect on the lipid chains. This effect was not observed for the mutants. A comparative study of a number of type 1 viral fusion peptides was performed by Taylor and Sansom (2010). In their study, the physical properties of the fusion peptide surface were related to the tilt angles determined both experimentally and by means of MD simulations. It was found that the relationship between the distribution of lipophilic potential over the peptide surface and the peptide geometry control the tilt angle of the peptide in a DMPC bilayer. The depth of penetration into the bilayer appears to be determined by the electrostatic potential and hydrogen bonding at the C-terminus.

A different approach to assess the effect of fusion peptides was taken by Fuhrmans, Knecht, and Marrink (2009). The use of a CG model allowed for a systematic exploration of the effect of the peptide on the lipid phase diagram. In a spontaneous aggregation approach, starting from random mixtures of DOPE lipids, HA fusion peptides and water, a new phase was found forming exclusively in the presence of the fusion peptide. This new phase is in fact morphologically equivalent to the single diamond cubic phase. In light of the role of the fusion peptide, it is insightful to describe the single diamond phase in terms of pores and stalks; one finds that there is a tight balance of stalks and pores with the network of stalks defining the pores and vice versa. The phenomenon of stalks and pores in close proximity is not uncommon and it was actually predicted to be energetically favorable to form a pore in the presence of a stalk in a field theoretic study (Katsov et al., 2006). Stalk–pore complexes are also observed as intermediate states during fusion along the stalk–pore complex pathway (see Section IV.B), pointing to a potential role of the HA fusion peptide. The peptides' boomerang structure appears to entail a preference to locate between the bases of two emerging stalks with its helical arms neatly lining the surface, possibly reducing the Gaussian curvature elastic energy associated with the stalk–pore structure. See Fig. 5A for a close-up of the location of the peptides at the membrane–water interface. In addition, the peptides seeming ability to stabilize stalks and pores might account for its observed effects on hemolysis and the stabilization of inverted cubic phases (Epand, 2003). Additional simulation studies (Fuhrmans & Marrink, 2011) probing the ability of mutants of the HA fusion peptide to perturb the lipid phase diagram show much weaker effects compared to the wild type. Again, this is attributed to the unique boomerang shape of the HA fusion peptide.

As mentioned before, full fusion assays are computationally demanding and therefore the simulation studies discussed in this section use indirect measures to assess the role of fusion peptides. Nevertheless, simulation studies suggest that the fusion peptides may be important in a number of ways. First, the presence of the peptides at the membrane–water interface may lower the hydration repulsion, and

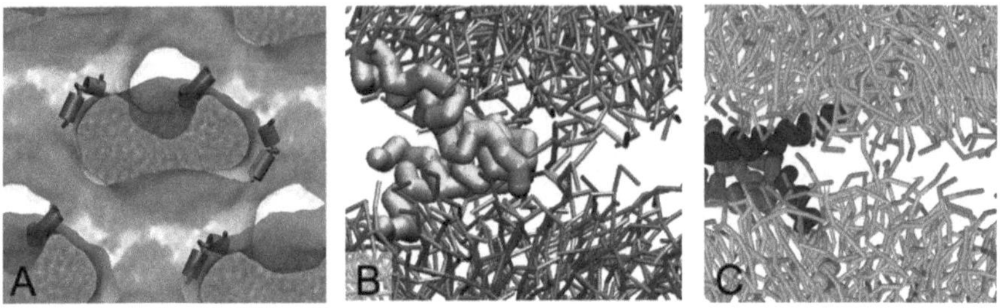

FIGURE 5 Protein-mediated membrane fusion intermediates. (A) Stalk–pore complex stabilized by HA fusion peptides (Fuhrmans et al., 2009). The snapshot shows a cut through the stalks, looking on top of the adjacent pores. The lipid–water interface is shown as a surface, the lipid tail beads as spheres, and the peptides as rods. (B) Fusion of liposomes triggered by lung surfactant protein B (Baoukina & Tieleman, 2010). A snapshot of the initial stage of the fusion process is shown, in which the protein binds the two liposomes together allowing lipids to bridge. (C) Liposomal fusion by the SNARE complex (Risselada et al., 2011). The initial stage of the formation of a splayed lipid intermediate is shown. (See Color Insert.)

thus helps bringing the bilayers together. Second, the peptides induce local disorder and facilitate the protrusion of lipid tails. Assuming this to be the initial energy barrier, as discussed in Section V, the rate of stalk formation is increased. Third, the peptides cause an expansion of the interfacial area and therefore positive curvature stress. Positive curvature helps to form the final fusion pore, or to stabilize the stalk–pore complex along the fusion pathway. The simulation studies also show that the typical V-shape, or boomerang shape of the fusion peptide (most notably for the HA fusion peptide) is optimal in these respects, in line with experimental findings (e.g., Lai, Park, White, & Tamm, 2006). Interestingly, the simulations predict that the stalk state itself is not favoured by the fusion peptides, only its formation rate by lowering of the initial barrier. The effect of peptides on stabilization of inverted phases should then be understood in terms of the positive curvature component that is induced by the peptide, allowing phases with saddle-splay curvature such as the inverted cubic phases. Further evidence for the role of fusion peptides in the formation of pores is found in the combined experimental and simulation study performed by Donald et al (2011). Here it is shown that the fusion peptide of the parainfluenza virus 5 can form a stable hexameric trans-membrane bundle, at least partly filled with water. In fact the structure resembles that of a closed mechano-sensitive channel which may gate in response to membrane tension during the final stage of fusion, opening the fusion pore.

B. Protein-Induced Fusion

Both in viral fusion and in eukaryotic fusion pathways, complex protein machineries exist that regulate and control the fusion process. Given the already complex energy landscape of lipid-mediated fusion, elucidating the precise role of proteins is an enormous challenge. In general, the proteins will be involved in bringing the fusing membranes together, and by lowering the various intermediate states, to guide the system through a specific fusion pathway (Chernomordik & Kozlov, 2008; Jahn & Grubmuller, 2002). Simulation studies are now probing the interplay between lipids and proteins during fusion, both at a generic level (protein mimetics) and also recently using detailed models.

A study of Smeijers, Pieterse, Markvoort, and Hilbers (2006) explored the effect of small CG protein mimetics on the process of fusion. Especially interesting is their finding that a so-called scramblase, that is, a transmembrane protein with a hydrophilic strip which allows rapid lipid flip–flopping, strongly promotes vesicle fusion. The authors conclude that it does so by providing a way to relieve membrane tension that results from crowding of the outer monolayer due to growth of the stalk. After the fusion process has completed, the presence of the scramblase proteins also helps the vesicles to relax to a spherical shape. Additionally, one of the proteins promotes fusion by inducing pore formation. Incorporation of these proteins allows even flat membranes to fuse spontaneously. Another example of the possible mode of action of fusion proteins is provided by the simulations of Noguchi and Takasu (2002a). They showed that small nanoparticles that adhere to the membrane surface can promote fusion by bending of the stalk (i.e., facilitating the alternative pathways, see Section IV.B).

The first simulations of spontaneous fusion induced by realistic proteins are reported by Baoukina and Tieleman (2010). The fusion of small unilamellar vesicles mediated by lung surfactant protein B (SP-B) was studied, using the MD method and a CG force field in which near-atomic detail is preserved. The fusogenic activity of SP-B protein is known from *in vivo* studies (Oosterlaken-Dijksterhuis, van Eijk, van Golde, & Haagsman, 1992), and is believed to play a role in the lung surfactant secretory pathway (Ryan et al., 2005). In the simulations, an SP-B monomer triggers spontaneous fusion events by anchoring two vesicles and facilitating the formation of a lipid bridge between the proximal leaflets. A snapshot of this stage is shown in Fig. 5B. Once a lipid bridge is formed, fusion proceeds via the previously described stalk–HD–pore pathway (see Section IV.A). In the absence of protein, fusion of vesicles was not observed in either unbiased simulations or upon application of a restraining potential to maintain the vesicles in close proximity. The authors conclude that the particular shape of the SP-B protein is crucial, enabling it to bind to two vesicles at once and forcing their proximity. It is plausible that the protein surface subsequently acts as a hydrophobic scaffold, which is used by the lipid tails to adopt a splayed conformation triggering the stalk formation.

Key player in eukaryotic fusion is the SNARE (Soluble *N*-ethylmaleimide-sensitive fusion protein Attachment Protein (SNAP) REceptor) complex. Assembly of the SNARE proteins into a so-called "*trans*" complex likely bridges the fusing membranes, inducing their fusion. The core SNARE complex is a four-α-helix bundle, where one α-helix is contributed by syntaxin-1, one α-helix by synaptobrevin and two α-helices are contributed by SNAP-25. Based on the stability of the resultant *cis*-SNARE complex, it has been postulated that energy released during the assembly process serves as a means for overcoming the repulsive forces between the membranes. Because SNARE proteins are anchored in the membranes, and these anchors have been shown to act as force transducers (McNew et al., 2000), a possibility is that the SNARE complex induces a local tension in the membranes allowing them to fuse. Simulation studies of Knecht and Grubmuller (2003) showed that the transmembrane anchors can in principle provide the required mechanical stiffness for them to act as force transducers. The importance of global membrane tension was already shown in the simulation studies of Shillcock et al., discussed in Section IV. In an extension of that work, Shillcock and Lipowsky (2006) tested the hypothesis that SNARE proteins facilitate fusion via local tension. The authors considered a system consisting of a 28 nm diameter vesicle fusing with a planar membrane patch of 100 nm linear dimension. The tension in the planar membrane was generated by pulling six artificial barrel proteins, mimicking the action of the SNARE proteins. Although full fusion could indeed be observed, the work required to induce formation of the fusion pore was quite large, approximately 90 k_BT per protein. Another DPD simulation study (Wu & Guo, 2009) has elaborated on this theme by including the extracellular piece of the SNARE proteins as well as their transmembrane anchors. By explicitly adding attractive forces between the model proteins, the formation of the fusion complex was mimicked. Binding of the proteins resulted in bringing the membranes together, creating enough tension in the center of the protein ring to open a fusion pore.

The power of simulation studies is further evidenced from the recent work of Risselada et al. (2011). Here, the SNARE-mediated fusion of two vesicles was simulated, using the same near-atomistic CG model as in the study of lung surfactant protein discussed above. The set-up consisted of two small POPE liposomes with either one or two preassembled *trans* SNARE complexes, in close resemblance to *in vitro* fusion assays (Weber et al., 1998). The model of the SNARE complex was based on the X-ray resolved structure (Stein, Weber, Wahl, & Jahn, 2009). Fusion, defined as content mixing, was observed to occur spontaneously in 4 out of 10 simulations in case of 2 complexes, and in 2 out of 10 with only 1 complex present. Control simulations without proteins did not show any fusion events. In all succesful cases, the fusion process proceeds via four stages. In the first stage, the transmembrane SNARE receptors (TMRs) are able to destabilize individual lipids, leading to the formation of a stalk via a splayed-lipid intermediate as illustrated in Fig. 5C. In the second stage, the stalk expands in a linear fashion leading to the inverted-micellar intermediate (see Section IV.B). The third and fourth stage comprise the rupture of the membranes to form an asymmetric HD and full fusion pore. The simulations suggest that the formation of the fusion pore is mediated by the charged C-terminal of the TMR. Insertion of this charged end into the membrane interior is forced by the mechanical stress stored in the SNARE complex; by doing so the charge opens up the fusion pore to avoid becoming dehydrated. Remarkably, the final structure of the SNARE complex resembles the postfusion structure suggested from the X-ray data (Stein et al., 2009).

Taken together, the *in silico* experiments on protein-mediated membrane fusion point to two key events for which the proteins are important: lipid splaying and pore formation. Concerning lipid splaying, direct evidence for the inducement of splayed intermediates comes from the simulations on lung surfactant protein B- (Baoukina & Tieleman, 2010) and SNARE- (Risselada et al., 2011) mediated fusion. Indirect evidence is provided by the numerous studies on the lipid disturbing effects of fusion peptides in general, as discussed in Section VII.A. The role of fusion proteins in the pore formation is evident from the simulated SNARE-mediated fusion (Risselada et al., 2011), and, more indirectly, from the ability of fusion peptides to induce positive curvature and to stabilize stalk–pore complexes (Fuhrmans et al., 2009). Note that, both lipid splaying and pore formation are favored by the presence of membrane tension, which may also be induced by the fusion proteins (Shillcock & Lipowsky, 2006). Additionally, simulations show that protein mimetics could have an effect on, for example, stalk bending (Noguchi & Takasu, 2002a) or lipid flip–flopping (Smeijers, Pieterse, et al., 2006).

VII. OUTLOOK

The progress in particle-based simulation studies of membrane fusion has been quite substantial. *In silico* modeling of fusion is becoming a valuable tool nowadays, completing the large range of experimental methods. Simulation studies have proven most powerful in revealing detailed pathways for membrane fusion and fission. The many ways in which lipids can change their conformations to adopt to, or give rise to, the fusion intermediates is rather impressive. On top of this sits the whole protein fusion machinery, adding another layer of complexity. Simulation studies are now beginning to explore their role as well, with recent highlights the spontaneous protein-mediated fusion of vesicles observed in Baoukina and Tieleman (2010) and Risselada et al. (2011).

What is needed now is a more systematic exploration of fusion pathways as a function of system parameters such as lipid composition, membrane curvature, membrane separation, system size, tension, temperature, salt conditions, etc., in order to extract statistical information about which pathways are observed under which conditions, and what are the associated energies along the way. Especially, we anticipate a significant increase in studies on peptide and protein-mediated fusion, probing the fascinating lipid/protein interplay. With the enormous growth in number of processor nodes foreseen for the next 10 years, such systematic studies can become very powerful. As an example might serve the work of Kasson et al. (2006) who used distributed computing to explore the fusion of small vesicles over thousands of trajectories. Another development of interest is the use of multiscale modeling techniques (Ayton, Noid, & Voth, 2007; Nielsen, Bulo, Moore, & Ensing, 2010) in which detailed, atomistic resolution is combined with the effectiveness of CG models. One could think of, for example, fusion or fission of large vesicles in which only the contacting parts are simulated in full detail and the remainder is treated at a reduced level of resolution.

From the experimental side, the challenge is to detect some of the pathways or intermediates that are predicted *in silico*. Most notable in this respect are the pathways involving a linear expansion of the stalk, seen in many different simulation studies. Direct observation of the inverted-micellar intermediate would lend strong support for this mechanism. However, the absence of this intermediate does not disprove the linear stalk expansion pathway, as pores might form before the water droplet is trapped. Mixing of *trans* and *cis* lipids from the two fusing membranes is another signature of the alternative pathways, which proceed via an asymmetric hemifusion state. Although probably hard to detect experimentally, the initial fusion events are predicted to be dominated by the ability of single lipids to splay and cross the inter bilayer space, at least when fusion involves strongly dehydrated membranes. Indirect indications could come from the expected fusogenicity of polyunsaturated lipids when present in the outer leaflets, or the decreased ability to fuse for membranes enriched in cyclic lipids, for instance. Finally, the combined simulation data point to novel roles of fusion peptides, facilitating lipid tail splaying and promoting pore or pore–stalk complex formation by inducing positive curvature stress; knowledge which can aid in the interpretation of experimental measurements.

In conclusion, simulation studies on membrane fusion point to a large variety of fusion pathways, revealing a more complex energy landscape beyond the traditional stalk–HD–fusion pore mechanism. Truly, lipids are remarkably versatile molecules that give rise to some amazing acrobatics in the fusion arena.

Acknowledgments

We acknowledge Yuliya Smirnova, Jelger Risselada, Svetlana Baoukina, Marc Fuhrmans, Sander Smeijers, Koen Pieterse, Peter Kasson, Kai Yang, and Mohamed Laradji for kindly providing snapshots for the figures, and Lars Schäfer and Sander Smeijers for proofreading of the manuscript.

References

Arkhipov, A., Yin, Y., & Schulten, K. (2008). Four-scale description of membrane sculpting by bar domains. *Biophysical Journal, 95*, 2806–2821.

Ayton, G. S., Blood, P. D., & Voth, G. A. (2007a). Membrane remodeling from N-BAR domain interactions: Insights from multi-scale simulation. *Biophysical Journal, 92*, 3595–3602.

Ayton, G. S., Noid, W. G., & Voth, G. A. (2007b). Multiscale modeling of biomolecular systems: In serial and in parallel. *Current Opinion in Structural Biology, 17*, 192–198.

Baoukina, S., & Tieleman, D. P. (2010). Direct simulation of protein-mediated vesicle fusion: Lung surfactant protein B. *Biophysical Journal, 99*, 2134–2142.

Baumgart, T., Hess, S. T., & Webb, W. W. (2003). Imaging coexisting fluid domains in biomembrane models coupling curvature and line tension. *Nature, 425*, 821–824.

Blochliger, E., Blocher, M., Walde, P., & Luisi, P. L. (1998). Matrix effect in the size distribution of fatty acid vesicles. *Journal of Physical Chemistry B, 102*, 10383–10390.

Blood, P. D., & Voth, G. A. (2006). Direct observation of Bin/amphiphysin/Rvs (BAR) domain-induced membrane curvature by means of molecular dynamics simulations. *Proceedings of the National Academy of Sciences of the United States of America, 103*, 15068–15072.

Chernomordik, L. V., & Kozlov, M. M. (2003). Protein-lipid interplay in fusion and fission of biological membranes. *Annual Review Biochemistry, 72*, 175–207.

Chernomordik, L. V., & Kozlov, M. M. (2008). Mechanics of membrane fusion. *Nature Structural & Molecular Biology, 15*, 675–683.

Cooke, I. R., Kremer, K., & Deserno, M. (2005). Tunable generic model for fluid bilayer membranes. *Physical Review E, 72*, 011506.

Donald, J. E., Zhang, Y., Fiorin, G., Carnevale, V., Slochower, D. R., Gai, F., Klein, M. L., & DeGrado, W. F. (2011). Transmembrane orientation and possible role of the fusiogenic peptide from parainfluenza virus 5 (PIV5) in promoting fusion. *Proceedings of the National Academy of Sciences of the United States of America, 108*, 3958–3963.

Epand, R. M. (2003). Fusion peptides and the mechanism of viral fusion. *Biochimica et Biophysica Acta-Biomembranes, 1614*, 116–121.

Eun, C., & Berkowitz, M. L. (2009). Origin of the hydration force: Water-dedicated interaction between two hydrophilic plates. *Journal of Physical Chemistry B, 113*, 13222–13228.

Finkelstein, A., Zimmerberg, J., & Cohen, F. S. (1986). Osmotic swelling of vesicles: Its role in the fusion of vesicles with planar phospholipid bilayer membranes and its possible role in exocytosis. *Annual Review of Physiology, 48*, 163–174.

Frolov, V. A., Dunina-Barkovskaya, A. Y., Samsonov, A. V., & Zimmerberg, J. (2003). Membrane permeability changes at early stages of influenza hemagglutinin-mediated fusion. *Biophysical Journal, 85*, 1725–1733.

Fuchslin, R. M., Maeke, T., & McCaskill, J. S. (2009). Spatially resolved simulations of membrane reactions and dynamics: Multipolar reaction DPD. *European Physical Journal E, 29*, 431–448.

Fuhrmans, M., Knecht, V., & Marrink, S. J. (2009). A single bicontinuous cubic phase induced by fusion peptides. *Journal of the American Chemical Society, 131*, 9166–9167.

Fuhrmans, M., & Marrink, S. J. (2011). In preparation.

Gao, L. H., Lipowsky, R., & Shillcock, J. (2008). Tension-induced vesicle fusion: Pathways and pore dynamics. *Soft Matter, 4*, 1208–1214.

Gentilcore, A. N., Michaud-Agrawal, N., Crozier, P. S., Stevens, M. J., & Woolf, T. B. (2010). Examining the origins of the hydration force between lipid bilayers using all-atom simulations. *Journal of Membrane Biology, 235*, 1–15.

Glaser, P. E., & Gross, R. W. (1994). Plasmenylethanolamine facilitates rapid membrane fusion: A stopped-flow kinetic investigation correlating the propensity of a major plasma membrane constituent to adopt an HII phase with its ability to promote membrane fusion. *Biochemistry, 33*, 5805–5812.

Grafmuller, A., Shillcock, J., & Lipowsky, R. (2007). Pathway of membrane fusion with two tension-dependent energy barriers. *Physical Review Letters, 98*, 218101.

Grafmuller, A., Shillcock, J., & Lipowsky, R. (2009). The fusion of membranes and vesicles: Pathway and energy barriers from dissipative particle dynamics. *Biophysical Journal, 96*, 2658–2675.

Gurtovenko, A. A., Anwar, J., & Vattulainen, I. (2010). Defect-mediated trafficking across cell membranes: Insights from in silico modeling. *Chemical Reviews, 110*, 6077–6103.

Han, X., Bushweller, J. H., Cafiso, D. S., & Tamm, L. K. (2001). Membrane structure and fusion-triggering conformational change of the fusion domain from influenza hemagglutinin. *Nature Structural Biology, 8*, 715–720.

Hong, B. B., Qiu, F., Zhang, H. D., & Yang, Y. L. (2007). Budding dynamics of individual domains in multicomponent membranes simulated by N-varied dissipative particle dynamics. *Journal of Physical Chemistry B, 111*, 5837–5849.

Huang, Q., Chen, C. L., & Herrmann, A. (2004). Bilayer conformation of fusion peptide of influenza virus hemagglutin: A molecular dynamics simulation study. *Biophysical Journal, 87*, 14–22.

Illya, G., Lipowsky, R., & Shillcock, J. C. (2006). Two-component membrane material properties and domain formation from dissipative particle dynamics. *Journal of Chemical Physics, 125*, 114710.

Issa, Z. K., Manke, C. W., Jena, B. P., & Potoff, J. J. (2010). Ca^{2+} bridging of apposed phospholipid bilayers. *Journal of Physical Chemistry B, 114*, 13249–13254.

Jahn, R., & Grubmuller, H. (2002). Membrane fusion. *Current Opinion in Cell Biology, 14*, 488–495.

Kamath, S., & Wong, T. C. (2002). Membrane structure of the human immunodeficiency virus gp41 fusion domain by molecular dynamics simulation. *Biophysical Journal, 83*, 135–143.

Kasson, P. M., Kelley, N. W., Singhal, N., Vrljic, M., Brunger, A. T., & Pande, V. S. (2006). Ensemble molecular dynamics yields submillisecond kinetics and intermediates of membrane fusion. *Proceedings of the National Academy of Sciences of the United States of America, 103*, 11916–11921.

Kasson, P. M., Lindahl, E., & Pande, V. S. (2010). Atomic-resolution simulations predict a transition state for vesicle fusion defined by contact of a few lipid tails. *Plos Computational Biology, 6*, e1000829.

Kasson, P. M., & Pande, V. S. (2007). Control of membrane fusion mechanism by lipid composition: Predictions from ensemble molecular dynamics. *Plos Computational Biology, 3*, 2228–2238.

Katsov, K., Muller, M., & Schick, M. (2004). Field theoretic study of bilayer membrane fusion. I. Hemifusion mechanism. *Biophysical Journal, 87*, 3277–3290.

Katsov, K., Muller, M., & Schick, M. (2006). Field theoretic study of bilayer membrane fusion: II. Mechanism of a stalk-hole complex. *Biophysical Journal, 90*, 915–926.

Kinnunen, P. K. J., & Holopainen, J. M. (2000). Mechanisms of initiation of membrane fusion: Role of lipids. *Bioscience Reports, 20*, 465–482.

Knecht, V., & Grubmuller, H. (2003). Mechanical coupling via the membrane fusion SNARE protein syntaxin 1A: A molecular dynamics study. *Biophysical Journal, 84*, 1527–1547.

Knecht, V., Mark, A. E., & Marrink, S. J. (2006). Phase behavior of fatty acids/lipid/water mixture studied in atomic detail. *Journal of the American Chemical Society, 128*, 2030–2034.

Knecht, V., & Marrink, S. J. (2007). Molecular dynamics simulations of lipid vesicle fusion in atomic detail. *Biophysical Journal, 92*, 4254–4261.

Kozlov, M. M., & Markin, V. S. (1983). Possible mechanism of membrane-fusion. *Biofizika, 28*, 242–247.

Kozlovsky, Y., Chernomordik, L. V., & Kozlov, M. M. (2002). Lipid intermediates in membrane fusion: Formation, structure, and decay of hemifusion diaphragm. *Biophysical Journal, 83*, 2634–2651.

Kozlovsky, Y., & Kozlov, M. M. (2003). Membrane fission: Model for intermediate structures. *Biophysical Journal, 85*, 85–96.

Kumar, P. B. S., Gompper, G., & Lipowsky, R. (2001). Budding dynamics of multicomponent membranes. *Physical Review Letters, 86*, 3911–3914.

Kuzmin, P. I., Zimmerberg, J., Chizmadzhev, Y. A., & Cohen, F. S. (2001). A quantitative model for membrane fusion based on low-energy intermediates. *Proceedings of the National Academy of Sciences of the United States of America, 98*, 7235–7240.

Lague, P., Roux, B., & Pastor, R. W. (2005). Molecular dynamics simulations of the influenza hemagglutinin fusion peptide in micelles and bilayers: Conformational analysis of peptide and lipids. *Journal of Molecular Biology, 354*, 1129–1141.

Lai, A. L., Park, H., White, J. M., & Tamm, L. K. (2006). Fusion peptide of influenza hemagglutinin requires a fixed angle boomerang structure for activity. *Journal of Biological Chemistry, 281*, 5760–5770.

Laradji, M., & Kumar, P. B. S. (2004). Dynamics of domain growth in self-assembled fluid vesicles. *Physical Review Letters, 93*, 198105.

Laradji, M., & Kumar, P. B. S. (2005). Domain growth, budding, and fission in phase-separating self-assembled fluid bilayers. *Journal of Chemical Physics, 123*, 224902.

Laradji, M., & Kumar, P. B. S. (2006). Anomalously slow domain growth in fluid membranes with asymmetric transbilayer lipid distribution. *Physical Review E, 73*, 040901.

Lentz, B. R., Malinin, V., Haque, M. E., & Evans, K. (2000). Protein machines and lipid assemblies: Current views of cell membrane fusion. *Current Opinion in Structural Biology, 10*, 607–615.

Leontiadou, H., Mark, A. E., & Marrink, S. J. (2004). Molecular dynamics simulations of hydrophilic pores in lipid bilayers. *Biophysical Journal, 86*, 2156–2164.

Li, D. W., & Liu, X. Y. (2005). Examination of membrane fusion by dissipative particle dynamics simulation and comparison with continuum elastic models. *Journal of Chemical Physics, 122*, 174909.

Li, J. Y., Das, P., & Zhou, R. H. (2010). Single mutation effects on conformational change and membrane deformation of influenza hemagglutinin fusion peptides. *Journal of Physical Chemistry B, 114*, 8799–8806.

Li, X. J., Liu, Y., Wang, L., Deng, M. G., & Liang, H. J. (2009). Fusion and fission pathways of vesicles from amphiphilic triblock copolymers: A dissipative particle dynamics simulation study. *Physical Chemistry Chemical Physics, 11*, 4051–4059.

Lipowsky, R. (1992). Budding of membranes induced by intramembrane domains. *Journal de Physique II, 2*, 1825–1840.

Markin, V. S., & Albanesi, J. P. (2002). Membrane fusion: Stalk model revisited. *Biophysical Journal, 82*, 693–712.

Markvoort, A. J., Pfleger, N., Staffhorst, R., Hilbers, P. A. J., van Santen, R. A., Killian, J. A., et al., (2010). Self-reproduction of fatty acid vesicles: A combined experimental and simulation study. *Biophysical Journal, 99*, 1520–1528.

Markvoort, A. J., Smeijers, A. F., Pieterse, K., van Santen, R. A., & Hilbers, P. A. J. (2007). Lipid-based mechanisms for vesicle fission. *Journal of Physical Chemistry B, 111*, 5719–5725.

Markvoort, A. J., van Santen, R. A., & Hilbers, P. A. J. (2006). Vesicle shapes from molecular dynamics simulations. *Journal of Physical Chemistry B, 110*, 22780–22785.

Marrink, S. J., De Vries, A. H., & Tieleman, D. P. (2009). Lipids on the move: Simulations of membrane pores, domains, stalks and curves. *Biochimica et Biophysica Acta-Biomembranes, 1788*, 149–168.

Marrink, S. J., Fuhrmans, M., Risselada, H. J., & Periole, X. (2008). The Martini forcefield. *In* "Coarse graining of condensed phase and biomolecular systems" (G. A. Voth ed.), CRC Press, Boca Raton.

Marrink, S. J., Lindahl, E., Edholm, O., & Mark, A. E. (2001). Simulation of the spontaneous aggregation of phospholipids into bilayers. *Journal of the American Chemical Society, 123*, 8638–8639.

Marrink, S. J., & Mark, A. E. (2003). The mechanism of vesicle fusion as revealed by molecular dynamics simulations. *Journal of the American Chemical Society, 125*, 11144–11145.

Marrink, S. J., & Mark, A. E. (2004). Molecular view of hexagonal phase formation in phospholipid membranes. *Biophysical Journal, 87*, 3894–3900.

Miao, L., Seifert, U., Wortis, M., & Döbereiner, H. (1994). Budding transitions of fluid-bilayer vesicles: The effect of area-difference elasticity. *Physical Review E, 49*, 5389–5407.

McNew, J. A., Weber, T., Parlati, F., Johnston, R. J., Melia, T. J., Sollner, T. H., et al., (2000). Close is not enough: SNARE-dependent membrane fusion requires an active mechanism that transduces force to membrane anchors. *Journal of Cell Biology, 150*, 105–117.

Mirjanian, D., Dickey, A. N., Hoh, J. H., Woolf, T. B., & Stevens, M. J. (2010). Splaying of aliphatic tails plays a central role in barrier crossing during liposome fusion. *Journal of Physical Chemistry B, 114*, 11061–11068.

Muller, M., Katsov, K., & Schick, M. (2002). New mechanism of membrane fusion. *Journal of Chemical Physics, 116*, 2342–2345.

Muller, M., Katsov, K., & Schick, M. (2003). A new mechanism of model membrane fusion determined from Monte Carlo simulation. *Biophysical Journal, 85*, 1611–1623.

Muller, M., Katsov, K., & Schick, M. (2006). Biological and synthetic membranes: What can be learned from a coarse-grained description? *Physics Reports-Review Section of Physics Letters, 434*, 113–176.

Nielsen, S. O., Bulo, R. E., Moore, P. B., & Ensing, B. (2010). Recent progress in adaptive multiscale molecular dynamics simulations of soft matter. *Physical Chemistry Chemical Physics, 12*, 12401–12414.

Nishizawa, M., & Nishizawa, K. (2010a). Molecular dynamics simulation analyses of viral fusion peptides in membranes prone to phase transition: Effects on membrane curvature, phase behavior and lipid-water interface destabilization. *Journal of Biophysical Chemistry, 1*, 19–32.

Nishizawa, M., & Nishizawa, K. (2010b). Curvature-driven lipid sorting: Coarse-grained dynamics simulations of a membrane mimicking a hemifusion intermediate. *Journal of Biophysical Chemistry, 1*, 86–95.

Noguchi, H., & Takasu, M. (2001). Fusion pathways of vesicles: A Brownian dynamics simulation. *Journal of Chemical Physics, 115*, 9547–9551.

Noguchi, H., & Takasu, M. (2002a). Adhesion of nanoparticles to vesicles: A Brownian dynamics simulation. *Biophysical Journal, 83*, 299–308.

Noguchi, H., & Takasu, M. (2002b). Structural changes of pulled vesicles: A Brownian dynamics simulation. *Physical Review E, 65*, 051907.

Norizoe, Y., Daoulas, K. C., & Muller, M. (2010). Measuring excess free energies of self-assembled membrane structures. *Faraday Discussions, 144*, 369–391.

Ohki, S. (1984). Effects of divalent cations, temperature, osmotic pressure gradient, and vesicle curvature on phosphatidylserine vesicle fusion. *Journal of Membrane Biology, 77*, 265–275.

Ohta-Iino, S., Pasenkiewicz-Gierula, M., Takaoka, Y., Miyagawa, H., Kitamura, K., & Kusumi, A. (2001). Fast lipid disorientation at the onset of membrane fusion revealed by molecular dynamics simulations. *Biophysical Journal, 81*, 217–224.

Oosterlaken-Dijksterhuis, M. A., van Eijk, M., van Golde, L. M. G., & Haagsman, H. P. (1992). Lipid mixing is mediated by the hydrophobic surfactant protein Sp-B but not by Sp-C. *Biochimica et Biophysica Acta, 1110*, 45–50.

Papahadjopoulos, D., Nir, S., & Duzgunes, N. (1990). Molecular mechanisms of calcium-induced membrane-fusion. *Journal of Bioenergetics and Biomembranes, 22*, 157–179.

Pool, R., & Bolhuis, P. G. (2006). Prediction of an autocatalytic replication mechanism for micelle formation. *Physical Review Letters, 97*, 018302.

Pool, R., & Bolhuis, P. G. (2007). Sampling the kinetic pathways of a micelle fusion and fission transition. *Journal of Chemical Physics, 126*, 244703.

Ramachandran, S., Kumar, P. B. S., & Laradji, M. (2008). Lipid flip-flop driven mechanical and morphological changes in model membranes. *Journal of Chemical Physics, 129*, 125104.

Ramachandran, S., Laradji, M., & Kumar, P. B. S. (2009). Lateral organization of lipids in multi-component liposomes. *Journal of the Physical Society of Japan, 78*, 041006.

Relini, A., Cassinadri, D., Mirghani, Z., Brandt, O., Gambacorta, A., Trincone, A., et al., (1994). Calcium-induced interaction and fusion of archaeobacterial lipid vesicles: A fluorescence study. *Biochimica et Biophysica Acta-Biomembranes, 1194*, 17–24.

Reynwar, B. J., Illya, G., Harmandaris, V. A., Muller, M. M., Kremer, K., & Deserno, M. (2007). Aggregation and vesiculation of membrane proteins by curvature-mediated interactions. *Nature, 447*, 461–464.

Risselada, H. J., Knutzer, C., & Grubmuller, H. (2011). Caught in the act: Visualization of SNARE-mediated fusion events in molecular detail. *Chem Bio Chem, 12*, 1049–1055.

Risselada, H. J., & Marrink, S. J. (2008). The molecular face of lipid rafts in model membranes. *Proceedings of the National Academy of Sciences of the United States of America, 105*, 17367–17372.

Ryan, M. A., Qi, X. Y., Serrano, A. G., Ikegami, M., Perez-Gil, J., Johansson, J., et al., (2005). Mapping and analysis of the lytic and fusogenic domains of surfactant protein B. *Biochemistry, 44*, 861–872.

Sammalkorpi, M., Karttunen, M., & Haataja, M. (2008). Micelle fission through surface instability and formation of an interdigitating stalk. *Journal of the American Chemical Society, 130*, 17977–17980.

Sapir, A., Avinoam, O., Podbilewicz, B., & Chernomordik, L. V. (2008). Viral and developmental cell fusion mechanisms: Conservation and divergence. *Developmental Cell, 14*, 11–21.

Seifert, U., Berndl, K., & Lipowsky, R. (1991). Shape transformations of vesicles: Phase diagram for spontaneous-curvature and bilayer-coupling models. *Physical Review A, 44*, 1182–1202.

Shillcock, J. C., & Lipowsky, R. (2005). Tension-induced fusion of bilayer membranes and vesicles. *Nature Materials, 4*, 225–228.

Shillcock, J. C., & Lipowsky, R. (2006). The computational route from bilayer membranes to vesicle fusion. *Journal of Physics-Condensed Matter, 18*, S1191–S1219.

Smeijers, A. F., Markvoort, A. J., Pieterse, K., & Hilbers, P. A. J. (2006a). A detailed look at vesicle fusion. *Journal of Physical Chemistry B, 110*, 13212–13219.

Smeijers, A. F., Pieterse, K., Markvoort, A. J., & Hilbers, P. A. J. (2006b). Coarse-grained trans-membrane proteins: Hydrophobic matching, aggregation, and their effect on fusion. *Journal of Physical Chemistry B, 110*, 13614–13623.

Smirnova, YG., Marrink, S. J., Lipowsky, R., & Knecht, V. (2010). Solvent-exposed tails as prestalk transition states for membrane fusion at low hydration. *Journal of the American Chemical Society, 132*, 6710–6718.

Stein, A., Weber, G., Wahl, M. C., & Jahn, R. (2009). Helical extension of the neuronal SNARE complex into the membrane. *Nature, 460*, 525–528.

Stevens, M. J. (2005). Complementary matching in domain formation within lipid bilayers. *Journal of the American Chemical Society, 127*, 15330–15331.

Stevens, M. J., Hoh, J. H., & Woolf, T. B. (2003). Insights into the molecular mechanism of membrane fusion from simulation: Evidence for the association of splayed tails. *Physical Review Letters, 91*, 188102.

Stillwell, W., & Wassall, S. R. (2003). Docosahexaenoic acid: Membrane properties of a unique fatty acid. *Chemistry and Physics of Lipids, 126*, 1–27.

Tamm, L. K., & Han, X. (2000). Viral fusion peptides: A tool set to disrupt and connect biological membranes. *Bioscience Reports, 20*, 501–518.

Taylor, A., & Sansom, M. S. P (2010). Studies on viral fusion peptides: The distribution of lipophilic and electrostatic potential over the peptide determines the angle of insertion into a membrane. *European Biophysics Journal with Biophysics Letters, 39*, 1537–1545.

Tieleman, D. P., & Bentz, J. (2002). Molecular dynamics simulation of the evolution of hydrophobic defects in one monolayer of a phosphatidylcholine bilayer: Relevance for membrane fusion mechanisms. *Biophysical Journal, 83*, 1501–1510.

Tieleman, D. P., Leontiadou, H., Mark, A. E., & Marrink, S. J. (2003). Simulation of pore formation in lipid bilayers by mechanical stress and electric fields. *Journal of the American Chemical Society, 125*, 6382–6383.

Tolpekina, T. V., den Otter, W. K., & Briels, W. J. (2004a). Nucleation free energy of pore formation in an amphiphilic bilayer studied by molecular dynamics simulations. *Journal of Chemical Physics, 121*, 12060–12066.

Tolpekina, T. V., den Otter, W. K., & Briels, W. J. (2004b). Simulations of stable pores in membranes: System size dependence and line tension. *Journal of Chemical Physics, 121*, 8014–8020.

Vaccaro, L., Cross, K. J., Kleinjung, J., Straus, S. K., Thomas, D. J., Wharton, S. A., et al., (2005). Plasticity of influenza haemagglutinin fusion peptides and their interaction with lipid bilayers. *Biophysical Journal, 88*, 25–36.

Volynsky, P. E., Polyansky, A. A., Simakov, N. A., Arseniev, A. S., & Efremov, R. G. (2005). Effect of lipid composition on the "membrane response" induced by a fusion peptide. *Biochemistry, 44*, 14626–14637.

Weber, T., Zemelman, B. V., McNew, J. A., Westermann, B., Gmachl, M., Parlati, F., et al., (1998). Snarepins: Minimal machinery for membrane fusion. *Cell, 92*, 759–772.

Wong-Ekkabut, J., Xu, Z. T., Triampo, W., Tang, I. M., Tieleman, D. P., & Monticelli, L. (2007). Effect of lipid peroxidation on the properties of lipid bilayers: A molecular dynamics study. *Biophysical Journal, 93*, 4225–4236.

Wu, S. G., & Guo, H. X. (2009). Simulation study of protein-mediated vesicle fusion. *Journal of Physical Chemistry B, 113*, 589–591.

Yamamoto, S., & Hyodo, S. (2003). Budding and fission dynamics of two-component vesicles. *Journal of Chemical Physics, 118*, 7937–7943.

Yang, K., & Ma, Y. Q. (2009). Computer simulations of vesicle fission induced by external amphipathic inclusions. *Journal of Physical Chemistry B, 113*, 1048–1057.

Yang, K., Shao, X., & Ma, Y. Q. (2009). Shape deformation and fission route of the lipid domain in a multicomponent vesicle. *Physical Review E, 79*, 051924.

Yang, L., & Huang, H. W. (2002). Observation of a membrane fusion intermediate structure. *Science, 297*, 1877–1879.

Zheng, C., Liu, P., Li, J., & Zhang, Y. W. (2010). Phase diagrams for multi-component membrane vesicles: A coarse-grained modeling study. *Langmuir, 26*, 12659–12666.

CHAPTER 12

An Alternate Path for Fusion and its Exploration by Field-Theoretic Means

Marcus Müller[1] and Michael Schick[2]

[1]Institut für Theoretische Physik, Georg-August-Universität, Göttingen, Germany
[2]Department of Physics, University of Washington, Seattle, Washington, USA

I. OVERVIEW

We review a pathway for fusion that is an alternative to the standard one, and that differs markedly from it. It became apparent in simulations, and was intensively studied subsequently by field-theoretic methods. Results of those studies are reviewed. They show that the alternative pathway is favored over the standard one under almost all circumstances. Furthermore, it appears to be more in accord with many other molecular simulations. The alternate pathway allows for the possibility that transient leakage occur during fusion, or that lysis pre-empt it. We emphasize that fusion is a relatively rare event, as it must be in order that bilayers be useful, stable, entities.

II. INTRODUCTION

The purpose of this chapter is twofold. First, we will review a pathway for the fusion of two lipid bilayers, a pathway that is markedly different from the

Current Topics in Membranes, Volume 68 0065-230X/10 $35.00

DOI: 10.1016/B978-0-12-385891-7.00012-X

standard stalk–hemifusion–fusion mechanism (Hui, Stewart, Boni, & Yeagle, 1981; Kozlov & Markin, 1983) discussed in other chapters of this book. One of the consequences of this alternate pathway, which distinguishes it from the standard one, is the possibility, not necessity, of leakage during the fusion process. We review the various experiments and simulations that support this scenario. Second, we will present the results from theoretical calculations employing field-theoretic methods that show this pathway can present lower free-energy barriers to fusion than does the standard mechanism. Third, in an outlook, we will discuss computational techniques to calculate the free energy of metastable fusion intermediates by computer simulations and indicate potential strategies for predicting the fusion pathway.

We shall focus on the fusion of two apposed lipid bilayers. There is little that we need to say about fusion *per se* except that there is no reason to believe that it will occur unless the bilayers to be fused are placed under tension; that is, their free energy per unit area is made nonzero. The tension is brought about by some external agents, fusogens like SNARE complexes or hemagglutinin, which expend energy to bring the bilayers together. This eliminates some of the water between them, which increases the free energy per unit area. The response of the two apposed lipid membranes to this nonzero tension is to reduce the bilayer area of the system. It can do this either by fusion or by rupture (Engel & Walter, 2008).

The understanding of the fusion of model lipid membranes in the absence of proteins highlights the universal aspects of this collective transformation that changes the membrane topology. If it is the case that fusion without proteins proceeds by one of the alternate mechanisms we describe below, which have the possibility that the fusion process can be leaky, and if the fusion of biological bilayers with their associated proteins be not leaky, as is generally thought, then the work discussed here highlights a role for the associated proteins in addition to that of providing the necessary energy to bring the bilayers together. This role, which is not understood, is to ensure that, of all the possible paths that could be taken, some that are leaky and others that are not, only the latter paths are selected.

The description of fusion—the pathway and the concomitant metastable states and saddle points—by an effective interface Hamiltonian has been seminal (Kozlov & Markin, 1983) and guided the interpretation of many experiments. In this approach a lipid membrane, or the individual leaflets, is described by an infinitely thin sheet that is characterized by a small number of elastic properties—their spontaneous curvature and bending moduli. These coarse-grained parameters encode the architecture of the constituent lipids and the way they pack in the bilayer. An accurate description of the free energy along the fusion pathway within this framework, however, remains a challenge because (i) the pathway itself is not predicted but is an input in the calculation and (ii) the transformation from two apposed bilayer membranes to a fusion pore or from two vesicles to a single vesicle involves a discontinuous change of topology and passes through highly curved intermediates, which are difficult to describe within an effective interface Hamiltonian. The former caveat prevents a systematic exploration of possible fusion pathways. The latter challenge can be illustrated by the following consideration: The free energy along the fusion pathway from two vesicles to a single vesicle is continuous. The effective interface Hamiltonian, however, contains a term proportional to the integral of the Gaussian curvature of each leaflet. By virtue of the Gauss–Bonnet theorem, this contribution does not depend on the detailed shape of the membrane but only on the Euler characteristics of the monolayers. At some point along this fusion path, this Euler characteristic abruptly changes from the value appropriate to two intact vesicles, $\chi E = 2 \times (2 + 2)$, to a value $\chi E = 2 \times (2 + 1)$, which characterizes a stalk, to its final value, $\chi E = 2 \times 2$, that of a single vesicle. In the framework of an effective interface Hamiltonian, this discontinuous change has to be exactly compensated by the other curvature terms, describing highly curved and strongly interacting monolayers. This is a formidable task because the effective interface Hamiltonian is based on a curvature expansion around a flat reference state that is only accurate if the radius of curvature is large compared to the thickness of the bilayer. Moreover, the strong coupling between the monolayers in a complex and highly bent geometry are unknown. This difficulty results in a significant variation of the free-energy barriers depending on the specific form of the effective interface Hamiltonian (Kozlovsky & Kozlov, 2002).

Molecular simulations can partially overcome both difficulties, and provide unbiased insights into the fusion pathway and describe geometrically complex, highly bent structures while retaining a molecular description of the bilayer's constituents. Such coarse-grained molecular simulations of collective phenomena in lipid membranes have been reviewed in (Marrink, de Vries, & Tieleman, 2009; Müller, Katsov, & Schick, 2003a; Müller, Katsov, & Schick, 2006; Murtola, Bunker, Vattulainen, Deserno, & Karttunen, 2009; Nielsen, Lopez, Srinivas, & Klein, 2004; Shelley, Shelley, Reeder, Bandyopadhyay, & Klein, 2001; Venturoli, Sperotto, Kranenburg, & Smit, 2006). Molecular simulations, however, also face significant challenges: (i) A systematic exploration of the fusion process within a chemically realistic, atomistic model is not yet possible because the time and length scale of the fusion process—tens of nanometers and microseconds—are difficult to assess computationally. Further, a systematic study requires the observation of many fusion events for several reasons: there might be several, competing fusion pathways (Müller, Katsov, & Schick, 2003b); the fusion process exhibits a significant spread in duration (Kasson & Pande, 2007; Shillcock & Lipowsky, 2005); there are many molecular parameters, and one would like to understand the way in which the fusion process depends upon them. Particularly because of the several molecular parameters, one has to clarify the relation between the coarse-grained model and the experimental system to be simulated. (ii) Key characteristics of the fusion pathway are the metastable intermediates and saddle points and their concomitant free energies. Since the free energy cannot be simply expressed as a function of the particle coordinates, special sampling techniques have to be devised to extract transition paths (Dellago & Bolhuis, 2009) or to compute the free energy from molecular simulations (Müller & de Pablo, 2006). The calculation of the free energies of self-assembling systems has become possible only very recently (Detcheverry, Pike, Nealey, Müller, & de Pablo, 2009; Lennon, Katsov, & Fredrickson, 2008; Martinez-Veracoechea & Escobedo, 2006; Müller & Daoulas, 2008).

Because of the above difficulties, we have adopted a two-track strategy. First, molecular simulations are used to identify the fusion pathway. Second, field-theoretic techniques are employed to obtain the free energy of the intermediate structures that are observed, and to explore systematically the role of tension and molecular architecture on the fusion process.

III. A DIFFERENT PATHWAY TO FUSION

The fusion pathway described in this section was discovered simultaneously by two different groups, each carrying out coarse-grained simulations on bilayers of amphiphiles. Noguchi and Takasu (Noguchi & Takasu, 2001) utilized a Brownian dynamics simulation on a model of amphiphiles consisting of one hydrophilic head bead and two hydrophobic tails beads. The solvent was not modeled explicitly. Müller, Katsov, and Schick (2002) performed a Monte Carlo simulation on a system in which the amphiphiles more closely resembled block copolymers containing 32 segments of which 10 were hydrophilic and 22 hydrophobic. The solvent was represented by a homopolymer of 32 segments. Although the computational methods, the lipid architecture and interactions were rather different, the two groups observed the same mechanism.

Initially a stalk between the opposing bilayers is formed, as depicted in Fig. 1A. However instead of expanding axially, as in the standard mechanism, the stalk was observed to elongate and wander. The presence of the stalk actively destabilizes the bilayer by causing an increase in the rate of nucleation of holes in it. These holes appear close to the stalk, as can be observed in Fig. 2A and D, because the presence of the stalk reduces the line tension of the hole and thus the free energy needed to create them (Müller et al., 2003b). Due to the reduction in line tension in that region, in which the stalk abuts the hole, it is not surprising that the stalk elongates so as to encircle it (see Fig. 2B) producing a hemifusion

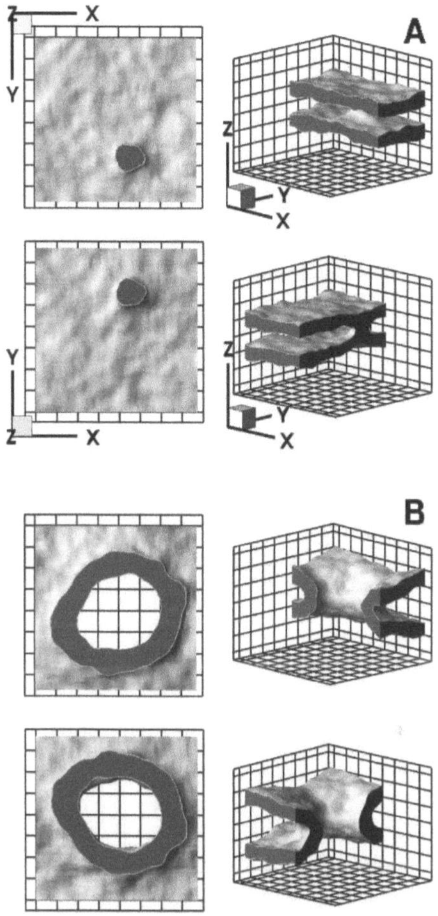

FIGURE 1 Representative snapshots of (A) the stalk intermediate and (B) the complete fusion pore from a coarse-grained simulation of model membranes. Each configuration is shown from four different viewpoints. The hydrophobic core is shown as dark gray, the hydrophilic–hydrophobic interface (defined as a surface on which densities of hydrophilic and hydrophobic segments are equal) is light gray. Hydrophilic segments are not shown for clarity. Top- and bottom-left subpanels have been generated by cutting the system along the middle $x - y$ plane, the top and bottom halves are viewed in the positive (up) and negative (down) z-direction correspondingly. Top-and bottom-right subpanels are side views with cuts made by $x - z$ and $y - z$ planes, respectively. 3D-orientation axis are the same for all snapshots and shown in panel (A). Reprinted from (Müller et al., 2003b), Copyright (2003), with permission from Elsevier.

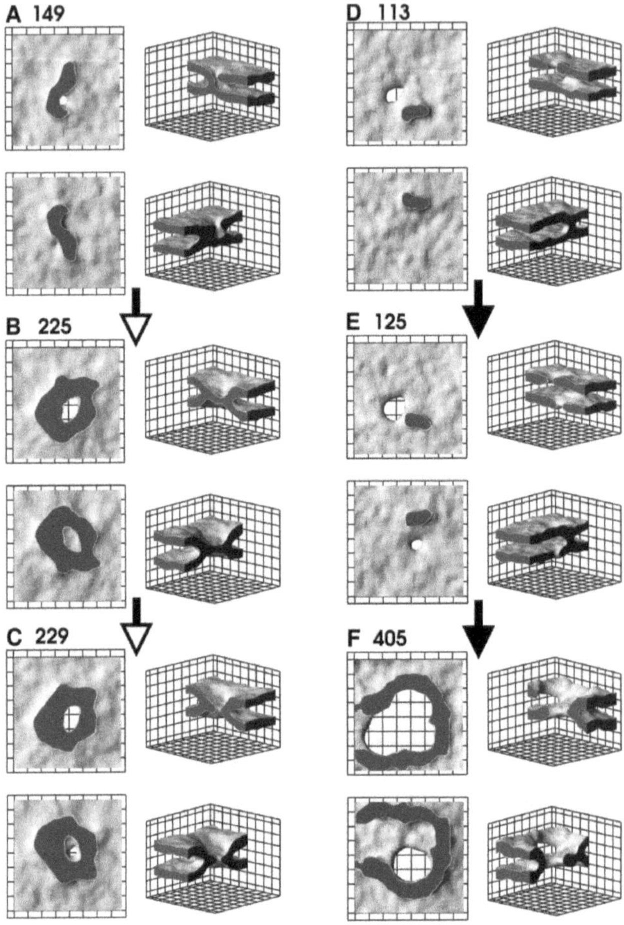

FIGURE 2 Two observed pathways of fusion process. The snapshots are taken from two representative simulation runs. Each configuration is numbered by the time (in multiples of 25,000 MCS of local Monte Carlo moves of the bond fluctuation model) at which it was observed. For comparison, 80,000 MCS are needed for a lipid to laterally diffuse a distance of the order of the radius of gyration R_g. Using the lateral self-diffusion coefficient to match the time scale between the coarse-grained simulation and an experiment, we find that fusion occurs on the microsecond scale (Müller et al., 2006). See Fig. 1 for explanation of the graphics shown. Reprinted from (Müller et al., 2003b), Copyright (2003), with permission from Elsevier.

diaphragm comprised of the one still-intact bilayer. Note that this is in marked contrast with the hemifusion diaphragm of the standard mechanism, which consists of the outer leaves only of the two fusing bilayers. The appearance of a second hole, one that pierces this diaphragm, completes the fusion process, as shown in Fig. 2C. One alternative scenario occurs if the rate at which holes form is sufficiently rapid that the second hole appears before the stalk completely encloses the first one. In this case, the fusion process is completed when the stalk encloses *both* the holes that form the fusion pore, as has almost occurred in Fig. 2F. Another alternative, analyzed by Katsov, Müller, and Schick (2006), occurs if the rate at which holes form is sufficiently slow so as to enable the stalk to elongate in a linear fashion and eventually close upon itself before either of the two holes form. In so doing, the stalk can entrap an amount of solvent and forms a structure, shown in Fig. 3, denoted in (Katsov et al., 2006) an inverted micellar intermediate (IMI) after a rather different, but topologically similar, structure discussed by Siegel (Siegel, 1993). A hole then appears in one of the bilayers, thereby forming a hemifusion diaphragm consisting of the remaining intact bilayer. Formation of a hole in this bilayer completes the fusion process.

Two consequences of the above fusion pathways, which do not occur in the standard one, were emphasized very early (Müller et al., 2003b). The first, which occurs in all three of the nonstandard pathways described above, is the mixing of lipids from different leaves of the *same* bilayer. The second is the possibility of leakage, which could occur in either of the pathways shown in Fig. 2. Both consequences are easily inferred from part (A) of that figure. Leakage would not occur, however, if the rate of hole formation was sufficiently slow so that an IMI formed.

Various aspects of the nonstandard pathway shown in Fig. 2A–C were later observed in molecular dynamics simulations of other coarse-grained models. The elongation of the initial stalk, the formation of a hole in one bilayer, and the formation of a hemifusion diaphragm consisting of one of the original bilayers

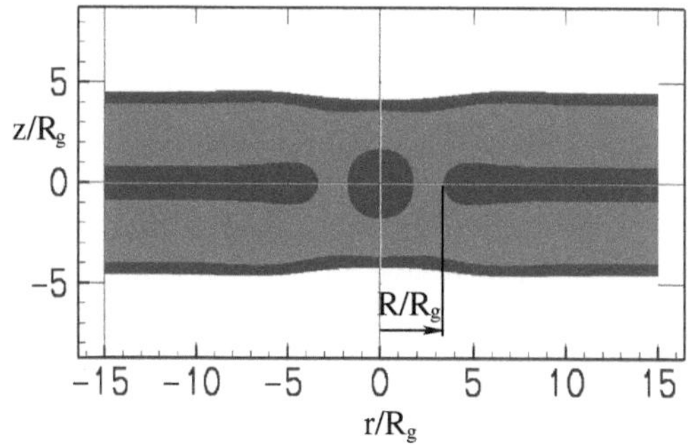

FIGURE 3 Density profile of an inverted micellar intermediate (IMI). The amphiphiles are characterized by $f = 0.3$. The radius of the IMI, in units of the radius of gyration, R/R_g is 3.4. The tension is zero. Only the majority component is shown at each point. Solvent segments are white. Hydrophilic and hydrophobic segments of the amphiphile are shaded dark and light, respectively. Reprinted from (Katsov et al., 2006), Copyright (2006), with permission from Elsevier.

was observed by Marrink and Mark (2003). However, they also observed fusion events in the same system that proceeded by the standard mechanism. In the same year, Stevens, Hoh, and Woolf (2003) observed the elongation and wandering of the initial stalk and the formation of an IMI. This structure was also observed more recently by Marrink, Fuhrmans, Risselada, and Periole (2008). Of great interest is the recent observation of the formation of an IMI before subsequent fusion in the simulation of a SNARE-mediated process (Risselada, Kutzner & Grubmüller, 2011).

What of the experimental evidence? Mixing of lipids originating in different leaves of the same bilayer is not easily observed, but there nevertheless is some evidence for it (Cevc & Richardsen, 1999; Evans & Lentz, 2002). Leakage, on the other hand, is much more readily detected, and there is much evidence in the fusion of model membranes (Bonnafous & Stegmann, 2000; Cevc & Richardsen, 1999; Evans & Lentz, 2002; Haque & Lentz, 2002; Haque, McIntosh, & Lentz, 2001; Lentz, Talbot, Lee, & Zheng, 1997; Shangguan, Alford, & Bentz, 1996; Weinreb & Lentz, 2007). There is also some evidence from the fusion of biological membranes for the occurrence of leakage during fusion (Aguilar, Engel, & Walter, 2007; Frolov, Dunina-Barkovskaya, Samsonov, & Zimmerberg, 2003; Merz & Wickner, 2004).

The different computer simulation studies employed very different molecular representations ranging from minimal models (Daoulas & Müller, 2010), which only represent the relevant interactions that are required to bring about the self-assembly into amphiphilic membranes, to sophisticated models, like the MARTINI force field (Marrink, Risselada, Yefimov, Tieleman, & de Vries, 2007), that have been carefully parameterized by a quantitative comparison to chemically realistic, atomistic models. The latter models are able to describe in molecular detail complex morphologies like the rhombohedral stalk structure, in the absence of peptides (Marrink & Mark, 2004), and in their presence (Fuhrmans, Knecht & Marrink, 2009). In spite of the marked differences in the interactions (e.g., solvent-free models (Noguchi & Takasu, 2001) vs. explicit representation of the solvent (Shillcock & Lipowsky, 2002)) and the details of the molecular architecture (e.g., linear chains of a small number of interaction centers (Gao, Shillcock, & Lipowsky, 2007) vs. a systematically derived model (Marrink & Mark, 2003) duly accounting for the molecular shape and its fluctuations), most of the studies find deviations from the standard stalk–hemifusion–fusion pathway (Gao, Lipowsky, & Shillcock, 2008; Grafmüller, Shillcock, & Lipowsky, 2009; Marrink & Mark, 2003; Müller et al., 2002; Müller et al., 2003b; Noguchi & Takasu, 2001; Smeijers, Marvoort, Pieterse, & Hilbers, 2006; Stevens et al., 2003). This observation suggests that the qualitative aspects of bilayer membrane fusion are rather universal and differ from the standard pathway.

The mere fact that the fusion process in the different coarse-grained models can be directly observed in molecular simulations without the application of special sampling techniques to overcome free-energy barriers provides a limit for the free energy barriers. When we describe fusion as a thermally activated process the time scale is set by $\tau_{\text{fusion}} \approx \tau_o \exp(\Delta G/k_B T)$, where the kinetic prefactor, τ_o, characterizes a microscopic time scale, $k_B T$ is the thermal energy, and ΔG is the largest barrier to be overcome in the course of the fusion process. Since molecular simulations can typically span about 6–7 orders in time and the smallest time scale of the simulation is certainly smaller than τ_o, the fusion barrier does not exceed $15k_B T$ in those simulations that directly observe fusion.

Moreover, most of the studies observe more than one fusion pathway indicating that there are multiple paths of transformation from two apposed bilayers to a fusion pore, which differ in their barriers by an amount comparable to $k_B T$. These findings suggests the following: (i) If the free energy barrier does not exceed $k_B T$ by very much, the fusion may not be conceived as a thermally activated process, that is, a description in terms of isolated, well-defined transition states along the fusion pathway may not be adequate and the full collective dynamics must be considered. (ii) Details of the molecular architecture of the lipids or the effect of fusion peptides may select one pathway over another. For instance, studies by Lipowsky and coworkers indicate that the fusion path significantly varies in response to modifications of the coarse-grained model (Gao et al., 2008; Grafmüller et al., 2009; Shillcock & Lipowsky, 2005). A systematic exploration of the free energy barriers as a function of the model parameters is warranted to provide some insights into the different observations.

IV. RESULTS FROM A FIELD-THEORETIC DESCRIPTION

This brief chapter is not the place to review self-consistent field theory and the means by which one derives a field-theoretic description from a particle-based one. The interested reader is referred to a few particularly clear sources: reviews by Schmid (Müller & Schmid, 2005; Schmid, 1998) and a text by Fredrickson (Fredrickson, 2006). We note that this mean-field approach has been successfully applied to investigate the behavior of block copolymers (Matsen, 2006, 2002) and provides accurate estimates for the free energy of stable and metastable structures in polymeric systems.

We present here some of the results that have been obtained from a field-theoretic description of the fusion of bilayers composed of flexible block copolymers. As stated earlier, the implicit assumption is that the fusion pathway is universal, that it is the same for these amphiphiles as for lipids. The fact that the same fusion pathway was observed in the simulation of two completely different systems of amphiphiles (Müller et al., 2002; Noguchi & Takasu, 2001) supports this view.

In our model of amphiphilic membranes, the block copolymers consist of a fraction f that is "hydrophilic" and a fraction $1-f$ that is "hydrophobic." The solvent consists of flexible homopolymers that are "hydrophilic"; for simplicity both molecular species have the same radius of gyration, R_g.

In order to relate the predictions of the coarse-grained model of diblock copolymers and homopolymers to a lipid system, one has to identify the relevant interactions and parameterize their strengths (Müller et al., 2003a; Müller et al., 2006). Systems of block copolymer in solvent and those of lipids in aqueous solution spontaneously self-assemble into bilayer membranes because of the repulsion between hydrophilic and hydrophobic entities and the connectivity of these entities along the backbone of the molecule. The hydrophobic thickness, d, of the membrane is dictated by the molecular extension, R_g, and defines the characteristic length scale. The relevant energy scale is set by the interfacial free energy, γ_{int}, of the hydrophilic and hydrophobic interface. In an aqueous lipid system, this quantity corresponds to the oil–water surface tension. In the self-consistent field theory the interfacial tension depends on the incompatibility, χN, between the hydrophilic and hydrophobic segments and the molecular density, which can be parameterized by the square root of the invariant chain length, $\sqrt{\bar{N}} \sim \rho_p R_g^3$, where R_g denotes the radius of gyration of the polymer and ρ_p is the number density of polymers. We emphasize that this molecular density sets the scale of the free energy, that is, all the different contributions to the free energy–those of the interfacial tension between hydrophilic and hydrophobic domains, of the bending rigidities of the monolayers, of the areal compressibility, and of the coupling between the monolayers–all depend on the number density in the same way. Because it sets the scale, it does not enter explicitly into the mean-field calculations, and we choose the value $\rho_p R_g^3 \approx 0.65$ in order to match the results of the theory with those obtained from our simulations of the bond fluctuation model (Müller et al., 2002). Furthermore, we set the incompatibility to $\chi N = 30$, which corresponds to a reasonably strong segregation between hydrophilic and hydrophobic domains. For these specific parameters the free energies of an aqueous lipid solution are about 2.5 times those of the coarse-grained model of the diblock copolymer–homopolymer mixture. This ratio is characteristic for coarse-grained models that have a hard-core repulsion. It is computationally advantageous because the free energy barriers in the simulation are smaller than in experimental lipid systems, therefore allowing the thermally activated fusion process to be observed spontaneously, and obviating the need to apply sophisticated sampling techniques. We note, however, that the molecular densities in experimental copolymer–homopolymer mixtures, or in the aqueous solutions of amphiphilic polymers that form polymersomes (Discher et al., 1999), are typically much larger than that we have chosen. Consequently, the barriers in experimental polymer systems are, in fact, larger than those in lipid systems.

Lastly, the molecular architecture of the amphiphiles is characterized by the fraction, f, of hydrophilic segments. By adjustment of the fraction f, the spontaneous curvature of these amphiphiles can be matched to those of a given lipid (Müller & Gompper, 2002). The product of the spontaneous curvature of a monolayer, c_0, and the membrane thickness, d, provides an appropriate dimensionless quantity to relate the coarse-grained model and an experimental system. For example, a value $f = 0.36$ corresponds to a spontaneous curvature appropriate to a 1:1 mixture of dioleoylphosphatidylethanolamine (DOPE) and dioleoylphosphatidylcholine (DOPC).

A. The Standard Fusion Pathway

The first fusion pathway to be examined by field-theoretic methods (Katsov, Müller, & Schick, 2004) was the standard one involving the formation of an axially symmetric stalk, which expands axially to a hemifusion diaphragm, as shown in Fig. 4.

The free energies of these configurations are shown in Fig. 5 as a function of their size for different architectures of the amphiphiles and for zero tension, in panel (a). The free energies increase linearly with radius for large radius due to the line tension of the rim of the diaphragm. Because the tension, or free energy per unit area, is zero, the decrease in bilayer area brought about by the hemifusion does not decrease the free energy of the system. In the inset, we indicate three configurations, which play a key role in the description of the fusion process via this pathway:

- The "transition" state between the unperturbed bilayers and the metastable stalk, S0.
- The metastable stalk itself, S1.
- The transition state that occurs as the initial small stalk is radially expanded into the hemifusion diaphragm, denoted S2.

The free energy of the stalk intermediate did not exceed $4k_BT$ for the block copolymer model, which was estimated to correspond to about $10k_BT$ for a relevant lipid system. For strong asymmetries, the free energy of a stalk, S1, becomes negative, that is, stalks between the two apposed bilayers are stable and the system will lower its free energy by forming multiple stalks. Under these conditions stalks will condense into a dense, two-dimensional liquid and, at high lateral densities, arrange on an hexagonal lattice forming a rhombohedral structure. This phase has been experimentally observed in lipid systems (Aeffner, Reusch, Weinhausen, & Salditt, 2009; Yang &

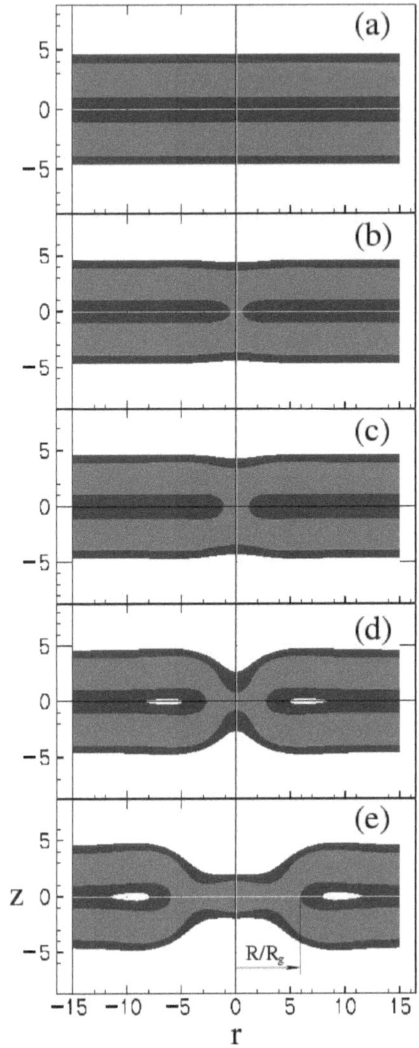

FIGURE 4 Density profiles of the stalk-like structures shown in the *r*, *z* plane of cylindrical coordinates. (a) Two bilayers in solvent. There is no stalk between them. (b) The limit of stability of the initial stalk, a state we label S0. (c) The metastable stalk itself, which we label S1. (d) The unstable transition state (saddle point) between the metastable stalk and the hemifusion diaphragm. This transition state is denoted S2. (e) A small hemifusion diaphragm. Reprinted from (Katsov et al., 2004), Copyright (2004), with permission from Elsevier.

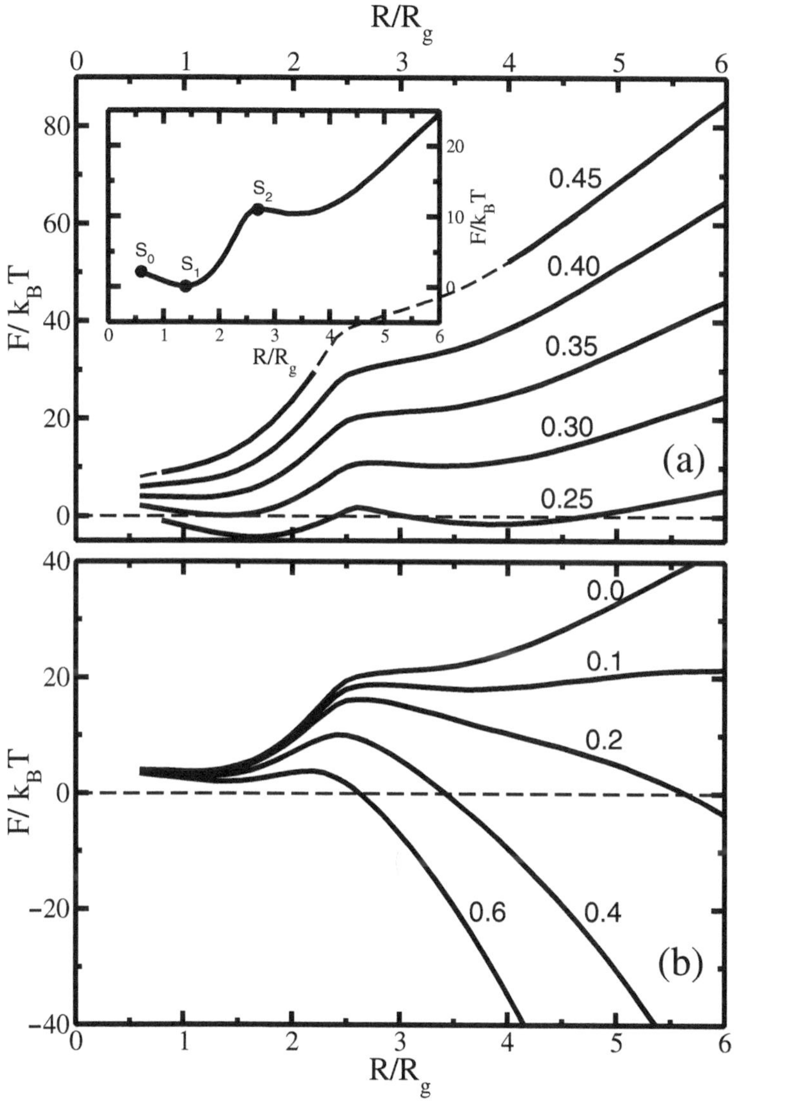

FIGURE 5 (a) The free energy, *F*, of the stalk-like structure connecting bilayers of fixed tension, zero, is shown for several different values of the amphiphile's hydrophilic fraction *f*. In the inset we identify the metastable stalk, S1, the limit of metastability of a stalk, S0, with a narrow cross section, and the transition state, S2, between the metastable stalk and an extended hemifusion diaphragm. The architectural parameter is $f = 0.30$ for this inset. No stable stalk solutions were found for $f = 0.45$ in the region shown with dashed lines. They were unstable to pore formation. (b) The free energy of the expanding stalk-like structure connecting bilayers of amphiphiles with fixed architectural parameter $f = 0.35$ is shown for several different bilayer tensions. These tensions, γ/γ_{int}, are shown next to each curve. Reprinted from (Katsov et al., 2004), Copyright (2004), with permission from Elsevier.

Huang, 2002) and block copolymers (Hajduk et al., 1997; Loo, Register, Adamson, & Ryan, 2005) illustrating again the universality of amphiphilic self-assembly. The (meta)stability of the stalk and its strong dependence on the molecular asymmetry predicted by Katsov et al. (Katsov et al., 2004) has also been corroborated by recent molecular simulation (Norizoe, Daoulas, & Müller, 2010).

In panel (b) the free energies are shown for a given architecture, $f = 0.35$, and different tensions. Now that the tensions are nonzero, the decrease in bilayer area brought about by hemifusion decreases the free energy of the system and dominates the line tension contribution for sufficiently large radii. As expected, the barrier associated with the expansion of the hemifusion diaphragm decreases with increasing tension (Monck, de Toledo, & Fernandez, 1990).

The free energy of a fusion pore was also calculated (Katsov et al., 2004). If it is assumed that a hole forms in the hemifusion diaphragm at the radius at which its free energy is equal to that of a pore with the same radius, then one finds that this occurs at a radius *larger* than that corresponding to the barrier associated with expansion of the hemifusion diaphragm. The conclusion from this study is that the major barrier to fusion along this pathway is that associated with the expansion of the hemifusion diaphragm itself, that is, the saddle point S2.

The results of the field-theoretic study of the standard, axially symmetric fusion pathway can be summarized in the fusion diagram, Fig. 6, which delineates the thermodynamic stability of stalks and its consequences for the standard fusion pathway. There is only a small window of molecular asymmetries, in which the stalk is a metastable structure. If the amphiphiles are too symmetric the stalk is not even metastable and, since all fusion pathways involve the stalk as an intermediate structure, fusion will be strongly suppressed. As the amphiphiles become more asymmetric, the excess free energy will decrease and eventually become negative. In the latter case, a rhombohedral stalk phase is predicted in agreement with experiments (Aeffner et al., 2009; Hajduk et al., 1997; Loo et al., 2005; Yang & Huang, 2002). For even more asymmetric amphiphiles, the stalk ceases to be metastable and spontaneously extends linearly marking the spinodal toward the inverted hexagonal phase.

B. The Alternate Fusion Pathway

The alternate path to fusion was subsequently examined by field-theoretic means (Katsov et al., 2006). This was more difficult than in the standard pathway because the stalk and the holes will not necessarily make an axially symmetric system. In principle, the free energy along the alternate path could be determined by three-dimensional self-consistent field calculations but it has

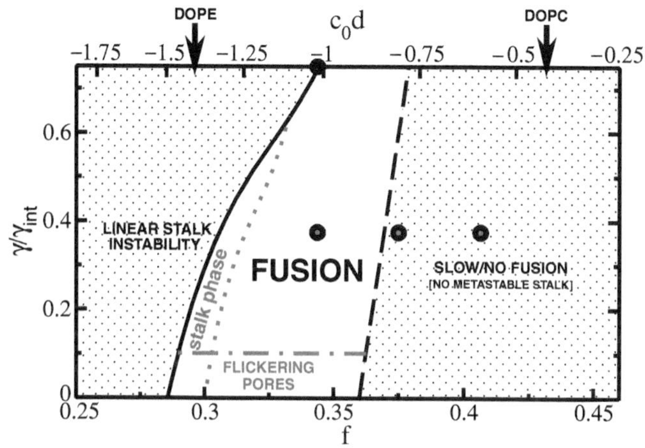

FIGURE 6 Stability diagram of stalks as a function of the hydrophilic fraction, f, and the bilayer tension, γ. Circles show points at which previous independent simulations were performed by us. Successful fusion can occur within the unshaded region. As the tension, γ, decreases to zero, the barrier to expansion of the pore increases without limit as does the time for fusion. As the right-hand boundary is approached, the stalk loses its metastability, causing fusion to be extremely slow. As the left-hand boundary is approached, the barriers to fusion are reduced, as is the time for fusion, but the process is eventually pre-emptied due to the stability either of radial stalks (forming the rhombohedral stalk phase), or linear stalks (forming the inverted hexagonal phase). Reprinted from (Katsov et al., 2004), Copyright (2004), with permission from Elsevier.

proven difficult to achieve the required numerical accuracy. Thus, the free energy has been estimated by patching together the results of axially symmetric calculations.

The state of the system immediately before the formation of the first hole was considered to be a stalk that had bent into an arc, which is a fraction α of a complete circle of radius R, as shown in Fig. 7a. Note that $\alpha = 0$ corresponds to the original stalk while $\alpha = 1$ corresponds to the IMI of Fig. 3. Just after the formation of the first hole, the system was modeled as the same arc as above, but now with a hole next to it as in Fig. 7b.

The free energy landscape in the $\alpha - R$ plane, defined as the minimum of the free energy of the configurations with and without a hole, is shown in Fig. 8. It exhibits two valleys. The first, near $\alpha = 0$, corresponds to a barely elongated stalk. The second, near $\alpha = 1$, corresponds to a hole nearly completely encircled by the stalk. A ridge separates these two valleys, and the minimum of this ridge defines the barrier between them. Of course for fusion to be completed, a hole must form in the hemifusion diaphragm to make a pore. But just as in the standard pathway, it was found that this final step did not make a

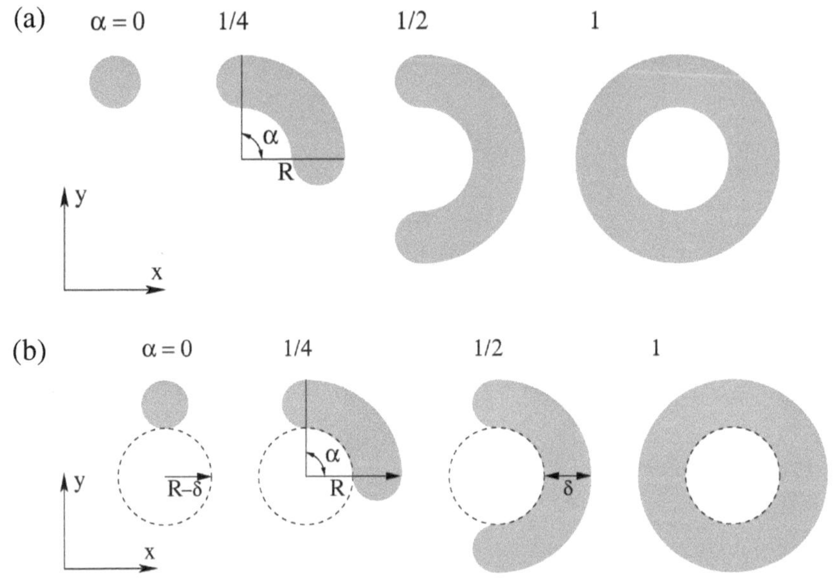

FIGURE 7 (a) Parameterization of the elongated stalk. Gray schematically shows location of the hydrophobic segments in the plane of symmetry between fusing bilayers. Values of the fractional arc angle α, defined in the range [0,1], are given at the top of each stalk configuration. (b) Parameterization of the stalk–hole complex. Gray schematically shows location of the hydrophobic segments in the plane of symmetry between fusing bilayers. Projection of the edge of a hole in one of the membranes is shown with dashed line. The radius of this hole is $R - \delta$. The other membrane does not have a hole. The hydrophobic thickness of the bilayer is δ. Values of the fractional arc angle α, defined in the range [0,1], are given at the top of each stalk configuration. Reprinted from (Katsov et al., 2006), Copyright (2006), with permission from Elsevier.

larger barrier than that already found; that is, in the alternate mechanism, the main barrier to fusion is to go from the initial state with an elongated stalk, perhaps even an IMI, to the hemifusion diaphragm consisting of the one intact bilayer.

C. Comparison of the Two Pathways

Katsov et al. (Katsov et al., 2006) could now compare the barriers to fusion along the standard and the alternate pathways as a function of the tension in the

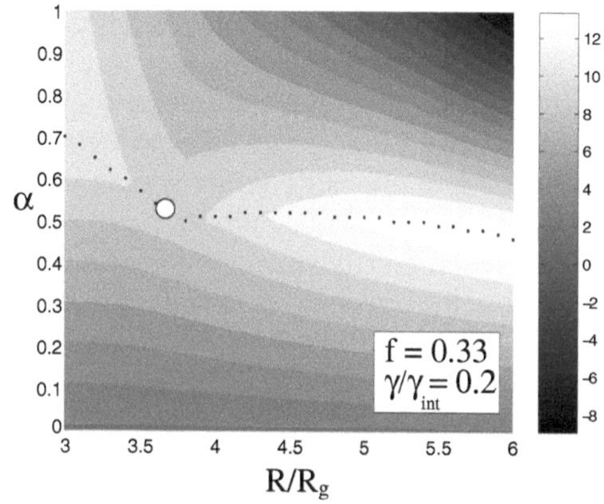

FIGURE 8 Free energy landscape (in units of $k_B T$) of the fusion process, plotted as a function of the radius, R (in units of the size of an amphiphile, R_g) and circumference fraction, α. The hydrophilic fraction is $f = 0.33$ and the tension $\gamma/\gamma_{int} = 0.2$. The dotted line shows a ridge of possible transition states, separating two valleys. The region close to the $\alpha = 0$ line corresponds to a barely elongated stalk intermediate. The other valley, close to $\alpha = 1$ states, corresponds to a hole almost completely surrounded by an elongated stalk. The saddle point on the ridge, denoted by an open dot, corresponds to the optimal (lowest free energy) transition state. Reprinted from (Katsov et al., 2006), Copyright (2006), with permission from Elsevier.

bilayers and the architecture of the amphiphiles. They found that the barrier to fusion was almost always lower along the new pathway, a difference that varied from about 3 to $8k_B T$ in the polymer system, equivalent to 7.5 to $12k_b T$ in the lipid system.

Several conclusions were drawn from this study. Among them were as follows:

- The amount of leakage, which can occur during fusion, is correlated with the amphiphile architecture; the more the architecture tends toward that of amphiphiles that form inverted-hexagonal phases, the less leakage is expected.
- The IMI is a metastable intermediate whose free energy is not large, a conclusion in contrast to those from phenomenological theories (Siegel, 1993). Were the IMI the fusion intermediate, there would be no leakage.
- The new mechanism is favored whenever the free energy of the linearly extending stalk is small. This is the case when the architecture tends toward that of amphiphiles which form inverted-hexagonal phases.
- The new mechanism is favored when the free energy to form a hole is not too large.

This last point is interesting. Clearly for a bilayer to be stable or a vesicle to be biologically useful, the probability of hole formation due to thermal fluctuations must be small. But for fusion to occur, the integrity of the vesicle must be disrupted and the probability for one or two holes to form must be appreciable. The conundrum is how these opposing requirements can be reconciled. As emphasized early on (Katsov et al., 2006; Müller et al., 2003b), the probability of hole formation is increased greatly when the formation takes place next to a stalk-like structure because the probability depends exponentially on the square of the line tension of the hole's rim, and this tension is reduced by the proximity of the stalk-like structure.

D. *Effect of Different Compositions of the Bilayer Leaves*

It is well known that the compositions of the inner and outer leaves of the plasma membrane can be quite different from one another. In human red blood cell membranes, for example, most of the cholinephospholipids, sphingomyelin (SM), and phosphatidylcholine (PC) are found in the outer, ectoplasmic, leaf, and most of the aminophospholipids, phosphatidylethanolamine (PE), and phosphatidylserine (PS) are found in the inner, cytoplasmic, leaf (Devaux, 1991; Rothman & Lenard, 1977). In particular the mole percent of PC in the outer/inner leaf is 22/8, of SM is 20/5, of PS is 0/10, and of PE is 8/27 (Rothman & Lenard, 1977). To maintain this imbalance costs energy (Seigneuret & Devaux, 1984), therefore it is reasonable to assume that it plays some physiological function. One suggestion is that this imbalance promotes fusion in intracellular events (Bailey & Cullis, 1994; Chernomordik, Kozlov, & Zimmerberg, 1995; Eastman, Hope, Wong, & Cullis, 1992). The reasoning is as follows. Of the four major lipid groups cited above, three of them, SM, PC, and PS (Hope & Cullis, 1980) form bilayers under physiological conditions. They make up 65% of the total bilayer, but 84% of the outer, trans, leaf. PE, however, does not form lamellae, but rather an inverted hexagonal phase (Gruner, 1989). It has often been noted (Siegel, 1986) that regions of this nonbilayer phase resemble the nonbilayer configurations posited to occur in fusion. Furthermore, PE resides predominantly in the inner, cytoplasmic, leaf of the plasma membrane. While it makes up only 35% of the total bilayer composition of human red blood cell membranes, it comprises 54% of the inner leaf. It is presumed to also reside predominantly in the outer leaf of a bilayer vesicle within the cell, as the outer leaf of such a vesicle would make contact with the inner leaf of the plasma membrane during fusion of the vesicle and plasma membrane, and thereby have the opportunity to exchange lipid content. But it is precisely the inner leaf of the plasma membrane and the outer leaf of a vesicle that would be closest to one another during fusion (i.e., would be the *cis* leaves), and would undergo the largest deviation from a planar configuration. Hence the enhanced concentration of PE in these leaves would presumably promote fusion.

The effect of this asymmetry has been investigated by field-theoretic methods (Lee & Schick, 2007b) and by simulation (Kasson & Pande, 2007). In the field-theoretic study, a system was considered in which the bilayers consisted of two kinds of amphiphiles; one with a fraction of hydrophilic groups $f = 0.4$ that gives the molecule a spontaneous curvature like that of DOPC, a molecule that forms lamellar phases, and the other with $f = 0.294$, approximately the value characterizing DOPE, a molecule that forms an inverted hexagonal phase. The hexagonal-former was first added to both leaves equally. The barrier to fusion along the standard pathway fell from an estimated 60 $k_B T$ with no hexagonal formers to $28 k_B T$ when the hexagonal formers occupied a fraction 0.35 of the volume as PE does in the plasma membrane of human red blood cells. The effect of the addition was more rapid initially, presumably because the first amount added goes to those place at which the stress can be relieved by these amphiphiles. The calculation showed that these hexagonal formers preferentially went to the rim of the hemifusion diaphragm. This nonlinear effect of the nonlamellar formers is in accord with experiment (Eastman et al., 1992). The effect on the fusion barrier in the alternate mechanism was more modest; from $21 k_B T$ to $17 k_B T$. The effect of the free energy of the stalk itself was non-negligible but small, a result that contrasts strongly with that from phenomenological theories (Kozlovsky & Kozlov, 2002). Of particular interest is that the change in the time of stalk formation due to the addition of hexagonal-former as determined from simulation (Kasson & Pande, 2007) could be used to extract the change in energy of the stalk with this addition. The result agreed with that obtained from self-consistent field theory to within a factor of two. This represents "... the first direct quantitative link between continuum-theory and molecular-dynamics predictions for membrane fusion" (Kasson & Pande, 2007).

Field theory was then used to calculate the effect of redistributing the hexagonal-formers between the two leaves while keeping their total amount fixed at 0.35 of the total (Lee & Schick, 2007b). The barrier to fusion in the standard mechanism was reduced from $28 k_B T$ to about $13 k_B T$ as the fraction in the inner leaf rose from 0.35 to 0.43, primarily due to the reduction in the energy of the hemifusion diaphragm. In the alternate pathway, the barrier was reduced from $17 k_B T$ to about $10 k_B T$. Again the energy of the stalk itself was not affected very much, decreasing from $13 k_B T$ to about $8 k_B T$. The most important result to be drawn from this is that in both mechanisms the major barrier to fusion is reduced by the introduction and asymmetric distribution of hexagonal-formers. It is now not so much larger than the barrier to make the initial stalk. Indeed in the alternate mechanism, the difference in the two barrier heights is quite small. This indicates that once the fusion process is initiated, it should proceed quite rapidly.

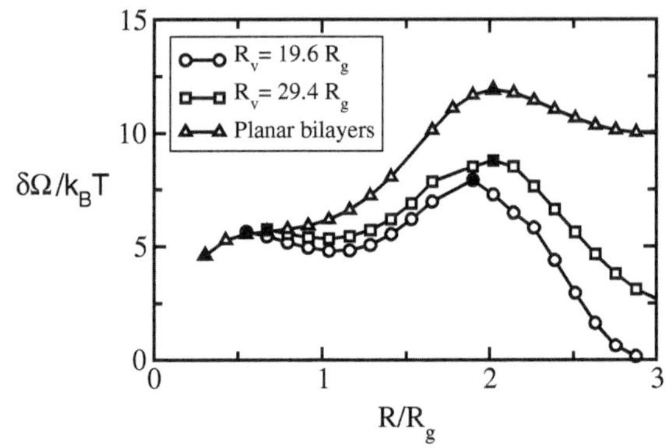

FIGURE 9 Excess free energy of stalk–hemifusion intermediates for two-component bilayers comprised of lamellar-forming $f_1 = 0.4$ and hexagonal-forming $f_2 = 0.294$ AB diblock copolymers under zero tension. The upper curve (triangles) is for two planar bilayers, the middle curve (squares) is for a planar bilayer and a vesicle of $R_v = 29.4R_g$, and the bottom (circles) is for a planar bilayer and a vesicle of $R_v = 19.6R_g$. The distance between the two bilayers is $1.96R_g$. Filled symbols indicate the first and second barriers for each system. Insets below the curves show density plots of head groups of a stalk of $R = 0.67R_g$ at which the first barrier occurs, a hemifusion diaphragm of $R = 2.0R_g$ at which the second barrier occurs, and a hemifusion diaphragm of the largest radius shown in the plot ($R = 2.9R_g$) between a planar bilayer and a vesicle of $R_v = 29.4R_g$. Reprinted from (Lee & Schick, 2008), Copyright (2008), with permission from Elsevier.

The knowledge that the barrier to the stalk formation was almost the dominant barrier prompted other studies (Lee & Schick, 2007a; Lee & Schick, 2008) to see whether there were conditions under which making the stalk was, in fact, the larger of two barriers. Indeed it was found that in the fusion of sufficiently small vesicles, the larger barrier was the initial one to make the stalk itself. This is seen in Fig. 9. If this result is correct, then it yields the picture of fusion of small synaptic vesicles, on demand; that is, the vesicle can be tethered at a sufficient distance from the plasma membrane so that fusion cannot proceed, and then be brought to it where the vesicle and membrane will fuse rapidly.

V. OUTLOOK

Results of the field-theoretic calculations point to several areas for further study. Some pertain to the field-theoretic methods we have employed, while others point more generally to the very modeling of the fusion process itself. We list here four such areas.

1. *Expansion of the coarse-grained models described by the field theory*: Within the coarse-grained model, the amphiphiles have been represented by linear, Gaussian, chain molecules with a hydrophilic fraction, *f*. The

product of the spontaneous curvature of a monolayer and the bilayer thickness has been used to relate the hydrophilic fraction, *f*, of our coarse-grained model to experimental systems. It is gratifying that the narrow window of metastability of a stalk that is predicted by the coarse-grained model agrees with the range in which fusion is indeed experimentally observed. However, the Gaussian chain model allows for large fluctuations of the molecular shape, which is indeed typical of long, flexible polymers. Biological lipids, however, are rather short and stiff molecules, characterized by a "mean molecular shape" that fluctuates much less than that of polymers. Rather than requiring a deformation of molecular shape, the bilayer packing can be thought of as requiring a packing of molecules with a rather well-defined geometry (Israelachvili, 1991). Although continuum Gaussian chains are often employed because they permit efficient and accurate computations, it is well known that the field-theoretic framework can be applied to other, more realistic, models as well. Molecular stiffness, for example, is described by the use of the worm-like chain continuum model (Morse & Fredrickson, 1994; Schmid & Müller, 1995). More general discrete architectures can be considered, although the calculation of the partition function now requires a direct enumeration of a very large number of molecular configurations. This is a procedure that has been championed by Szleifer and coworkers (Ben-Shaul & Szleifer, 1985; Szleifer & Carignano, 1996), and by the Wageningen group (Leermakers, Rabinovitch, & Balabaev, 2003; Meijer, Leermakers, & Lyklema, 1999). It would be very interesting to study the excess free energy of stalks between apposed membranes that are comprised of amphiphiles with more realistic molecular architectures, for example, double-tailed lipids (Li & Schick, 2000; Smirnova, Marrink, Lipowsky, & Knecht, 2010; Stevens et al., 2003).

2. *Determination of the accuracy of self-consistent field theory*: Since the free energy of the metastable states in the biologically relevant parameter regime are comparable to the thermal energy, $k_B T$, it is certainly warranted to assess the accuracy of the mean-field approximation invoked in the self-consistent field calculations. Recently, three different methods have been devised to compute the free energy of self-assembled structure without employing a mean-field approximation. Lennon, Katsov, and Fredrickson (Lennon et al., 2008) used the same continuum Gaussian chain model in conjunction with field-theoretic simulations and obtained the exact free energy by studying the fluctuations of the collective densities around their mean-field values. The advantage of this field-theoretic approach is that one can directly calculate the free energy of the same model as the one used in the self-consistent field calculations, and thereby directly determine the accuracy of the mean-field approximation. A second route is to calculate

directly the free energy of a particle-based model. One strategy to do this, consists in simply computing the pressure, p, in a grand canonical ensemble utilizing the relation between it and the grand potential $\Omega = -pV$. Another possibility is to compute the chemical potential in the isothermal–isobaric ensemble (Detcheverry et al., 2009; Martinez-Veracoechea & Escobedo, 2006). Both calculations are made more tractable by employing coarse-grained models that do not feature a hard-core repulsion (Daoulas & Müller, 2010; Shillcock & Lipowsky, 2002; Venturoli et al., 2006). A third strategy consists in reversibly transforming the structures of interest (e.g., two apposed bilayers with and without a stalk) into each other with the help of an external field (Müller & Daoulas, 2008). This technique has recently been applied to calculate the excess free energy of a stalk within a coarse-grained solvent-free model. The results corroborate that the excess free energy of a stalk is on the order of $10k_BT$, and strongly decreases as the composition of the amphiphile becomes more asymmetric. Finally, the change in free energy due to a change of lipid architecture can be obtained via thermodynamic integration using all-chemical mutations of one lipid species into the other in the semigrand canonical ensemble (Norizoe et al., 2010).

3. *Concise description of the fusion pathway*: Our field-theoretic calculations relied on a pathway that became evident from direct observation of the fusion process in molecular simulations. However, the systematic exploration of fusion pathways by such simulations remains prohibitively expensive even within the framework of minimal, coarse-grained models. The basic difficulty stems from two sources: (i) the intrinsically slow dynamics of the self-assembling systems and (ii) the absence of a simple reaction coordinate that parameterizes the transformation path. The former effect is related to the difference between the weak, nonbonded, forces that bring about the self-assembly itself, and the comparably stronger, bonded forces that dictate the molecular architecture and set the smallest time scale of the computational model. The latter source of difficulty is related to the problem of defining a simple, collective quantity that parameterizes the complex molecular rearrangement during the fusion process.

   This problem can be illustrated by the very basic process of stalk formation between two apposed bilayers, which is a cornerstone of the fusion process. Previous field-theoretic methods utilized the radius of the stalk as a reaction coordinate, constraining that radius to have a given value. The excess free energy of the stalk exhibits a minimum as a function of its radius, and this minimum identifies the metastable stalk, S1 (cf. Fig. 5). Upon decreasing the radius even further, one observes that the free-energy increases and that, at a certain well-defined value, the constrained stalk ceases to be even metastable, S0. This point marks the spinodal limit of stability of the constrained stalk and provides an indication of the scale of the barrier toward stalk formation.

Employing molecular simulations utilizing the coarse-grained MARTINI model of double-tailed lipids (Marrink et al., 2007), Smirnova et al. (2010) dragged the tail of a lipid from the bilayer in which it had been embedded to the apposing bilayer. They used the coordinate, z, of the center of mass of the tail as a reaction coordinate. The resulting splayed configuration in which a lipid has one tail in each of the apposing bilayers had previously been implicated in the formation of a stalk (Stevens et al., 2003). Smirnova and coworkers indeed observed that the formation of a stalk can be induced by this configuration of the lipid tails. They calculated the potential of mean force as a function of the position, z, and the minimal distance, u, between the two hydrophobic portions of the bilayer. They estimated both the excess free energy of the metastable stalk as well as the free energy barrier of stalk formation. While this study highlights the role of specific conformations of a single lipid in the formation of a stalk, the two reaction coordinates, z and u, do not allow for a *reversible* formation of a stalk. Therefore, the significance of the free energies extracted from this process needs to be clarified.

One promising route to describe and explore the pathways to fusion consists in applying string methods (Maragliano, Fischer, Vanden-Eijnden, & Ciccotti, 2006), transition path sampling, or other advanced simulation techniques useful for rare events (Dellago & Bolhuis, 2009). Recently, the former strategy has been used to identify transition states in the nucleation of ordered phases in block copolymer melts within self-consistent field theory, an application that revealed intricate, three-dimensional, geometrically complex, critical nuclei (Cheng, Lin, E, Zhang, & Shi, 2010; Lin, Cheng, E, Shi, & Zhang, 2010), as well as pore formation in membranes (Ting, Appelö, & Wang, 2011). An application of similar techniques to the field-theoretic study of fusion would potentially allow for a systematic and accurate exploration of possible pathways.

4. *Assessment of possible simultaneous fusion pathways*: Given that the free energies of metastable intermediates and of saddle points are comparable to the thermal energy, k_BT, the applicability of the concept of a single, or of only a few, well-defined transition states with specific free energies (Kasson & Pande, 2007; Knecht & Marrink, 2007; Marrink & Mark, 2003; Smeijers et al., 2006) remains to be assessed. The experimentally relevant quantity is the fusion rate rather than the free-energy barrier, and the rate depends upon the kinetics. A change in the parameters of the system, such as the tension, temperature, or lipid architecture, not only alters the free energy along the fusion path but it also affects the collective dynamics and thereby the fusion rate. Thus, the possibility that the kinetics contributes to the selection of the fusion path cannot be ignored.

VI. CONCLUSION

We have presented a pathway by which fusion can take place that is an alternative to the standard one discussed elsewhere in this book. We presented calculations of the free energies of fusion intermediates that indicate that the alternate pathway is more favorable than the standard one. In addition it seems to be much more in accord with simulations.

In this alternative scenario, there is little left of the elements of the standard pathway. There is a stalk, but in the fusion of a vesicle and a planar bilayer, there is no stalk formation on the axis of symmetry. Further there is no axially symmetric expansion of the stalk, and there is no hemifusion diaphragm consisting of the two outer leaves of the opposing bilayers.

The picture that emerges consists of several of the following elements:

- Fusion is a difficult process to initiate. This is probably a consequence of the necessary stability and integrity of biological membranes. Simply bringing membranes up to one another and removing locally the water between them seems to be insufficient to begin the fusion process. Vesicles simply deform under the enforced contact.
- Eventually a barrier to the formation of a stalk, a barrier whose magnitude is unknown at present, is breached and a stalk forms, presumably at a locus of increased deformation or of increased tension, such as the perimeter of the area of enforced contact.
- The stalk expands asymmetrically sewing together the two bilayers as it expands. It closes upon itself, perhaps by following the perimeter of the area of enforced contact.
- The nucleation of holes in either or both of bilayers is promoted by the presence of the stalk and they can appear before the stalk closes upon itself. If they do, then transient leakage, or even lysis, can occur.
- If holes do not appear before the stalk has closed upon itself, a circumstance made plausible by the inherent stability of the bilayer, an IMI is formed in which the two bilayers are simply sewn together by the elongated stalk. Again the stalk promotes the nucleation of holes in the bilayers. After a hole appears in one bilayer, a hemifusion diaphragm consisting of the other intact bilayer remains. Hole formation in that bilayer completes the fusion process.

The above picture is still very incomplete. In particular it is unclear whether the pathway we have described is *the* pathway of fusion, or simply one of several that are utilized simultaneously. Much remains to be understood, and to be done, but the field is quite an active one. This book is a testament to that fact.

Acknowledgments

It is a great pleasure to thank Kirill Katsov and Jae Youn Lee for fruitful and enjoyable collaboration. This work was supported by the National Science Foundation under Grant No. DMR-0803956, the Volkswagen foundation, and the SFB 803 "Functionality controlled by organization in and between membranes."

References

Aeffner, S., Reusch, T., Weinhausen, B., & Salditt, T. (2009). Membrane fusion intermediates and the effect of cholesterol: An in-house X-ray scattering study. *European Physical Journal E, 30*, 205–214.

Aguilar, P., Engel, A., & Walter, P. (2007). The plasma membrane proteins Prm1 and Fig1 ascertain fidelity of membrane fusion during yeast mating. *Molecular Biology of the Cell, 18*, 547–556.

Bailey, A. L., & Cullis, P. R. (1994). Modulation of membrane fusion by asymmetric transbilayer distributions of amino lipids. *Biochemistry, 33*, 12573–12580.

Ben-Shaul, A., & Szleifer, I. (1985). Chain organization and thermodynamics in micelles and bilayers. 1. Theory. *Journal of Chemical Physics, 83*, 3597.

Bonnafous, P., & Stegmann, T. (2000). Membrane perturbation and fusion pore formation in influenza hemagglutinin-mediated membrane fusion. *Journal of Biological Chemistry, 275*, 6160–6166.

Cevc, G., & Richardsen, H. (1999). Lipid vesicles and membrane fusion. *Advanced Drug Delivery, 38*, 207–232.

Cheng, X. Y., Lin, L., E, W., Zhang, P. W., & Shi, A. C. (2010). Nucleation of ordered phases in block copolymers. *Physical Review Letters, 104*, 148301.

Chernomordik, L., Kozlov, M., & Zimmerberg, J. (1995). Lipids in biological membrane fusion. *Journal of Membrane Biology, 146*, 1–14.

Daoulas, K. C., & Müller, M. (2010). Comparison of simulations of lipid membranes with membranes of block copolymers. *Advances in Polymer Science (Polymer Membranes/Biomembranes), 224*, 197–233.

Dellago, C., & Bolhuis, P. G. (2009). Transition path sampling and other advanced simulation techniques for rare event. *Advances in Polymer Science, 221*, 167–233.

Detcheverry, F. A., Pike, D. Q., Nealey, P. F., Müller, M., & de Pablo, J. J. (2009). Monte Carlo simulation of coarse grain polymeric systems. *Physical Review Letters, 102*, 197801.

Devaux, P. (1991). Static and dynamic lipid asymmetry in cell membranes. *Biochemistry, 30*, 1163–1173.

Discher, B. D., Won, Y. -Y., Ege, D. S., Lee, J. C. -M., Bates, F. S., Discher, D. E., & Hammer, D. A. (1999). Polymersomes: Tough vesicles made from diblock copolymers. *Science, 284*, 1143–1146.

Eastman, S., Hope, M. J., Wong, K., & Cullis, P. R. (1992). Influence of phospholipid asymmetry on fusion between large unilamellar vesicles. *Biochemistry, 31*, 4262–4268.

Engel, A., & Walter, P. (2008). Membrane lysis during biological membrane fusion: Collateral damage by misregulated fusion machines. *Journal of Cell Biology, 183*, 181–186.

Evans, K. O., & Lentz, B. R. (2002). Kinetics of lipid rearrangements during poly(ethylene glycol)-mediated fusion of highly curved unilamellar vesicles. *Biochemistry, 41*, 1241–1249.

Fredrickson, G. (2006). The equilibrium theory of inhomogeneous polymers. Oxford University Press, Inc., Oxford.

Frolov, V. A., Dunina-Barkovskaya, A. Y., Samsonov, A. V., & Zimmerberg, J. (2003). Membrane permeability changes at early stages of influenza hemagglutinin-mediated fusion. *Biophysical Journal, 85*, 1725–1733.

Fuhrmans, M., Knecht, V., & Marrink, S. J. (2009). A single bicontinuous cubic phase induced by fusion peptides. *Journal of the American Chemical Society, 131*, 9166–9167.
Gao, L. H., Lipowsky, R., & Shillcock, J. (2008). Tension-induced vesicle fusion: Pathways and pore dynamics. *Soft Matter, 4*, 1208–1214.
Gao, L. H., Shillcock, J., & Lipowsky, R. (2007). Improved dissipative particle dynamics simulations of lipid bilayers. *Journal of Chemical Physics, 126*, 1208–1214.
Grafmüller, A., Shillcock, J., & Lipowsky, R. (2009). The fusion of membranes and vesicles: Pathway and energy barriers from dissipative particle dynamics. *Biophysical Journal, 96*, 2658–2675.
Gruner, S. M. (1989). Stability of lyotropic phases with curved interfaces. *Journal of Physical Chemistry, 93*, 1570–7562.
Hajduk, D. A., Takenouchi, H., Hillmyer, M. A., Bates, F. S., Vigild, M. E., & Almdal, K. (1997). Stability of the perforated layer (pl) phase in diblock copolymer melts. *Macromolecules, 30*, 3788–3795.
Haque, M. E., & Lentz, B. R. (2002). Influence of gp41 fusion peptide on the kinetics of poly (ethylene glycol)-mediated model membrane fusion. *Biochemistry, 41*, 10866–10876.
Haque, M. E., McIntosh, T., & Lentz, B. R. (2001). Influence of lipid composition on physical properties and PEG-mediated fusion of curved and uncurved model membrane vesicles: "Nature's Own" fusogenic lipid bilayer. *Biochemistry, 40*, 4340–4348.
Hope, M. J., & Cullis, P. R. (1980). Effects of divalent cations and pH on phosphatidylserine model membranes: A ^{31}P NMR study. *Biochemical and Biophysical Research Communications, 92*, 846–852.
Hui, S., Stewart, T., Boni, L., & Yeagle, P. (1981). Membrane fusion through point defects in bilayers. *Science, 212*, 921–923.
Israelachvili, J. N. (1991). Intermolecular and surfaces forces, 2nd ed. Academic Press, London.
Li, X. J., & Schick, M. (2000). Theory of polymorphism: Application to phosphatidylethanolamine and phosphatidylserine. *Biophysical Journal, 78*, 34–46.
Kasson, P., & Pande, V. S. (2007). Control of membrane fusion mechanism by lipid composition: Predictions from ensemble molecular dynamics. *PLOS Computational Biology, 3*, 2228–2238.
Katsov, K., Müller, M., & Schick, M. (2004). Field theoretic study of bilayer membrane fusion: I. Hemifusion mechanism. *Biophysical Journal, 87*, 3277–3290.
Katsov, K., Müller, M., & Schick, M. (2006). Field theoretic study of bilayer membrane fusion: II. Mechanism of a stalk-hole complex. *Biophysical Journal, 90*, 915–926.
Knecht, V., & Marrink, S. -J. (2007). Molecular dynamics simulation of lipid vesicle fusion in atomic detail. *Biophysical Journal, 92*, 4254–4261.
Kozlov, M. M., & Markin, V. S. (1983). Possible mechanism of membrane fusion. *Biofizika, 28*, 255–261.
Kozlovsky, Y., & Kozlov, M. M. (2002). Stalk model of membrane fusion: Solution of energy crisis. *Biophysical Journal, 82*, 882–895.
Lee, J. -Y., & Schick, M. (2007a). Dependence of the energies of fusion on the inter-membrane separation: Optimal and constrained. *Journal of Chemical Physics, 127*, 0751021–0751026.
Lee, J. -Y., & Schick, M. (2007b). Field theoretic study of bilayer membrane fusion III: Membranes with leaves of different composition. *Biophysical Journal, 92*, 3938–3948.
Lee, J. -Y., & Schick, M. (2008). Calculation of the free energy barriers to the fusion of small vesicles. *Biophysical Journal, 94*, 1699–1706.
Leermakers, F., Rabinovitch, A., & Balabaev, N. (2003). Self-consistent-field modelling of hydrated unsaturated bilayers in the liquid-crystal phase and comparison to molecular dynamics simulations. *Physical Review E, 67*, 011910-1–011910-17.

Lennon, E. M., Katsov, K., & Fredrickson, G. H. (2008). Free energy evaluation in field-theoretic polymer simulations. *Physical Review Letters, 101*, 138302.

Lentz, B. R., Talbot, W., Lee, J., & Zheng, L. -X. (1997). Transbilayer lipid redistribution accompanies poly(ethylene glycol) treatment of model membranes but is not induced by fusion. *Biochemistry, 36*, 2076–2083.

Lin, L., Cheng, X. Y., E, W., Shi, A. C., & Zhang, P. W. (2010). A numerical method for the study of nucleation of ordered phases. *Journal of Computational Physics, 229*, 1797–1809.

Loo, Y. L., Register, R. A., Adamson, D. H., & Ryan, A. J. (2005). A highly regular hexagonally perforated lamellar structure in a quiescent diblock copolymer. *Macromolecules, 38*, 4947–4949.

Maragliano, L., Fischer, A., Vanden-Eijnden, E., & Ciccotti, G. (2006). String method in collective variables: Minimum free energy paths and isocommittor surfaces. *Journal of Chemical Physics, 125*, 503–523.

Marrink, S., Fuhrmans, M., Risselada, H., & Periole, X. (2008). Coarse graining of condensed phase and biomolecular systems. CRC Press, Boca Raton, FL.

Marrink, S. J., de Vries, A. H., & Tieleman, D. P. (2009). Lipids on the move: Simulations of membrane pores, domains, stalks and curves. *Biochimica et Biophysica Acta-Biomembranes, 1788*, 149–168.

Marrink, S. J., & Mark, A. E. (2003). The mechanism of vesicle fusion as revealed by molecular dynamics simulations. *Journal of the American Chemical Society, 125*, 11144–11145.

Marrink, S. J., & Mark, A. E. (2004). Molecular view of hexagonal phase formation in phospholipid membranes. *Biophysical Journal, 87*, 3894–3900.

Marrink, S. J., Risselada, H. J., Yefimov, S., Tieleman, D. P., & de Vries, A. H. (2007). The MARTINI force field: Coarse grained model for biomolecular simulations. *Journal of Physical Chemistry B, 111*, 7812–7824.

Martinez-Veracoechea, F. J., & Escobedo, F. A. (2006). Simulation of the gyroid phase in off-lattice models of pure diblock copolymer melts. *Journal of Chemical Physics*, 125.

Matsen, M. W. (2006). In: G. Gompper, M. Schick (Eds.), Soft Matter (Vol. 1, p. 87). Weinheim: Wiley-VCH.

Matsen, M. W. (2002). The standard Gaussian model for block copolymer melts. *Journal of Physics: Condensed Matter, 14.*, R21–R47.

Meijer, L., Leermakers, F., & Lyklema, J. (1999). Self-consistent field modelling of complex molecules with united atom detail in inhomogeneous systems. Cyclic and branched foreign molecules in dimyristoylphosphatidylcholine membranes. *Journal of Chemical Physics, 110*, 6560–6578.

Merz, A. J., & Wickner, W. (2004). Trans-SNARE interactions elicit Ca^{2+} efflux from the yeast vacuole lumen. *Journal of Cell Biology, 164*, 195–206.

Monck, J. R., de Toledo, G., & Fernandez, J. M. (1990). Tension in granule secretory membranes causes extensive membrane transfer through the exocytotic fusion pore. *Proceedings of the National Academy of Sciences of the United States of America, 87*, 7804–7808.

Morse, D. C., & Fredrickson, G. H. (1994). Semiflexible polymers near interfaces. *Physical Review Letters, 73*, 3235.

Müller, M., & Daoulas, K. C. (2008). Calculating the free energy of self-assembled structures by thermodynamic integration. *Journal of Chemical Physics, 128*, 024903.

Müller, M., & de Pablo, J. J. (2006). Simulation techniques for calculating free energies. *Lecture Notes in Physics, 703*, 67–122.

Müller, M., & Gompper, G. (2002). Elastic properties of polymer interfaces: Aggregation of pure diblock, mixed diblock, and triblock copolymers. *Physical Review E, 66*, 041805,1–041805,13.

Müller, M., Katsov, K., & Schick, M. (2002). New mechanism of membrane fusion. *Journal of Chemical Physics, 116*, 2342–2345.

Müller, M., Katsov, K., & Schick, M. (2003a). Coarse-grained models and collective phenomena in membranes: Computer simulation of membrane fusion. *Journal of Polymer Science Part B: Polymer Physics, 41*, 1441–1450.

Müller, M., Katsov, K., & Schick, M. (2003b). A new mechanism of model membrane fusion determined from Monte Carlo simulation. *Biophysical Journal, 85*, 1611–1623.

Müller, M., Katsov, K., & Schick, M. (2006). Biological and synthetic membranes: What can be learned from a coarse-grained description? *Physics Reports, 434*, 113–176.

Müller, M., & Schmid, F. (2005). Incorporating fluctuations and dynamics in self-consistent field theories for polymer blends. *Advances in Polymer Science, 185*, 1–58.

Murtola, T., Bunker, A., Vattulainen, I., Deserno, M., & Karttunen, M. (2009). Multiscale modeling of emergent materials: Biological and soft matter. *Physical Chemistry Chemical Physics, 11*, 1869–1892.

Nielsen, S., Lopez, C., Srinivas, G., & Klein, M. L. (2004). Coarse grain models and the computer simulation of soft materials. *Journal of Physics: Condensed Matter, 16*, R481.

Noguchi, H., & Takasu, M. (2001). Fusion pathways of vesicles: A Brownian dynamics simulation. *Journal of Chemical Physics, 115*, 9547–9551.

Norizoe, Y., Daoulas, K. C., & Müller, M. (2010). Measuring excess free energies of self-assembled membrane structures. *Faraday Discussions, 144*, 369–391.

Risselada, H. J., Kutzner, C., & Grubmüller, H. (2011). Caught in the Act: Visualization of SNARE-Mediated Fusion Events in Molecular Detail. *ChemBioChem, 12*, 1049–1055.

Rothman, J. E., & Lenard, J. (1977). Membrane asymmetry. *Science, 195*, 743–753.

Schmid, F. (1998). Self-consistent field theories for complex fluids. *Journal of Physics: Condensed Matter, 10*, 8105–8138.

Schmid, F., & Müller, M. (1995). Quantitative comparison of self-consistent-field theories for polymers near interfaces with Monte-Carlo simulations. *Macromolecules, 28*, 8639–8645.

Seigneuret, M., & Devaux, P. F. (1984). Asymmetric distribution of spin-labeled phospholipids in the erythrocyte membrane: Relation to shape changes. *Proceedings of the National Academy of Sciences of the United States of America, 81*, 3751–3755.

Shangguan, T., Alford, D., & Bentz, J. (1996). Influenza virus-liposome lipid mixing is leaky and largely insensitive to the material properties of the target membrane. *Biochemistry, 25*, 4956–4965.

Shelley, J., Shelley, M., Reeder, R., Bandyopadhyay, S., & Klein, M. (2001). A coarse grain model for phospholipid simulations. *Journal of Physical Chemistry B, 105*, 4464–4470.

Shillcock, J. C., & Lipowsky, R. (2002). Equilibrium structure and lateral stress distribution of amphiphilic bilayers from dissipative particle dynamics simulations. *Journal of Chemical Physics, 117*, 5048–5061.

Shillcock, J. C., & Lipowsky, R. (2005). Tension-induced fusion of bilayer membranes and vesicles. *Nature Materials, 4*, 225–228.

Siegel, D. P. (1986). Inverted micellar intermediates and the transitions between lamellar, inverted hexagonal, and cubic lipid phases. II. Implications for membrane-membrane interactions and membrane fusion. *Biophysical Journal, 49*, 1171–1183.

Siegel, D. P. (1993). Energetics of intermediates in membrane fusion: Comparison of stalk and inverted micellar intermediate mechanisms. *Biophysical Journal, 65*, 2124–2140.

Smeijers, A. F., Marvoort, A. J., Pieterse, K., & Hilbers, P. A. J. (2006). A detailed look at vesicle fusion. *Journal of Physical Chemistry B, 110*, 13212–13219.

Smirnova, Y., Marrink, S. -J., Lipowsky, R., & Knecht, V. (2010). Solvent-exposed tails as prestalk transition states for membrane fusion at low hydration. *Journal of the American Chemical Society, 132*, 6710–6718.

Stevens, M. J., Hoh, J., & Woolf, T. (2003). Insights into the molecular mechanism of membrane fusion from simulation: Evidence for the association of splayed tails. *Physical Review Letters, 91*, 188102-1–188102-4.

Szleifer, I., & Carignano, M. A. (1996). Tethered polymer layers. *Advances in Chemical Physics, 94*, 165–260.

Ting, C. L., Appelö, D., & Wang, Z. -G. (2011). Minimum Energy Path to Membrane Pore Formation and Rupture. *Physical Review Letters, 106*, 168101.

Venturoli, M., Sperotto, M. M., Kranenburg, M., & Smit, B. (2006). Mesoscopic models of biological membranes. *Physics Reports, 437*, 1–54.

Weinreb, G., & Lentz, B. R. (2007). Analysis of membrane fusion as a two-stalk sequential process: Evaluation of the stalk model. *Biophysical Journal, 92*, 4012–4029.

Yang, L., & Huang, H. W. (2002). Observation of a membrane fusion intermediate structure. *Science, 297*, 1877–1879.

Index

D

I

T

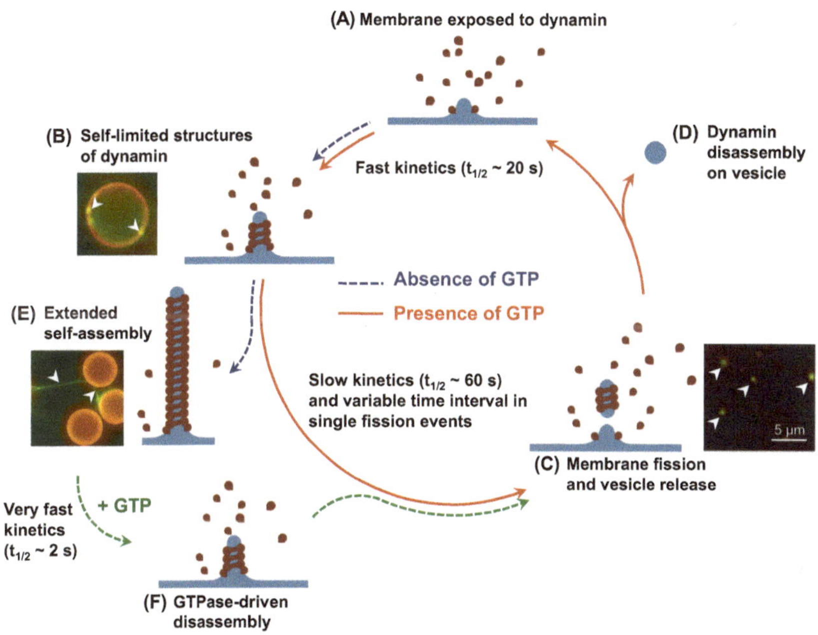

Plate 1 (Please refer to Chapter 2, Figure 1 on page 38 for figure legend).

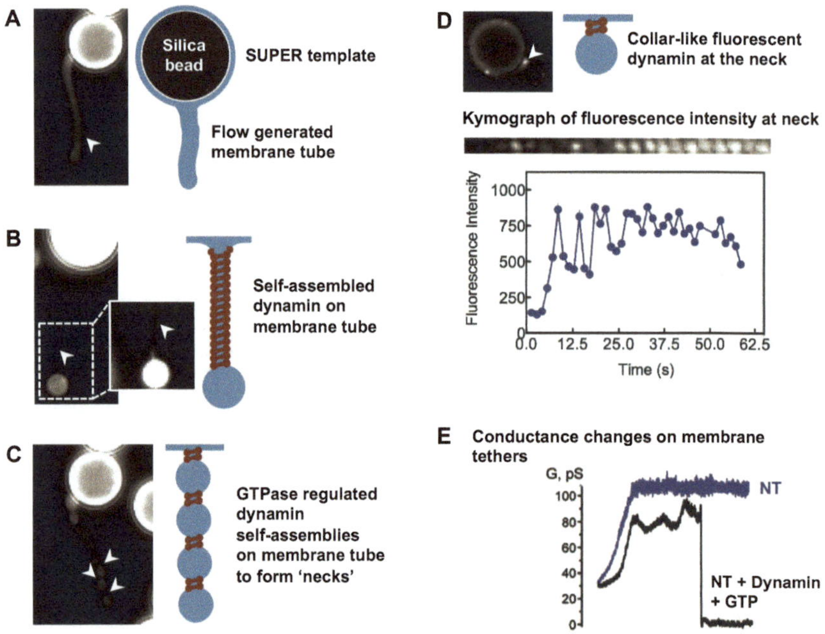

Plate 2 (Please refer to Chapter 2, Figure 2 on page 39 for figure legend).

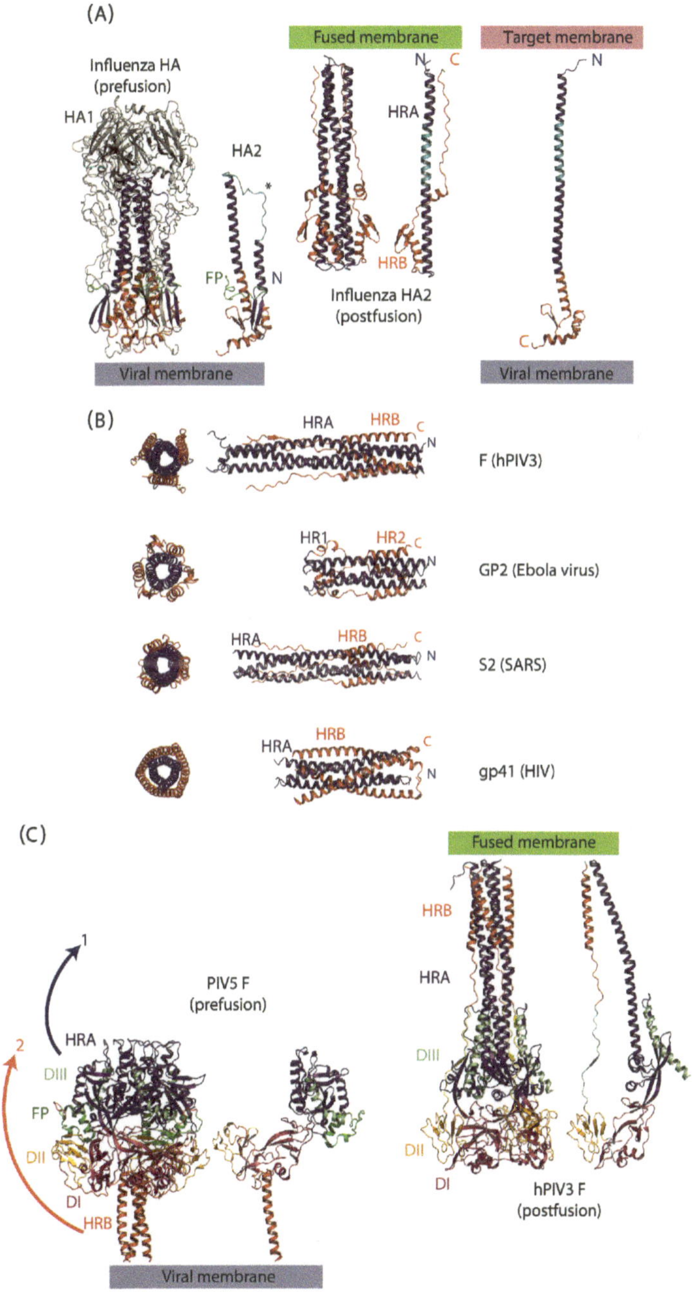

Plate 3 (Please refer to Chapter 3, Figure 1 on page 52 for figure legend).

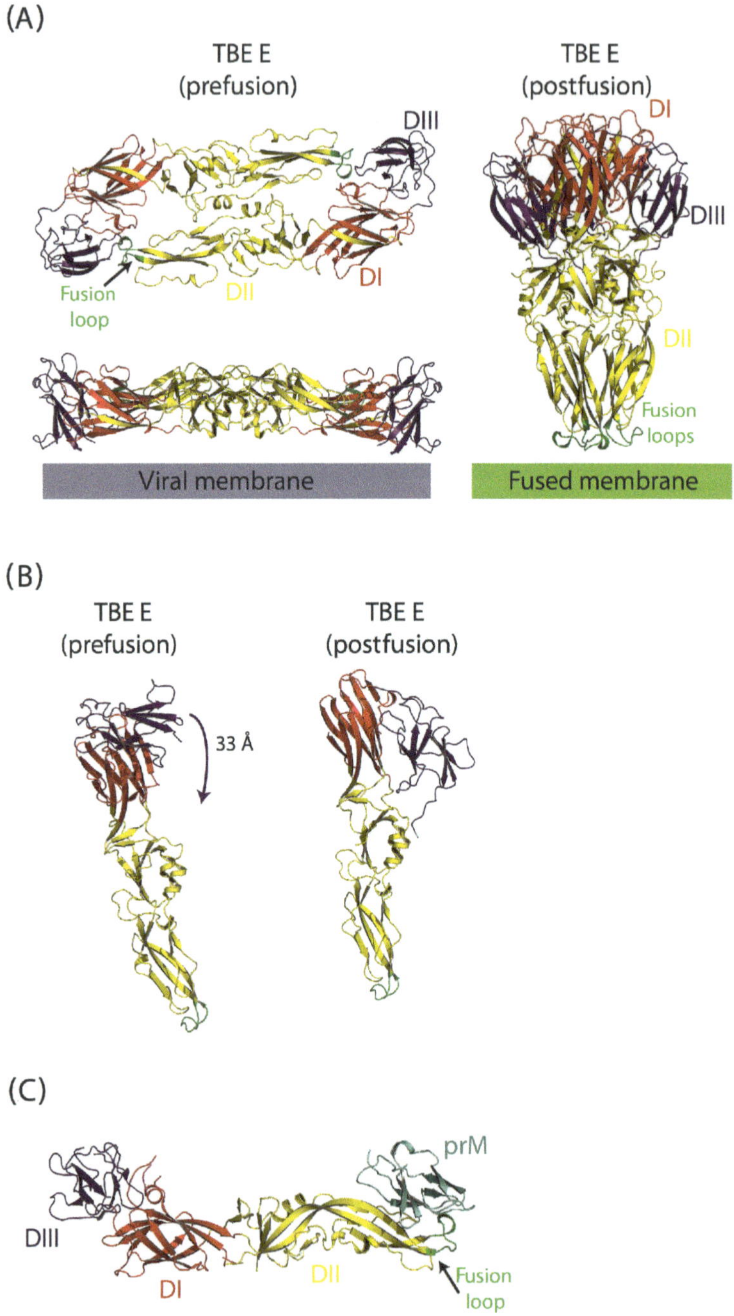

Plate 4 (Please refer to Chapter 3, Figure 2 on page 56 for figure legend).

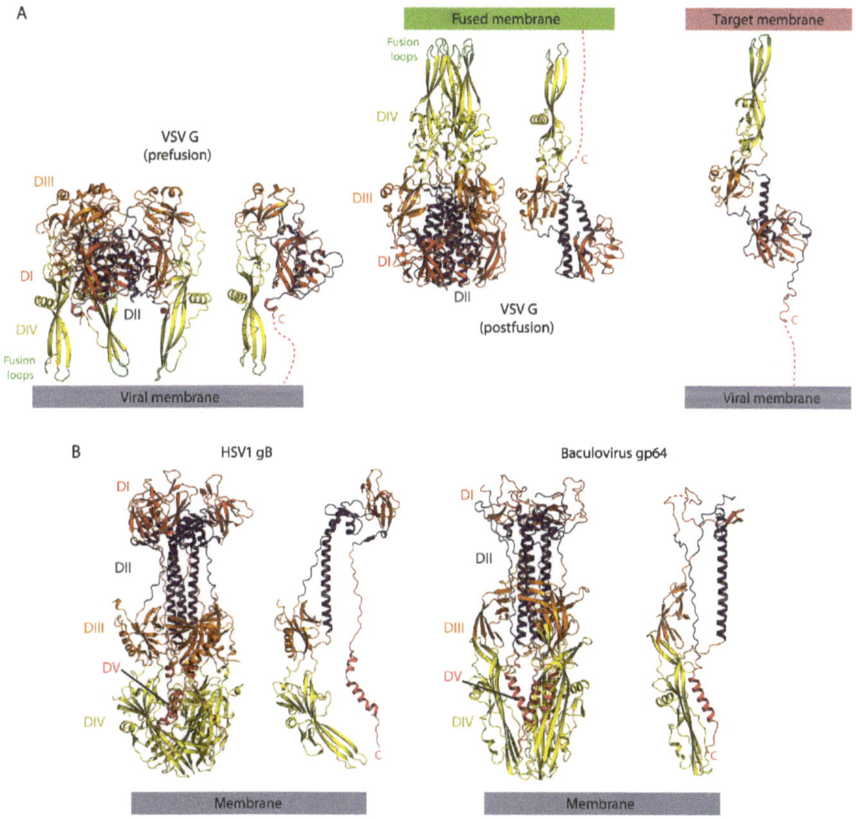

Plate 5 (Please refer to Chapter 3, Figure 3 on page 60 for figure legend).

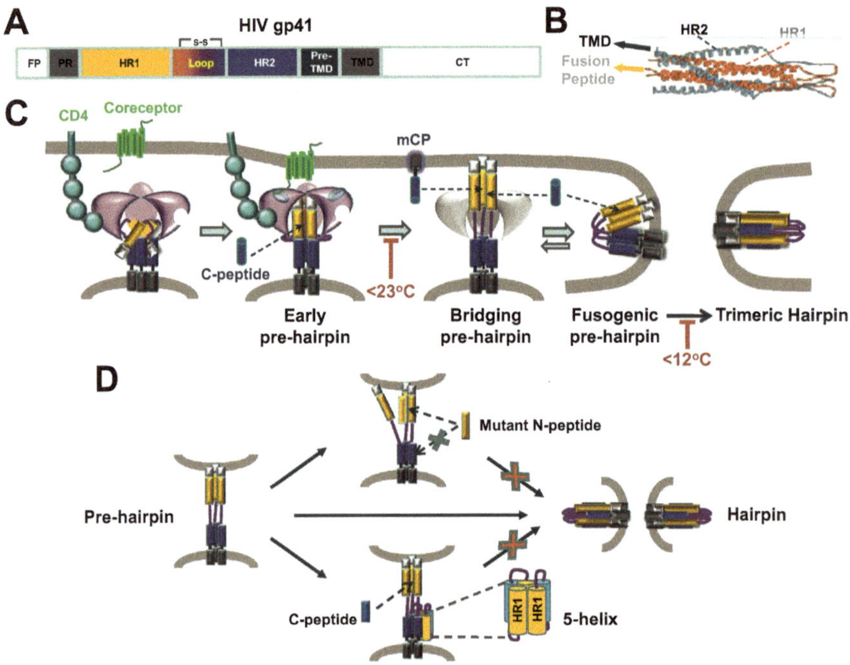

Plate 6 (Please refer to Chapter 4, Figure 2 on page 85 for figure legend).

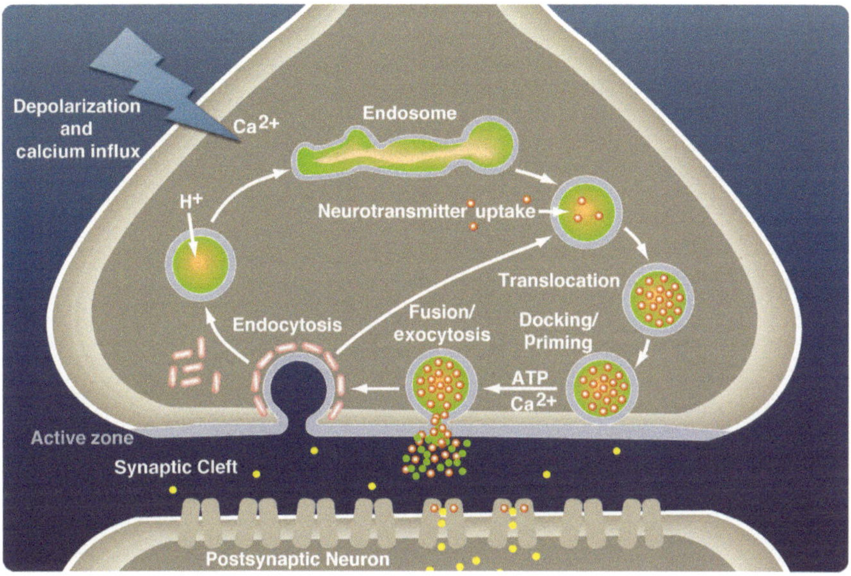

Plate 7 (Please refer to Chapter 6, Figure 1 on page 143 for figure legend).

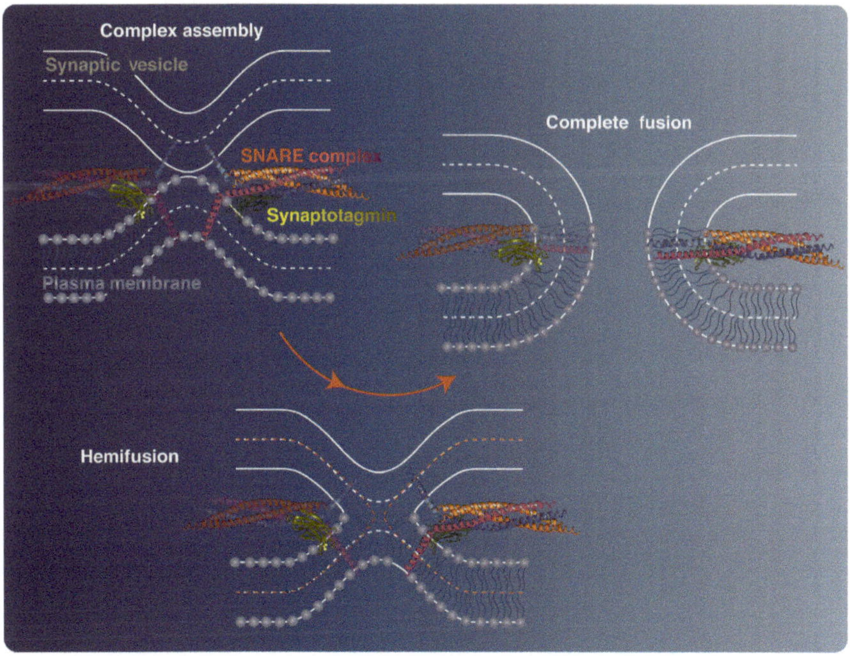

Plate 8 (Please refer to Chapter 6, Figure 2 on page 147 for figure legend).

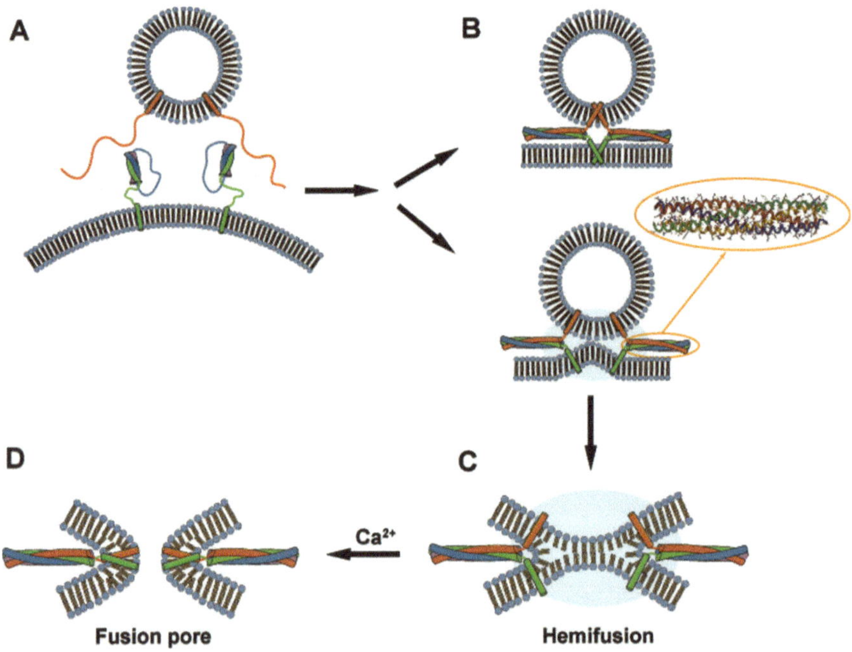

Plate 9 (Please refer to Chapter 7, Figure 1 on page 163 for figure legend).

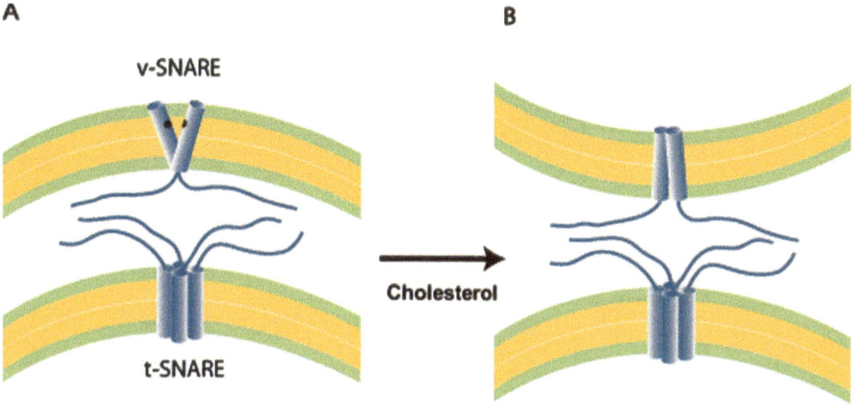

Plate 10 (Please refer to Chapter 7, Figure 2 on page 170 for figure legend).

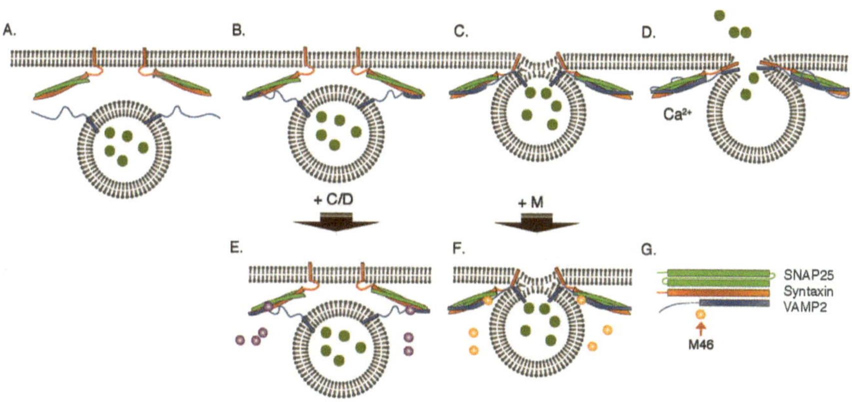

Plate 11 (Please refer to Chapter 7, Figure 3 on page 172 for figure legend).

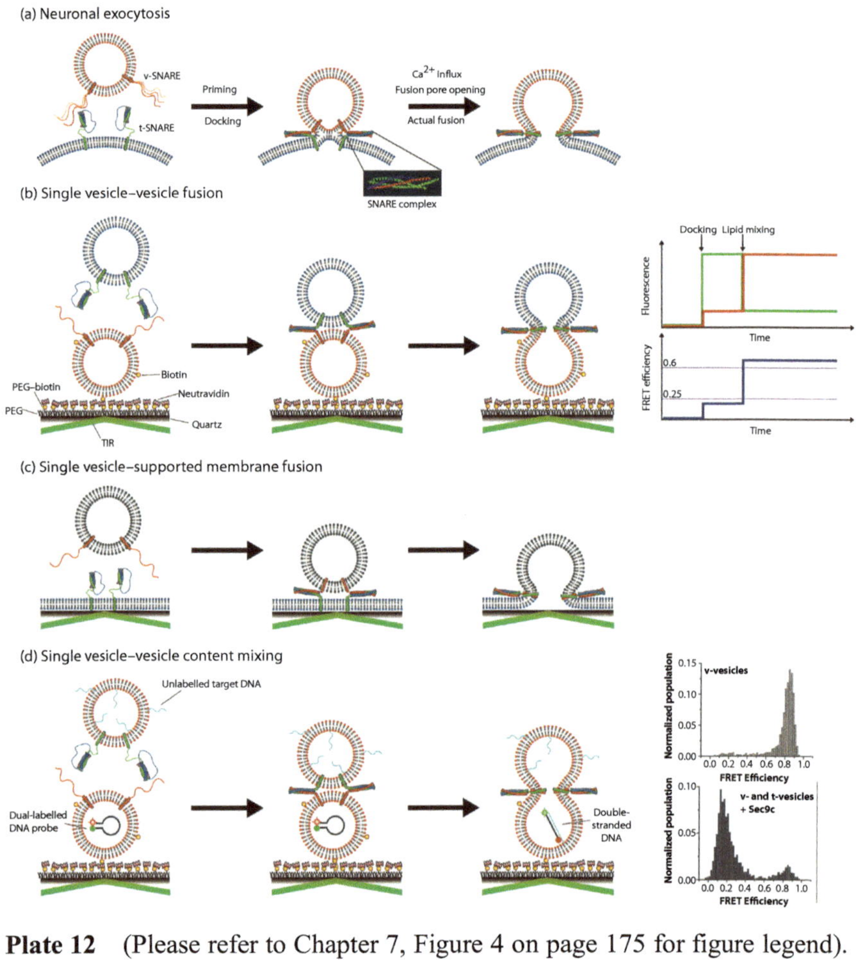

Plate 12 (Please refer to Chapter 7, Figure 4 on page 175 for figure legend).

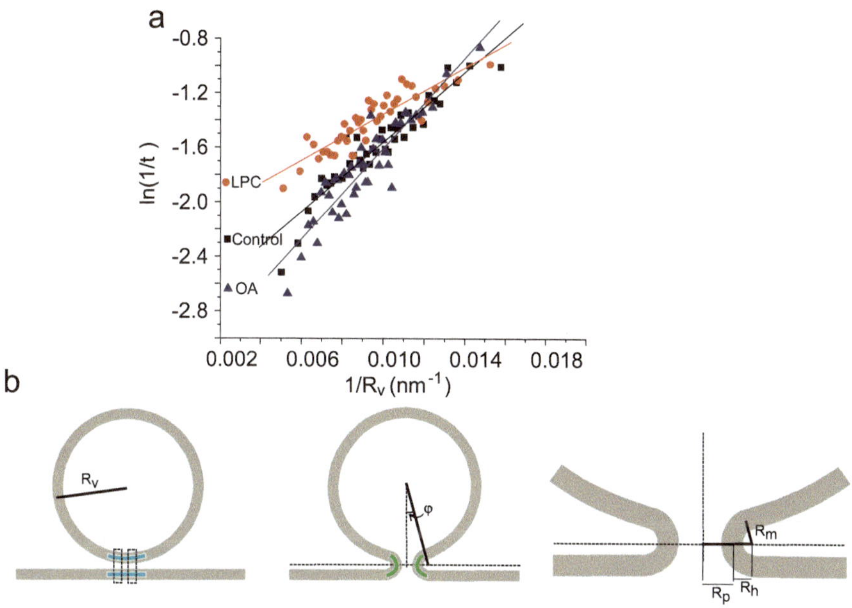

Plate 13 (Please refer to Chapter 8, Figure 3 on page 196 for figure legend).

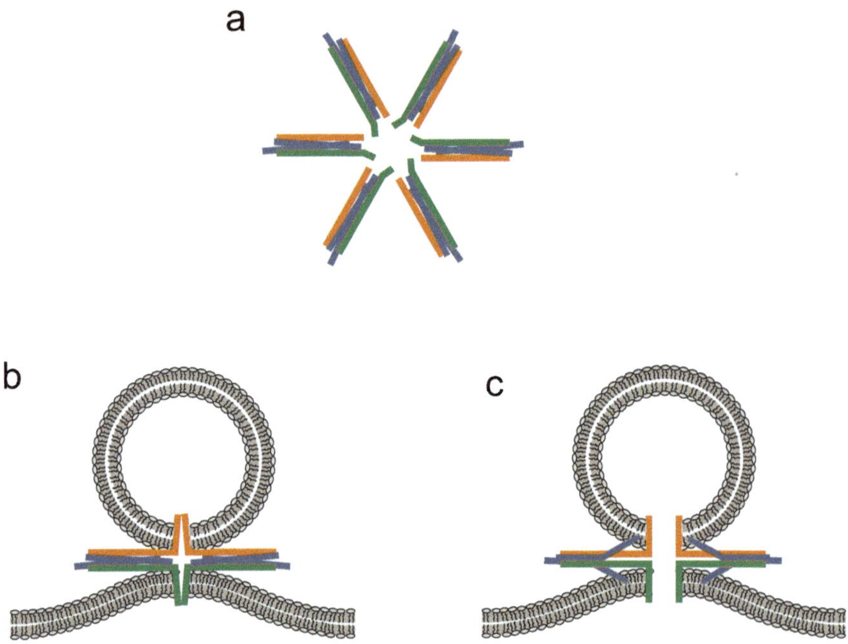

Plate 14 (Please refer to Chapter 8, Figure 4 on page 200 for figure legend).

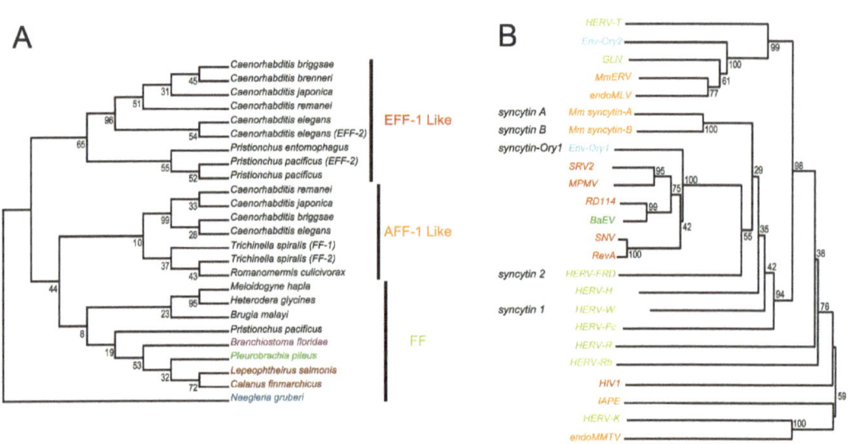

Plate 15 (Please refer to Chapter 9, Figure 1 on page 218 for figure legend).

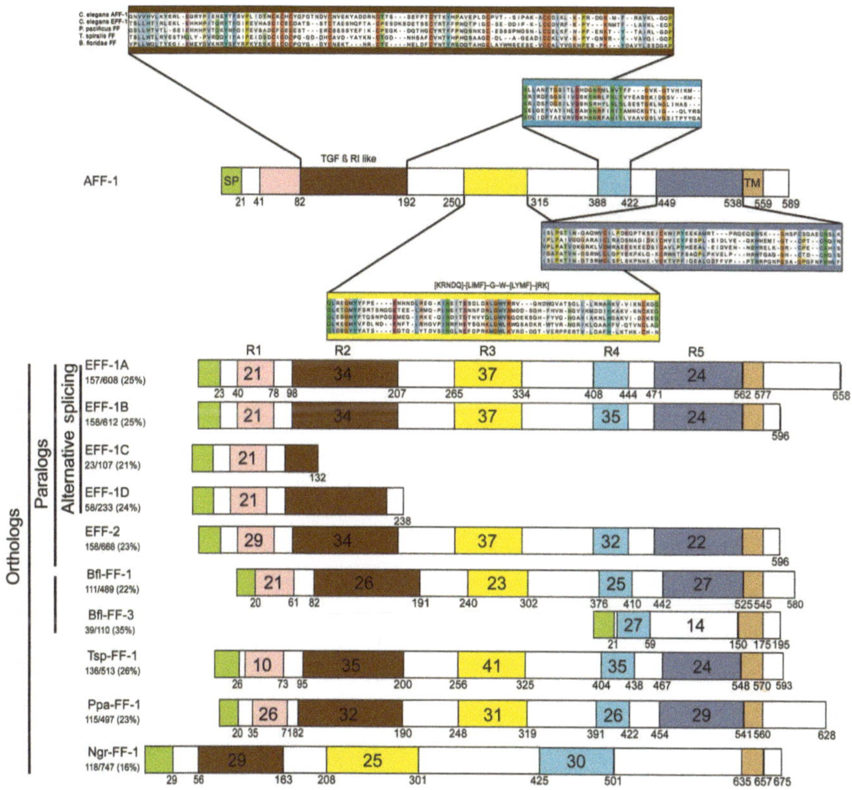

Plate 16 (Please refer to Chapter 9, Figure 2 on page 219 for figure legend).

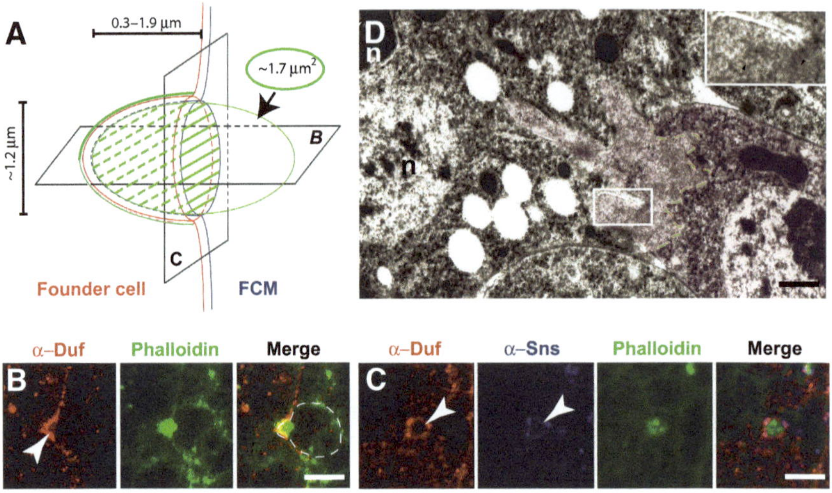

Plate 17 (Please refer to Chapter 10, Figure 2 on page 243 for figure legend).

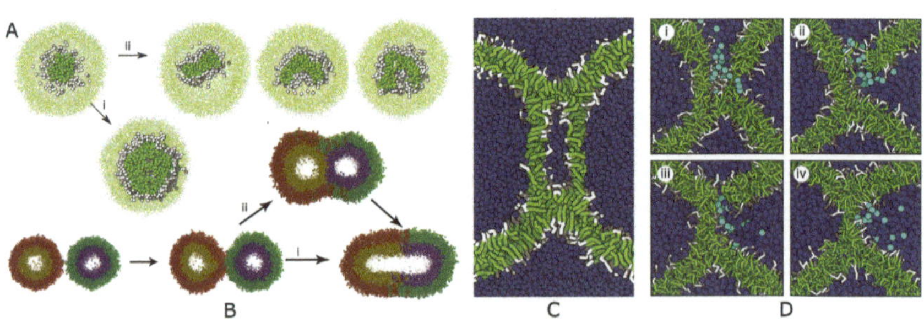

Plate 18 (Please refer to Chapter 11, Figure 1 on page 264 for figure legend).

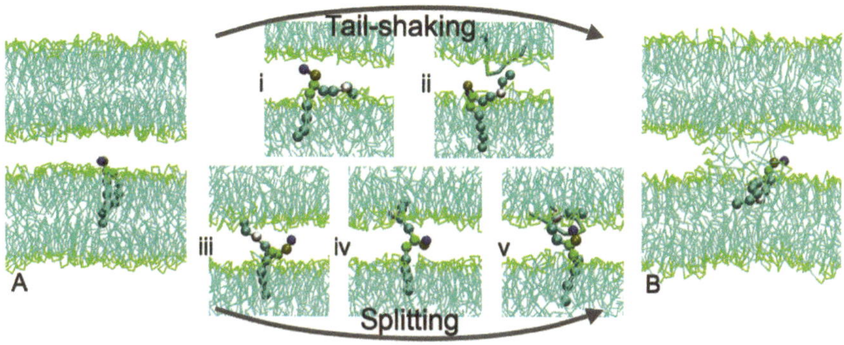

Plate 19 (Please refer to Chapter 11, Figure 2 on page 271 for figure legend).

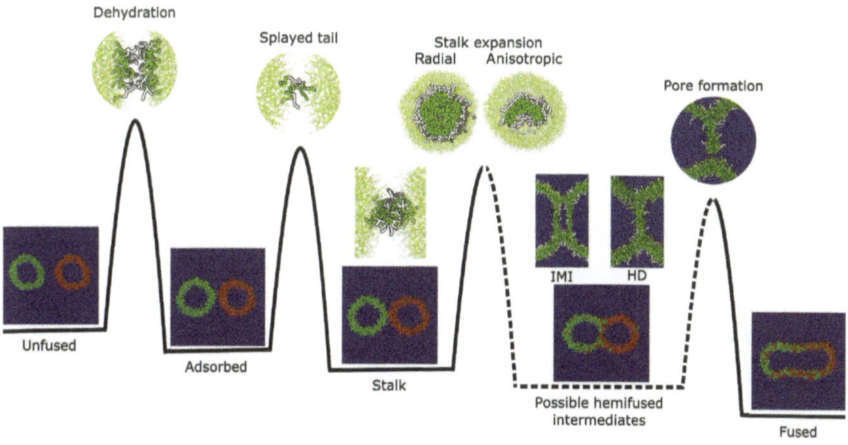

Plate 20 (Please refer to Chapter 11, Figure 3 on page 274 for figure legend).

Plate 21 (Please refer to Chapter 11, Figure 4 on page 278 for figure legend).

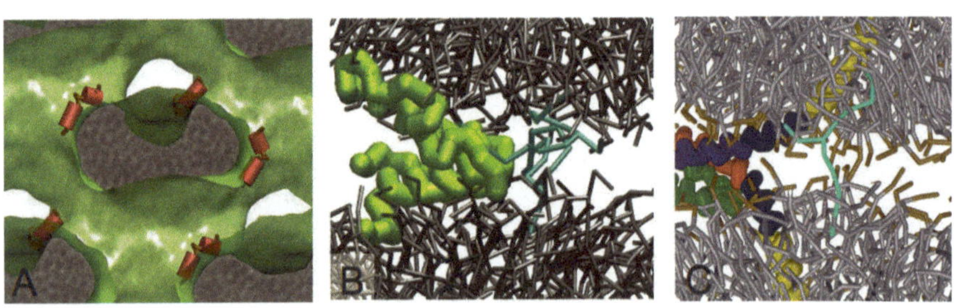

Plate 22 (Please refer to Chapter 11, Figure 5 on page 284 for figure legend).

MIX
Papier | Fördert gute Waldnutzung
FSC® C083411

Kolibri 360 GmbH
Ferdinand-Jühlke-Straße 7
99095 Erfurt, Deutschland
produktsicherheit@kolibri360.de